W9-CEF-585

Lipids in Freshwater Ecosystems

Springer

New York
Berlin
Heidelberg
Barcelona
Hong Kong
London
Milan
Paris
Singapore
Tokyo

Contents

is, they collect and integrate information from studies scattered in diverse journals, a pragmatic service in our era of burgeoning data and reports. More important perhaps, they provide critical evaluations and interpretations of existing data and explore interrelationships among biotic components of the ecosystem as a whole. Such analyses produce new insights that instill vigor and new directions into a discipline.

This text begins with summaries of some of the prevalent lipid compounds in freshwater ecosystems, produced within the aquatic systems and imported from the catchment areas. It then presents selected methodology for differentiation of lipid classes and fatty acids. Certain fatty acids have sufficient source specificity that they can be utilized as biochemical and trophic markers. With careful interpretation, such markers can provide qualitative insights into trophic relationships at higher levels in the food web.

Examination of environmental factors controlling synthesis of lipids in algae provides insights into their usefulness as indicators of the physiological status of the algae. Because zooplankton and probably most consumer organisms are capable of little de novo synthesis, lipids are primarily dietary in origin. Fluctuations in lipid content and quality in food sources, as well as feeding behaviors, are thus critically important to subsequent development, growth, and reproductive success of higher aquatic organisms. Differences in lipid requirements, utilization, and storage strategies among biota, particularly invertebrates and fish, can potentially influence evolutionary pathways, competitive interactions, and biodiversity.

The impacts of human activities on freshwater ecosystems are now so pervasive that they influence most biogeochemical cycling. Many nonpolar organic contaminants bioaccumulate in lipid compounds. Understanding of biochemical and environmental factors regulating bioaccumulation and cycling is therefore essential to freshwater management and remediation. Finally, hydrophobicity of lipids results in an accumulation of natural and anthropogenic compounds into lipid-rich tissues of organisms, water-surface microlayers, and foams. Numerous microbes and higher organisms have life stages that, during particular periods, inhabit the surface regions and thus are exposed to potentially elevated concentrations of lipid-associated contaminants. Study of this complex community is still in its infancy.

This book is a comprehensive summary of the status of our knowledge of the multifarious roles of lipids in freshwater ecosystems. Although gaps in our understanding still remain, the authors raise questions that will give impetus to accelerated experimental research in this exciting and essential area of inquiry.

Robert G. Wetzel
Bishop Professor of Biological Sciences
University of Alabama

Foreword

Although limnology is a young discipline, it has, over the past century, experienced marked growth. Its early descriptive period was a long one, given the enormous diversity of biota and environments in freshwater ecosystems. With the development of quantitative techniques came the ability to measure production rates and other parameters and to demonstrate the effects of nutrient limitation and predation on productivity and energy flow. As understanding of these phenomena grew, so too did our appreciation of the many complex chemical interactions among the biotic and habitat components of freshwater ecosystems.

A recent, exciting phase of limnology, which may be called *biochemical limnology,* is evolving rapidly. One of its many facets is the study of population and community dynamics at basic physiological levels. Examples are many. The integration of recent studies of food biochemistry with traditional studies of food quantity has begun to reveal the striking importance of food quality to reproduction and to the growth dynamics of many aquatic animals. Positive as well as negative alleleochemical interactions, already known in terrestrial ecosystems, are emerging as a major factor of many competitive interactions in fresh waters.

The role of dissolved organic matter, particularly humic and fulvic compounds of plant origin, in the aquatic ecosystem is complex. Not only do these compounds function as large stores of carbon and energy, but they are also metabolically interactive. For example, humic substances can complex with enzymes and other metabolic macromolecules and become stored in inactivated states for various periods of time. These complexes can then be displaced to other parts of the ecosystem, potentially to be reactivated at a later time. Thus, chemical communication among biotic components is certainly as prevalent in freshwater ecosystems as it is in the complex metabolic biochemistry of metazoans. If we are to manage freshwater ecosystems effectively, we must discover more about such interactive control mechanisms. Those controls are chemical; therefore, understanding the biology requires an understanding of the biochemistry.

This collection of work on lipids represents a synthesis of existing information on a diverse group of hydrophobic organic compounds of biological origin in freshwater ecosystems: their origins, functional couplings among biotic and abiotic processes, and fates. Syntheses serve many functions. If done well, as this one

Michael T. Arts
National Water Research Institute
Saskatoon, Saskatchewan S7N 3H5
Canada
michael.arts@ec.gc.ca

Bruce C. Wainman
McMaster Midwifery Education Program
McMaster University
St. Joseph's Hospital
Hamilton, Ontario L8N 4A6
Canada
wainmanb@ihs.csu.mcmaster.ca

Cover illustration: Schematic representation of the deposition pattern of lipid droplets (triacylglycerols) in a well-fed freshwater calanoid copepod.

Library of Congress Cataloging-in-Publication Data
Arts, Michael Theodore, 1958–
    Lipids in freshwater ecosystems / Michael T. Arts, Bruce C. Wainman.
      p. cm.
    Includes bibliographical references and index.
    ISBN 0-387-98505-0 (hardcover : alk. paper)
    1. Freshwater ecology. 2. Lipids—Research. I. Wainman, Bruce C.
II. Title.
QH541.5.F7A78   1998
577.6'14—dc21   98-13050

Printed on acid-free paper.

Production coordinated by Princeton Editorial Associates, Inc., and managed by Francine McNeill; manufacturing supervised by Nancy Wu.
Photocomposed copy prepared by Princeton Editorial Associates, Inc., Scottsdale, AZ, and Roosevelt, NJ, using the authors' WordPerfect files.
Printed and bound by Braun-Brumfield, Inc., Ann Arbor, MI.
Printed in the United States of America.

9 8 7 6 5 4 3 2 1

ISBN 0-387-98505-0 Springer-Verlag New York Berlin Heidelberg   SPIN 10556833

Michael T. Arts   Bruce C. Wainman

Editors

Lipids in Freshwater Ecosystems

Foreword by Robert G. Wetzel

With 63 Illustrations

Springer

Contributors

Robert G. Ackman
Canadian Institute of Fisheries
    Technology
Daltech, Dalhousie University
Halifax, Nova Scotia B3J 2X4
Canada
odorjr@tuns.ca

S. Marshall Adams
Environmental Sciences Division
Oak Ridge National Laboratory
Oak Ridge, TN 37831-6036
USA
sma@ornl.gov

Michael T. Arts
National Water Research Institute
Saskatoon, Saskatchewan S7N 3H5
Canada
michael.arts@ec.gc.ca

Joann F. Cavaletto
Great Lakes Environmental
    Research Laboratory
Ann Arbor, MI 48105
USA
cavaletto@glerl.noaa.gov

Daniel S. Cicerone
Environmental Sciences Division
Oak Ridge National Laboratory

Oak Ridge, TN 37831-6038
USA
efp@ornl.gov

Susan W. Fisher
Department of Entomology
Ohio State University
Columbus, OH 43210
USA
fisher.14@postbox.acs.ohio-state.edu

John A. Furgal
Department of Biology
University of Waterloo
Waterloo, Ontario N2L 3G1
Canada

Wayne S. Gardner
Marine Science Institute
University of Texas at Austin
Port Aransas, TX 78373
USA
gardner@utmsi.zo.utexas.edu

Clyde E. Goulden
Division of Environmental Research
Academy of Natural Sciences
Philadelphia, PA 19103
USA
cgoulden@mail.sas.upenn.edu

Peter F. Landrum
Great Lakes Environmental
    Research Laboratory
Ann Arbor, MI 48105
USA
landrum@glerl.noaa.gov

James N. McNair
Ecological Modeling Section
Patrick Center for
    Environmental Research
The Academy of Natural Sciences
Philadelphia, PA 19103-1195
USA
mcnair@acnatsci.org

Robert E. Moeller
Department of Earth and
    Environmental Sciences
Lehigh University
Bethlehem, PA 18015
USA
rem3@lehigh.edu

Guillermo E. Napolitano
Environmental Sciences Division
Oak Ridge National Laboratory
Oak Ridge, TN 37831-6351
USA
gnapolitano@mci2000.com

Yngvar Olsen
Trondhjem Biological Station
Norwegian University of Science
    and Technology
N-7034 Trondheim
Norway
Yngvar.Olsen@vm.ntnu.no

Christopher C. Parrish
Ocean Sciences Centre
Memorial University of
    Newfoundland
St. John's, Newfoundland A1C 5S7
Canada
cparrish@kean.ucs.mun.ca

Allen R. Place
Center of Marine Biotechnology
University of Maryland
Baltimore, MD 21202
USA
place@umbi.umd.edu

Hakumat Rai
Max Planck Institut für Limnologie
D-24302 Plön
Germany
rai@mpil-ploen.mpg.de

Ralph E.H. Smith
Department of Biology
University of Waterloo
Waterloo, Ontario N2L 3G1
Canada
rsmith@biology.watstar.uwaterloo.ca

Bruce C. Wainman
McMaster Midwifery Education
    Program
McMaster University
St. Joseph's Hospital
Hamilton, Ontario L8N 4A6
Canada
wainmanb@fhs.csu.mcmaster.ca

Introduction

Michael T. Arts and Bruce C. Wainman

This book has two major objectives to fulfill. One is to provide a general reference on research into lipids in aquatic ecosystems. The second is to provide a comprehensive and authoritative summary for scientists engaged in lipid research and, especially, those whom the complexity of lipid research has intimidated.

The study of lipids in aquatic ecosystems has matured to the point where it is necessary to establish a general reference and review book for those interested in aquatic lipids. The literature on lipids in freshwater aquatic organisms has expanded greatly over the past 10 years. For many aquatic species, there are now accurate descriptions of seasonal changes in lipid content, lipid class, and even fatty acid composition. There are also excellent review articles and even an entire volume of *Freshwater Biology* (vol. 38) dedicated to lipids. However, a current and readable critical summary is required for the busy researcher. This book should also allow researchers to become familiar with work outside their direct field and thus increase the number of aquatic biologists who can accurately analyze lipids and interpret the ecological significance of their work.

The second objective relates to the need to demystify lipid research. The main appeal, and the chief difficulty, of working with lipids is their structural and functional diversity. This diversity makes lipid research a continual challenge even for the experienced and somewhat daunting for researchers wanting to integrate lipid analysis into their work. The analytical difficulties may explain the abundance of research concerned with methodological problems overcome during a study. This is not a criticism of the work that is being done. Rather, it is a recognition of the fact that the analysis and interpretation of lipid are particularly difficult undertakings. It is the hope of the editors and contributors of this book that there is enough material within these covers to encourage and simplify the work of both new and newly interested researchers.

Although other groups of biomolecules are linked by structural and functional similarity, lipids are linked by one purely physical feature—their solubility in nonpolar solvents. This makes lipid extraction a relatively simple task but complicates lipid analysis because biomolecules must be separated on their physical properties.

In Chapter 1, "Determination of Total Lipid, Lipid Classes, and Fatty Acids in Aquatic Samples," Chris Parrish details simple and effective means for lipid extraction and storage. He then sets out detailed instructions for lipid class analysis using the Iatroscan thin-layer chromatography system, which is not available elsewhere in the literature.

The second chapter, "Fatty Acids as Trophic and Chemical Markers in Freshwater Ecosystems" by Guillermo Napolitano, begins with a clear explanation of lipid nomenclature. The body of the chapter is concerned with the characteristics of ideal lipid markers for trophic transfer studies. The author highlights several promising candidates for biomarkers in many types of aquatic ecosystems of both bacterial and algal origin.

The source of almost all the lipid for aquatic food webs is discussed in Chapter 3, "Irradiance and Lipid Production in Natural Algal Populations" by Bruce Wainman, Ralph Smith, Hakumat Rai, and John Furgal. This international effort was necessary to assemble a large matrix of lipid production, physical, and chemical data. These data are used to investigate seasonal fluctuations in lipid production and overnight reallocation of lipids. A highlight of this chapter is the use of mathematical models to calculate whole-lake lipid production on an areal basis.

Once the source of carbon has been discussed, Chapter 4 looks at the conversion of phytoplankton into zooplankton lipid. In Chapter 4, "Lipids in Freshwater Zooplankton: Selected Ecological and Physiological Aspects," Michael Arts explores several facets of lipid dynamics in zooplankton including the time course of lipid deposition and loss, the presentation of lipid contents/concentration, lipids as indices of stress, and the putative effects of ultraviolet B on zooplankton lipids. The chapter ends with a discourse on research needs and suggested future directions.

The thread of lipids in zooplankton continues in Chapter 5, in which Clyde Goulden, Robert Moeller, James McNair, and Allen Place examine "Lipid Dietary Dependencies in Zooplankton." Through carefully controlled field and lab experiments, these authors set out to to determine whether dietary lipids may limit lake zooplankton populations. Their conclusion that protein and lipids each play a role in limiting zooplankton populations depending on the circumstances is an important one.

The planktonic material that cannot be used by the zooplankton ends up in the benthic community. What the benthos does with this material is the subject of Chapter 6, "Seasonal Dynamics of Lipids in Freshwater Benthic Invertebrates," by Joann Cavaletto and Wayne Gardner. This encyclopedic recounting of the lipid content and lipid class composition of benthic invertebrates ranges from the perpetually lean Tubificidae to the obese *Diporeia*. This chapter illustrates how seasonal lipid values and lipid class analyses provide powerful means for evaluating the physiological health of invertebrates and highlights the variety of lipid storage strategies.

Just as lipids play a critical role in the health of the benthic and planktonic invertebrates, they are also critical in the metabolism of fishes. Marshall Adams, in Chapter 7, "Ecological Role of Lipids in the Health and Success of Fish Populations," illustrates the absolute requirements of fish for adequate supplies for lipids in overwintering, reproductive performance, and fry survival.

The critical role of lipids in fry survival is the driving force behind the lipid research reported in Yngvar Olsen's chapter, "Lipids and Essential Fatty Acids in Aquatic Food Webs: What Can Freshwater Ecologists Learn from Mariculture?" This chapter illustrates the tremendous work that has been done by mariculturists to understand the fatty acid requirements of many economically important fish species and how it may be applicable to freshwater work. The author manages to carry the story of essential fatty acids requirements of zooplankton and fish fry through the aquatic food web to humans.

Lipophilic contaminants closely follow the movement of lipids along the food web. Peter Landrum and Susan Fisher detail this process in their chapter, "Influence of Lipids on the Bioaccumulation and Trophic Transfer of Organic Contaminants in Aquatic Organisms." In this chapter, they explain from first principles how the lipids influence contaminant transfer at all stages of bioaccumulation.

The essential contradiction of this book on aquatic lipids is that lipids by definition do not dissolve in water. The mutual insolubility gives rise to much of the water-surface microlayer and the foam, the subject of Guillermo Napolitano and Daniel Cicerone's contribution entitled "Lipids in Water-Surface Microlayers and Foams." This interface forms an essential link between the worlds above and below the surface of the water, and the chemical processes that go on there have a large impact on the composition and abundance of chemicals that enter the water. The complex chemistry of the interface itself poses unique challenges to the organisms that reside or pass through it.

The remarkably diverse physical nature and physiological role of lipids are the topic of the final chapter, "Comparison of Lipids in Marine and Freshwater Organisms." In this chapter, Robert Ackman draws on years of experience in lipid research and cites many diverse examples in which lipids play a part in the marine and freshwater ecosystems.

The editors and contributors of this book are indebted to the many other individuals who made this book possible. In particular, we extend our heartfelt thanks to our reviewers:

- Gunnel Ahlgren, Upsala University, Institute of Limnology, Upsala, Sweden
- William W. Christie, The Scottish Crop Research Institute, Dundee, Scotland
- Maureen H. Conte, Woods Hole Oceanographic Institution, Massachusetts, USA
- Les Dickson, National Hydrology Research Institute, Saskatoon, Saskatchewan, Canada
- Gerhard Kattner, Alfred Wegener Institute for Polar and Marine Research, Bremerhaven, Germany
- Christopher Langdon, Hatfield Marine Science Center, Newport, Oregon, USA
- David R.S. Lean, University of Ottawa, Ottawa, Canada
- Joseph B. Rasmussen, McGill University, Montreal, Canada
- Barbara Santer, Max Planck Institute for Limnology, Plön, Germany
- John R. Sargent, NERC Unit of Aquatic Biochemistry, University of Stirling, Stirling, UK
- Peter A. Thompson, CSIRO Marine Laboratory, Western Australia, Australia

1

Determination of Total Lipid, Lipid Classes, and Fatty Acids in Aquatic Samples

Christopher C. Parrish

1.1. Introduction

The hydrophobic nature of lipids provides a convenient means of separating them from other compounds in an aqueous sample matrix. Extraction in nonpolar solvents is universally employed and is the basis of the operational definition of lipids. This approach is used routinely in algal biosynthetic studies in which the fate of a radiolabeled precursor is followed into the lipid pool. By adding ^{14}C bicarbonate to a sample from the field (Wainman and Lean, 1992) or a culture (Rai, 1995) and then later extracting the sample with a water-immiscible organic solvent, the "lipid fraction of carbon fixation" (LFCF) can be determined (Wainman and Lean, 1992). By performing a chromatographic separation before counting, this procedure can be further refined to determine the subclasses in which the ^{14}C ends up (Smith and D'Souza, 1993). Subfractionation is important when a differentiation between allocation to lipid storage and membrane synthesis is required. The radiolabeling approach is convenient, sensitive, and not prone to contamination. However, many ecological studies are not amenable to this approach, and so chemical analysis of the constituents of lipid extracts has to be performed.

A lipid extract may contain as many as 16 different subclasses of both biogenic and anthropogenic origin (Parrish, 1988). Figure 1.1 shows some of the more important biogenic classes. Triacylglycerols and phospholipids are biochemically related, as they both possess a glycerol backbone to which two or three fatty acids are esterified, and they also share common precursors: diacylglycerols derived from phosphatidic acid. Triacylglycerols are a very important energy storage substance, whereas phospholipids are essential components of membranes. Sterols share with phospholipids a structural function in membranes, but in terms of polarity they are grouped with triacylglycerols in the neutral lipids. Phospholipids are grouped with the polar lipids including glycolipids. Glycolipids contain one or more molecules of a sugar and are found in bacteria, plants, and animals. Glycoglycerolipids of the form shown in Figure 1.1 are found predominantly in the plant kingdom in association with chloroplasts.

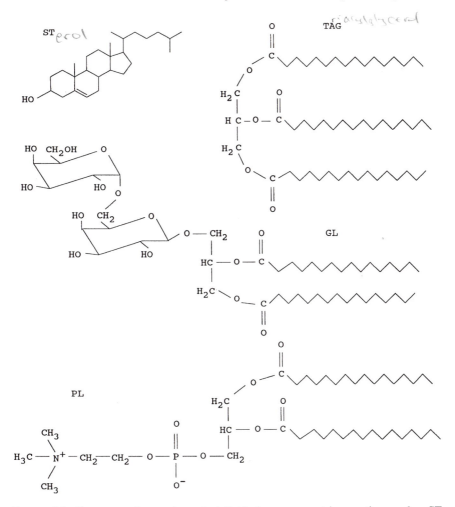

FIGURE 1.1. Structures of some important lipid classes present in aquatic samples. ST, sterol, represented by cholesterol; TAG, triacylglycerol, represented by tripalmitin; GL, glycolipid, represented by digalactosyl diacylglycerol; PL, phospholipid, represented by phosphatidyl choline.

Each class in a lipid extract can contain many compounds of similar polarity but sometimes with important structural differences. Most biogenic lipid classes contain the acyl group ($R-\overset{|}{C}=O$) because a fatty acid is incorporated into the molecule. Some fatty acids occur in the free form as shown in Figure 1.2, but most occur esterified in acyl lipids. These fatty acids can be released from the parent classes for analysis. In Figure 1.1, the triacylglycerol, glycolipid, and phospholipid are examples of acyl lipid classes, and if the fatty acids were cleaved from the glycerol backbone, they would all be the same: a 16-carbon fatty acid with no double bonds. This acid has the trivial name *palmitic acid* and the systematic

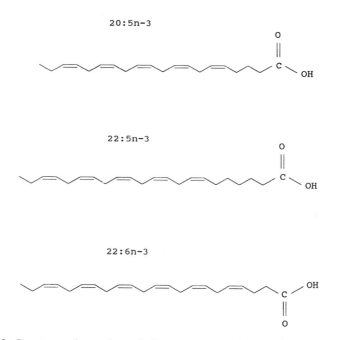

FIGURE 1.2. Structures of some long-chain polyunsaturated fatty acids. Eicosapentaenoic acid (20:5n-3), docosapentaenoic acid (22:5n-3), and docosahexaenoic acid (22:6n-3) are all related biochemically because of the location of the first double bond from the methyl end of the chain.

name n-*hexadecanoic acid;* the convenient shorthand notation for it is 16:0. This fatty acid is often one of the major fatty acids in lipid extracts, but most interest is focused on the polyunsaturated fatty acids with two or more double bonds. These fatty acids may be present in lower amounts but serve very important physical and metabolic functions in cells. Some of these fatty acids are essential to the normal functioning of the cell, and in the animal kingdom, these fatty acids or the correct precursors have to be acquired in the diet. Long-chain "ω3" and "ω6" fatty acids fall into this group. Figure 1.2 shows three related polyunsaturated fatty acids. These long-chain fatty acids are written as the ratio of carbon atoms to double bonds. The "n-3" is another way of indicating ω3, and it signifies the position of the first double bond from the methyl end of the molecule. Animals can elongate and desaturate fatty acids, but they are unable to place a double bond in the ω3 or ω6 position. Thus, freshwater fish have to be supplied with 18:2ω6 and 18:3ω3 in their diets (Henderson, 1996). The 18:2ω6 is converted to 20:4ω6, whereas the 18:3ω3 is converted to 20:5ω3 and 22:6ω3, with 22:5ω3 as an intermediate (Fig. 1.2).

Lipid classes and fatty acids have been determined in a wide variety of fresh-water studies. Recently, lipid classes have been determined in operationally defined dissolved and particulate matter (Yunker et al., 1994; Meyers and Eadie,

1993; Parrish et al., 1992a), algae (Dembitsky and Rozentsvet, 1996; Napolitano, 1994), sponges (Dembitsky et al., 1994a), crustaceans (Butler, 1994; Dembitsky et al., 1994b; Wainman et al., 1993; Arts et al., 1992), and fish (Verreth et al., 1994). In the past 2 years, there have also been studies of fatty acids in algae (Napolitano et al., 1995; Napolitano, 1994), sponges (Dembitsky et al., 1994a), crustacea (Dembitsky et al., 1994b; Rezanka and Dembitsky, 1994), and fish (Andrade et al., 1995; Fodor et al., 1995; Harrell and Woods, 1995; Rahman et al., 1995). Other solvent extractable compounds have also been investigated recently in freshwater samples including hydrocarbons (Yunker et al., 1994; Meyers and Eadie, 1993; Parrish et al., 1992a), alcohols (Yunker et al., 1994; Meyers and Eadie, 1993), and organochlorine contaminants (Renaud et al., 1995a,b). These studies suggest that there are nearly as many different ways of determining lipids in aquatic samples as there are laboratories that routinely perform these measurements. The purpose of this chapter is to indicate procedures that we have used successfully over the past several years for a large number and a wide variety of aquatic samples.

1.2. Results and Discussion

1.2.1. Sampling and Storage

As soon as a sample is taken, the lipid content and composition begin to change. These changes are catalyzed by enzymes, heat, oxygen, and light. Thus, samples should be kept as cold as possible and processed as soon as possible. Sampling of tissues or filtration of seston should be performed as soon as it can be done cleanly. To avoid contamination, it is important to avoid contact with any oils or grease at this stage. After processing, the tissue sample or filter should be immediately placed in a clean glass vial containing chloroform. At this point, it is important to avoid contact of the solvent with plastic or rubber to prevent contamination. Samples are handled with metal forceps or glass pipettes, and they must be stored in vials that have previously been washed with good-quality chloroform (e.g., high-performance liquid chromatography [HPLC] grade) and that have caps with Teflon liners. After immersing the sample in the chloroform, the airspace above the chloroform should be flushed with a clean, relatively unreactive gas (e.g., nitrogen). The vial is then capped, and Teflon tape is wrapped around the cap. Under these conditions, the sample can be stored in a freezer ($< -20°C$) for months without appreciable alteration in the lipid content or composition (Sasaki and Capuzzo, 1984).

After analysis, the integrity of the sample can be assessed from the free fatty acid content (Parrish, 1988) or the free fatty acid plus alcohol content (Parrish et al., 1995). Ideally, fresh samples should be analyzed for comparison with stored ones so that normal levels of these hydrolysis products can be determined. However, an interesting paper suggests that free fatty acid levels in marine diatoms should, in fact, be close to zero and that problems are encountered at the sample collection and extraction stage (Berge et al., 1995). They advocate the use of boiling water to deactivate lipolytic enzymes.

1.2.2. Lipid Extraction

Most laboratories use some variation of the Folch et al. (1957) extraction procedure involving chloroform and methanol (Fig. 1.3). The sample is homogenized in a 20-fold volume of chloroform:methanol (2:1) and then washed with water to remove nonlipid contaminants. Excessive washing should be avoided, and use of salt in the wash may influence the chromatographic properties of some lipid classes (Nelson, 1991). The Bligh and Dyer (1959) method is a very popular simplification of the method of Folch et al. (1957), which is to be recommended for large samples (>5 g). For very small freshwater samples (<20 mg), a microprocedure developed by Gardner et al. (1985) has been used successfully by other workers as well (Butler, 1994; Wainman et al., 1993; Arts and Evans, 1991). A variation on the Bligh and Dyer method that is currently popular for freshwater fish samples (Andrade et al., 1995; Rahman et al., 1995) is that of Kinsella et al. (1977). The modifications are mainly in terms of scale, phase separation, and sample preservation. Kinsella et al. (1977) published the first comprehensive analyses of lipid and fatty acid content of a variety of freshwater fish.

In most extraction procedures, the proportions of sample, water, and solvents are critical for maximal recovery of lipids and minimal contamination by nonlipids. In our version of the Folch et al. (1957) procedure (Fig. 1.3), it is important to extract the sample in a solvent-to-sample ratio of at least 3 ml per 150 mg dry weight of actual sample. The volume includes the chloroform and the methanol, and the weight does not include any filters or water associated with the filter or sample. It is also important that in the water washing procedure, the ratio of chloroform:methanol:water is 8:4:3. This amount of water includes that in the sample and filter if present (in our case, 0.5 ml) and is equivalent to a solvent-to-water ratio of 4:1. We typically put our samples on precombusted 4.25-cm Whatman GF/C glass microfiber filters for ease of handling. This also makes samples easier to grind with a rod.

There is growing concern over the potential health hazard associated with the use of chloroform, and so there are other solvent mixtures available (Ackman, 1993; Nelson, 1991; Shaikh, 1986). However, chloroform:methanol is still considered to be the most efficient extraction mixture for general use. Chen et al. (1981) have indicated that dicholormethane (methylene chloride) may be substituted for chloroform in the Folch et al. procedure. It is less toxic and more volatile, making it easier to use when solvent removal is required, but some lipids (e.g., tribehenin [1,2,3-tridocosanoylglycerol]) are less soluble in it.

1.2.3. Determination of Total Lipid

Total lipid can be determined gravimetrically, colorimetrically, or by summing individually determined lipid classes. In each case, the extract or a known portion

Place wet sample in centrifuge tube containing 2 ml of chloroform. Flush with nitrogen gas. At this point, samples can be left at $< -20°C$ for up to 1 year.

↓

Add 1 ml of ice-cold methanol.

↓

Grind the filter into a pulp quickly with a Teflon or metal-ended rod. Wash the rod with 1 ml of chloroform:methanol 2:1 and then with 0.5 ml of chloroform-extracted water. Recap the tube and sonicate the mixture in an ice bath for 4 minutes.

↓

If the sample is in a test tube that can be centrifuged, then it is centrifuged for 2–3 minutes at >1,000 rpm ($125 \times g$). If the test tube is too large to be centrifuged, then flush the sample with nitrogen, cap, and return to the freezer until the sample separates, usually overnight.

↓

Remove the organic layer (bottom layer) by using the double pipetting technique, which involves placing a long Pasteur pipette inside a short one. Remove all of the organic layer and transfer it into the prerinsed vials.

↓

Wash the pipette used to remove the organic layer into the vial containing the organic layer with 3×1 ml of ice-cold chloroform. Then wash the shorter pipette into the tube containing the aqueous layer with 3×1 ml of ice-cold chloroform.

↓

Sonicate and centrifuge the sample again and double pipette when separated, using new pipettes each time. Repeat at least three times, or until no color remains in the organic layer, and pool all organic layers. Between washes of the pulp in the tube, start to evaporate the solvent in the vial under nitrogen.

↓

Concentrate down to volume under a gentle stream of nitrogen.

↓

If the sample is not sufficiently ground up, the entire procedure may be repeated.

FIGURE 1.3. Lipid extraction procedure for aquatic samples of 10–150 mg dry weight, based on Folch et al. (1957). Seston samples and samples of small fauna are most easily handled on a precombusted (450°C) glass-fiber filter. One 10-ml glass centrifuge tube and one 15-ml glass vial each with a Teflon-lined cap are required for each extraction. All tubes and Teflon-lined caps should be rinsed three times with HPLC-grade methanol and then three times with HPLC-grade chloroform by sealing the solvent in the tube and shaking and then discarding.

of the extract is used, and precautions have to be taken against including in the determination any nonlipid material that may have become entrained. To avoid overestimating with gravimetry, care has to be taken to wash the extract properly, to evaporate all solvent from the extract, and to minimize any contact with oxygen, especially for highly unsaturated samples. Such samples may gain as much as 3.5% in mass due to adsorption of oxygen (Kaitaranta and Ke, 1981). Other precautions for gravimetric analyses are detailed in Wood (1991). Usually in ecological studies, only small samples are available, so that microprocedures such as those described by Gardner et al. (1985) are very useful.

Gravimetry is the most direct method for determining total lipids, provided there is sufficient material present. If a spectrophotometer is available, colorimetric methods can be more convenient and more sensitive (Parsons et al., 1989; Barnes and Blackstock, 1973; Marsh and Weinstein, 1966). These methods tend to rely on reactions with certain compounds or functional groups (e.g., carbon–carbon double bonds: C=C) within the sample matrix and are thus prone to biasing. In addition, the rate of reaction with those functional groups can vary according to the type of compound in which they are found (Ahlgren and Merino, 1991). Intercalibration with gravimetry for different sample types has been recommended (Barnes and Blackstock, 1973). As with gravimetry, colorimetric methods can also suffer from the possibility of inclusion of nonlipid compounds in the estimation. The only certain way of avoiding this is to perform a chromatographic separation. In the process of separating the lipids from nonlipid contaminants, additional information on the nature of the lipids in the sample is also usually obtained.

1.2.4. Determination of Lipid Classes

The heterogeneous nature of lipids means that much information can be gained by quantifying individual classes (Delbeke et al., 1995; Parrish, 1988). Although there are some very convenient colorimetric kits available for plasma samples (bioMerieux, France, or Boehringer Mannheim, Germany), these rarely are directly applicable to tissue or seston samples (Barnes and Blackstock, 1973). For these samples, the lipids usually have to be chromatographically subdivided into classes. For this purpose, thin-layer chromatography (TLC) or HPLC can be used.

In TLC, a lipid extract is spotted at the bottom of a glass plate or quartz rod that is covered in silica gel, and the lipids are developed with solvents of differing polarities. After separation on TLC plates, the lipid classes can be visualized and then scraped off for further analysis (Christie, 1989; Shaikh, 1986). However, quantification of the separated material directly on the silica gel is much more convenient and less prone to problems with contamination or recovery. Lipid class quantification can be performed directly on silica gel-coated rods (Parrish, 1987) or plates (Olsen and Henderson, 1989; Conte and Bishop, 1988); however, we prefer to use rod TLC (Fig. 1.4) because in the Chromarod-Iatroscan system there is a partial scanning facility that permits extensive analysis of a single sample on a single rod (Fig. 1.5). By separating out all the lipid classes in this way, one obtains

Sequence Leading to the First Chromatogram (HC to KET)

a. Blank scan the rods three times in the Iatroscan.
b. Apply samples and standards with an Autospotter or a Hamilton syringe fitted with a repeating dispenser.
c. Focus twice (three times if sample is very concentrated) in acetone to produce a narrow band of lipid material near the lower end of the rods.
d. Dry and condition in a constant-humidity chamber (over saturated $CaCl_2$) for 5 minutes.
e. Develop twice in hexane:diethyl ether:formic acid (98.95:1:0.05). The first development is for 25 minutes; the rods are dried in the constant humidity chamber for 5 minutes and redeveloped for 20 minutes.
f. Dry for 5 minutes in the Iatroscan.
g. Scan to the lowest point behind the KET peak (pps 25 on the Iatroscan).

↓

Sequence Leading to the Second Chromatogram (TAG to DAG)

a. Condition for 5 minutes.
b. Develop for 40 minutes in hexane:diethyl ether:formic acid (79:20:1).
c. Dry and scan to lowest point behind the DAG peak (pps 11 on the Iatroscan).

↓

Sequence Leading to the Third Chromatogram (AMPL and PL)

a. Condition for 5 minutes.
b. Develop twice for 15 minutes in 100% acetone.
c. Condition for 5 minutes.
d. Develop twice for 10 minutes in chloroform:methanol:chloroform extracted water (5:4:1).
e. Dry and scan entire length of rods.

FIGURE 1.4. Developing and conditioning sequences used routinely for the separation of aquatic lipid classes on Chromarods, based on Parrish (1987). Lipid class abbreviations are explained in the legend to Figure 1.5.

much greater confidence in the identities of individual peaks and any nonlipid material remains at the origin. For freshwater samples, the sum of the Iatroscan-measured lipid classes has been found to be 84–87% of the gravimetric lipid weight (Vanderploeg et al., 1992; Parrish, 1987). Gravimetric values tend to be higher, probably because the Iatroscan will determine only nonvolatile lipids and because there is always the possibility of the inclusion of undetected nonlipid material in gravimetric determinations.

Occasionally, peak splitting may be observed in the wax ester, triacylglycerol, or free fatty acid region (Fig. 1.5) of the chromatograms. This is due to the presence of lipid molecular species with widely differing degrees of unsaturation (Parrish et al., 1992b). The second part of a split triacylglycerol peak could be

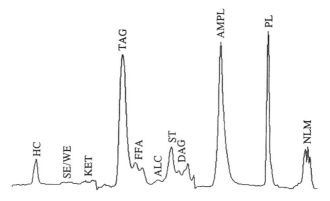

FIGURE 1.5. TLC/FID chromatogram of a net-tow sample (20-μm mesh) from a stream on Random Island running into Trinity Bay, Newfoundland. The sample was taken in May, and it consisted mainly of the pennate diatoms *Asterionella* sp., *Fragilaria* sp., and *Tabellaria* sp. Iatroscan conditions: air flow: 2.0 L/min, H_2: 190 ml/min, Chromarod developments as in Figure 1.4. The chromatogram is a composite of three separate scans. Peak identities: HC, hydrocarbon; SE/WE, steryl and wax esters; KET, ketone; TAG, triacylglycerol; FFA, free fatty acid; ALC, alcohol; ST, sterol; DAG, diacylglycerol; AMPL, acetone-mobile polar lipids; PL, phospholipid; NLM, nonlipid material.

mistaken for a free fatty acid peak, or the first part of a split free fatty acid peak (Fig. 1.5) could be mistaken for a polyunsaturated triacylglycerol peak. Under these circumstances, the peak's identity should be verified with a second development system (Fig. 1.4) of hexane/diethyl ether/formic acid (79.9:20:0.1). This solvent system should probably not, however, be used routinely, as it is not good for separating sterol and diacylglycerol peaks. In hexane/diethyl ether/formic acid (79:20:1), the 1,2-isomer of diacylglycerol runs behind sterol, whereas the less common 1,3-isomer runs slightly ahead of sterol. In hexane/diethyl ether/formic acid (79.9:20:0.1), the 1,3-isomer runs with alcohol and the 1,2-isomer runs with sterol. A single 40-minute development in hexane/diethyl ether/formic acid (97:2:1) may be even more useful for looking specifically at free fatty acid and triacylglycerol peak splitting (Parrish et al., 1992b), but information on sterols and more polar lipid classes is lost with this approach.

Although the Chromarod-Iatroscan procedure is sensitive, with a detection limit of about 50 ng, individual animals from water samples sometimes have to be pooled to obtain sufficient material. In ecological studies, it can be important to assess individual variability in lipid storage. For copepods, this can be assessed using an optical-digital procedure (Arts and Evans, 1991). For small quantities of algal cells, neutral lipid can be determined using Nile Red (Cooksey et al., 1987).

For specific lipid components, especially separations of molecular species within classes, HPLC can be very useful (Ratnayake and Ackman, 1989). Until recently, HPLC has suffered from the lack of a sensitive universal detector (Ratnayake and Ackman, 1989), but with increasing refinement of the evaporative light-scattering detector and its application to complex lipid samples (Christie, 1996; Heinz, 1996), this may soon be no longer true.

1.2.5. *Determination of Fatty Acids and Carbon Number Profiles*

There are about 15–20 major fatty acids that can be readily recognized in most samples. They differ in chain length and degree of unsaturation; certain ones among them can be used as biomarkers of microorganisms in ecological studies (Ahlgren et al., 1992; Sargent et al., 1987), and others are studied for the essential nutritional role they play (Henderson, 1996).

Fatty acids are esterified in most biogenic lipid classes. They are present in those that are termed *acyl lipids* but not in others such as hydrocarbons, ketones, alcohols, and sterols. The fatty acids are released from the acyl lipid classes and are re-esterified to methyl esters to make them amenable to gas chromatographic (GC) analysis (Fig. 1.6). Boron trifluoride has been recommended as a catalyst for the formation of fatty acid methyl esters by the American Oil Chemists' Society (AOCS, 1989) in its official method, but other derivatization procedures are available (Liu, 1994; Christie, 1989). Problems can occur with BF_3 if abnormally high concentrations are used (Morrison and Smith, 1964).

To perform fatty acid analyses, a variety of GCs are available; however, we have generally used a Varian 3400 GC equipped with an autoinjector. Likewise, a variety of capillary columns is available. The AOCS official method for marine oils states that the column should be at least 25 m long, with a 0.20–0.35-mm internal diameter, and it should be made of flexible fused silica. The method recommends Carbowax-20M or an equivalent such as SUPELCOWAX-10 for the liquid phase coating the inner wall of the column; however, in my laboratory an Omegawax 320 column (30 m, 0.32 mm i.d., 0.25-μm film thickness; Supelco, Inc.) has generally been used. This column was introduced in 1990 specifically for use with official methods for polyunsaturated fatty acids. Recently, Supelco has introduced a slightly lower polarity column, SPB-PUFA, which may be useful for specific analyses or as a confirmational tool for analyses performed on other columns. Similarly, Hewlett-Packard has introduced a midpolarity-phase column (HP-225) that may be especially useful for rapid separation of hydroxy acids.

By using an Omegawax 320 column in our Varian GC, good analyses have been obtained (Fig. 1.7), provided special attention is paid to solvent plug size in the autoinjector and to oxygen in the carrier gas. For fatty acid analyses, the optimum plug size was found to be 0.8 μl, and the best carrier gas was hydrogen that had been passed through an oxygen trap (S. Budge, Memorial University of Newfoundland, personal communication). With the use of hydrogen as a carrier gas, extra attention has to be paid to leaks and to laboratory ventilation.

Standards are available from several sources for peak identification and quantitation. Polyunsaturated fatty acid mixtures from Supelco have been used for peak identification in my laboratory. A very useful reference to aid in this process is that of Ackman (1986). The areas under the peaks in the chromatograms should

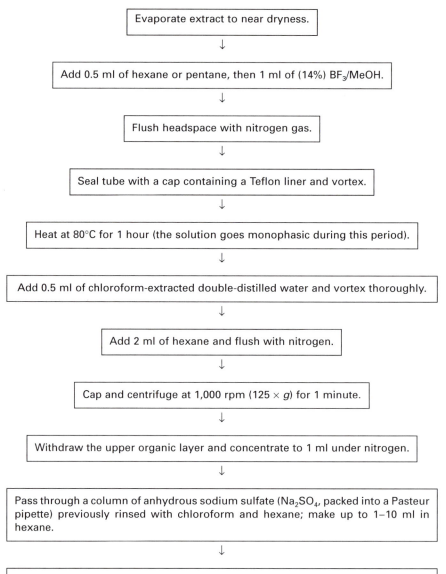

FIGURE 1.6. Fatty acid derivatization procedure for 2–20 mg of extract, based on Morrison and Smith (1964). For smaller amounts of extract, the volumes should be decreased proportionately.

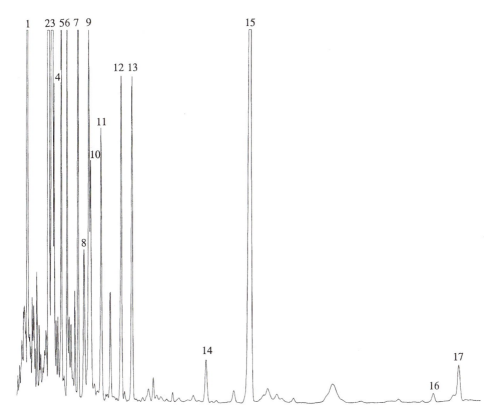

FIGURE 1.7. GC chromatogram of a net-tow sample (20-μm mesh) from a stream on Random Island running into Trinity Bay, Newfoundland. GC conditions: varian injector temperature: 250°C; column: Omegawax 320; column temperature program: 65°C for 30 sec, 40°C/min ramp to 195°C, hold for 15 min, 2°C/min ramp to 225°C, hold for 3.25 min; detector temperature: 260°C; gas flows: air 300 ml/min, H_2 (combustion) 32 ml/min, He (make-up) 18 ml/min, H_2 (carrier gas) 1.9 ml/min. Peak identities: (1) 14:0, (2) 16:0, (3) 16:1ω7, (4) 16:1ω5, (5) 16:2ω4, (6) 16:3ω4, (7) 16:4ω1, (8) 18:0, (9) 18:1ω9, (10) 18:1ω7, (11) 18:2ω6, (12) 18:3ω3, (13) 18:4ω3, (14) 20:4ω6, (15) 20:5ω3, (16) 22:5ω3, and (17) 22:6ω3.

be linearly proportional to the weight of material eluting from the column, provided GC conditions have been optimized according to the manufacturer's specifications and provided the column is not overloaded (Christie, 1989). It is important, however, to check that the GC is, in fact, operating correctly by using quantitative standards. Fatty acid methyl ester mixtures from Alltech Associates, Inc. (Deerfield, IL) can be used for this purpose. A standard with equal proportions of saturated fatty acids of different chain lengths (C_{13}–C_{21}) and another with equal proportions of fatty acids of the same chain length (C_{20}) but with varying degrees of unsaturation (20:0–20:4) are used in my laboratory. If the GC is operating properly, the peak areas of each component in each of these standards should be close in any one injection. Some of the components of these quantitative standards are also useful for identification purposes.

The reproducibility of repeated determinations of sample peak areas should be regularly checked. For major peaks (>5% of the total fatty acids), Christie (1989) indicates the coefficient of variation (CV) should be <10%. For replicate injections of the same extract, a CV <5% for the major peaks is routinely expected in my laboratory. However, it is important to replicate not just at the analysis step. Samples are routinely extracted in triplicate for every lipid determination in my laboratory and a CV <10% is anticipated. Ideally, replication should occur at the sampling stage in which significantly larger CVs are expected. CVs >20% have been obtained in my laboratory for settling particulate matter from different collection tubes on the same sediment trap frame deployed in the water column.

A new gas chromatographic method for profiling marine samples has recently been developed at Memorial University of Newfoundland (Yang et al., 1996). In terms of the information provided, this method is located between detailed fatty acid analyses and the class analyses provided by the Chromarod-Iatroscan system. The carbon number profiling method groups compounds according to carbon number within each class. By summing the groups of molecular species within each class, total class amounts are obtained. This method uses a short (5.5 m) nonpolar capillary column (DB-5: J. & W. Scientific, Folsom, CA) that is programmed to be heated through a wide temperature range (62–340°C). It has been successfully applied to neutral lipids in a wide range of marine samples (Yang et al., 1996) but has yet to be applied to polar lipids. However, this approach is used routinely for phospholipids in plasma samples after dephophorylation with phospholipase C (Myher and Kuksis, 1984).

1.3. Conclusion

There are probably nearly as many different variations on standard extraction (e.g., Wainman and Lean, 1994; Ackman, 1993; Nelson, 1991; Shaikh, 1986) and derivatization techniques (e.g., Liu, 1994) as there are laboratories that routinely perform these procedures. The purpose of this chapter has been to indicate procedures that have been used successfully over the past several years in my laboratory for both freshwater and marine samples. These procedures have been intercalibrated among themselves as well as with those used routinely in other laboratories. For example, Chromarod-Iatroscan analyses have been compared both with gravimetry (Parrish, 1987; Gardner et al., 1985) and with GC analyses (Yang et al., 1996; Parrish et al., 1995). Such intercalibration is to be recommended, especially for new users of any analytical approaches. A useful first step is to repeat the chosen extraction or derivatization procedure on the sample and then to compare the results of the original determination with those of the repeat procedure. If the repeat analysis yields no more lipid than is present in a total system blank, then one knows that one is at least close to the operational definition of the determination. With careful attention to methodology, practice, and a good knowledge of the literature in the chosen area of research, confidence in the determination of lipids in aquatic samples can soon be obtained.

Acknowledgments. I thank Sue Budge, Sharon Kennedy, and Jeanette Wells for extracting and analyzing the freshwater samples and for carefully reading the manuscript. I am especially grateful for the contribution that Sue Budge has made, in the course of her Ph.D. thesis work, toward the procedures described here for the determination of fatty acid methyl esters. I also thank Dr. C. H. McKenzie for algal identifications. Drs. M. T. Arts, W. W. Christie, and B. C. Wainman provided critical reviews of the manuscript. This chapter is OSC contribution 275.

References

Ackman, R.G. Extraction and analysis of omega-3 fatty acids: procedures and pitfalls. In: Drevon, C.A.; Baksaas, I.; Krokan, H.E., eds. Omega-3 Fatty Acids: Metabolism and Biological Effects. Basel: Birkhauser Verlag; 1993:p. 11–20.

Ackman, R.G. WCOT (capillary) gas-liquid chromatography. In: Hamilton, R.J.; Rossell, J.B., eds. Analysis of Oils and Fats. London: Elsevier; 1986:p. 137–206.

Ahlgren, G.; Gustafsson, I-B.; Boberg, M. Fatty acid content and chemical composition of freshwater microalgae. J. Phycol. 28:37–50; 1992.

Ahlgren, G.; Merino, L. Lipid analysis of freshwater microalgae: a method study. Arch. Hydrobiol. 121:295–306; 1991.

American Oil Chemists' Society. A.O.C.S. official method Ce 1b-89. In: Official Methods and Recommended Practices of the American Oil Chemists' Society. Champaign, IL: The Society; 1989.

Andrade, A.D.; Rubira, A.F.; Matsushita, M.; Souza, N.E. ω3 Fatty acids in freshwater fish from south Brazil. J. Am. Oil Chem. Soc. 72:1207–1210; 1995.

Arts, M.T.; Evans, M.S. Optical-digital measurements of energy reserves in calanoid copepods: intersegmental distribution and seasonal patterns. Limnol. Oceanogr. 36:289–298; 1991.

Arts, M.T.; Evans, M.S.; Robarts, R.D. Seasonal patterns of total and energy reserve lipids of dominant zooplankton crustaceans from a hyper-eutrophic lake. Oecologia 90:560–571; 1992.

Barnes, H.; Blackstock, J. Estimation of lipids in marine animals and tissues: detailed investigation of the sulphophosphovanillin method for 'total' lipids. J. Exp. Mar. Biol. Ecol. 12:103–118; 1973.

Berge, J-P.; Gouygou, J-P.; Dubacq, J-P.; Durand, P. Reassessment of lipid composition of the diatom *Skeletonema costatum*. Phytochemistry 39:1017–1021; 1995.

Bligh, E.G.; Dyer, W.J. A rapid method of total lipid extraction and purification. Can. J. Biochem. Physiol. 37:911–917; 1959.

Butler, N.M. Lipid storage in *Diaptomus kenai* (Copepoda; Calanoida): effects of inter- and intraspecific variation in food quality. Hydrobiologia 274:9–16; 1994.

Chen, I.S.; Shen, C-S. J.; Sheppard, A.J. Comparison of methylene chloride and chloroform for the extraction of fats from food products. J. Am. Oil Chem. Soc. 58:599–601; 1981.

Christie, W.W. Separation of phospholipid classes by high-performance liquid chromatography. In: Christie, W.W., ed. Advances in Lipid Methodology—Three. Dundee, UK: Oily Press; 1996:p. 77–107.

Christie, W.W. Gas Chromatography and Lipids. A Practical Guide. Ayr, UK: Oily Press; 1989.

Conte, M.H.; Bishop, J.K.B. Nanogram quantification of nonpolar lipid classes in environmental samples by high performance thin layer chromatography. Lipids 23:493–500; 1988.

Cooksey, K.E.; Guckert, J.B.; Williams, S.A.; Callis, P.R. Fluorometric determination of the neutral lipid content of microalgal cells using Nile Red. J. Microbiol. Methods 6:333–345; 1987.

Delbeke, K.; Teklemariam, T.; de la Cruz, E.; Sorgeloos, P. Reducing variability in pollution data: the use of lipid classes for normalization of pollution data in marine biota. Int. J. Environ. Anal. Chem. 58:147–162; 1995.

Dembitsky, V.M.; Rozentsvet, O.A. Distribution of polar lipids in some marine, brackish and freshwater green macrophytes. Phytochemistry 41:483–488; 1996.

Dembitsky, V.M.; Rezanka, T.; Kashin, A.G. Comparative study of the endemic freshwater fauna of Lake Baikal—VI. Unusual fatty acid and lipid composition of the endemic sponge *Lubomirskia baicalensis* and its amphipod crustacean parasite *Brandtia (Spinacanthus) parasitica*. Comp. Biochem. Physiol. 109B:415–426; 1994a.

Dembitsky, V.M.; Kashin, A.G.; Rezanka, T. Comparative study of the endemic freshwater fauna of Lake Baikal—V. Phospholipid and fatty acid composition of the deep-water amphipod crustacean *Acanthogammarus (Brachyuropus) grewingkii*. Comp. Biochem. Physiol. 108B:443–448; 1994b.

Fodor, E.; Jones, R.H.; Buba, C.; Kitajka, K.; Dey, I.; Farkas, T. Molecular architecture and biophysical properties of phospholipids during thermal adaptation in fish: an experimental and model study. Lipids 30:1119–1126; 1995.

Folch, J.; Lees, M.; Sloane Stanley, G.H. A simple method for the isolation and purification of total lipids from animal tissues. J. Biol. Chem. 226:497–509; 1957.

Gardner, W.S.; Frez, W.A.; Cichocki, E.A.; Parrish, C.C. Micromethod for lipids in aquatic invertebrates. Limnol. Oceanogr. 30:1099–1105; 1985.

Harrell, R.M.; Woods, L.C., III. Comparative fatty acid composition of eggs from domesticated and wild striped bass (*Morone saxatilis*). Aquaculture 133:225–233; 1995.

Heinz, E. Plant glycolipids: structure, isolation and analysis. In: Christie, W.W., ed. Advances in Lipid Methodology—Three. Dundee, UK: Oily Press; 1996:p. 211–332.

Henderson, R.J. Fatty acid metabolism in freshwater fish with particular reference to polyunsaturated fatty acids. Arch. Anim. Nutr. 49:5–22; 1996.

Kaitaranta, J.K.; Ke, P.J. TLC-FID assessment of lipid oxidation as applied to fish lipids rich in triglycerides. J. Am. Oil Chem. Soc. 58:710–713; 1981.

Kinsella, J.E.; Shimp, J.L.; Mai, J.; Weihrauch, J. Fatty acid content and composition of freshwater finfish. J. Am. Oil Chem. Soc. 54:424–429; 1977.

Liu, K-S. Preparation of fatty acid methyl esters for gas-chromatographic analysis of lipids in biological materials. J. Am. Oil Chem. Soc. 71:1179–1187; 1994.

Marsh, J.B.; Weinstein, D.B. Simple charring method for determination of lipids. J. Lipid Res. 7:574–576; 1966.

Meyers, P.A.; Eadie, B.J. Sources, degradation and recycling of organic matter associated with sinking particles in Lake Michigan. Organic Geochem. 20:47–56; 1993.

Morrison, W.R.; Smith, L.M. Preparation of fatty acid methyl esters and dimethylacetals from lipids with boron fluoride-methanol. J. Lipid Res. 5:600–608; 1964.

Myher, J.J.; Kuksis, A. Determination of plasma total lipid profiles by capillary gas-liquid chromatography. J. Biochem. Biophys. Methods 10:13–23; 1984.

Napolitano, G.E. The relationship of lipids with light and chlorophyll measurements in freshwater algae and periphyton. J. Phycol. 30:943–950; 1994.

Napolitano, G.E.; Heras, H.; Stewart, A.J. Fatty acid composition of freshwater phytoplankton during a red tide event. Biochem. System. Ecol. 23:65–69; 1995.

Nelson, G.J. Isolation and purification of lipids from biological matrices. In: Perkins, E.G., ed. Analysis of Fats, Oils and Lipoproteins. Champaign, IL: American Oil Chemists' Society; 1991:p. 22–59.

Olsen, R.E.; Henderson, R.J. The rapid analysis of neutral and polar marine lipids using double-development HPTLC and scanning densitometry. J. Exp. Mar. Biol. Ecol. 129:189–197; 1989.

Parrish, C.C. Dissolved and particulate marine lipid classes: a review. Mar. Chem. 23:17–40; 1988.

Parrish, C.C. Separation of aquatic lipid classes by Chromarod thin-layer chromatography with measurement by Iatroscan flame ionization detection. Can. J. Fish. Aquat. Sci. 44:722–731; 1987.

Parrish, C.C.; McKenzie, C.H.; MacDonald, B.A.; Hatfield, E.A. Seasonal studies of seston lipids in relation to microplankton species composition and scallop growth in South Broad Cove, Newfoundland. Mar. Ecol. Prog. Ser. 129:151–164; 1995.

Parrish, C.C.; Eadie, B.J.; Gardner, W.S.; Cavaletto, J.F. Lipid class and alkane distribution in settling particles of the upper Laurentian Great Lakes. Organic Geochem. 18:33–40; 1992a.

Parrish, C.C.; Bodennec, G.; Gentien, P. Separation of polyunsaturated and saturated lipids from marine phytoplankton on silica gel coated Chromarods. J. Chromatogr. 607:97–104; 1992b.

Parsons, T.R.; Maita, Y.; Lalli, C.M. A Manual of Chemical and Biological Methods for Seawater Analysis. Oxford: Pergamon Press, 1989.

Rahman, S.A.; Huah, T.S.; Hassan, O.; Daud, N.M. Fatty acid composition of some Malaysian freshwater fish. Food Chem. 54:45–49; 1995.

Rai, H. The influence of photon flux density (PFD) on short term ^{14}C incorporation into proteins, carbohydrates and lipids in freshwater algae. Hydrobiologia 308:51–59; 1995.

Ratnayake, W.M.N.; Ackman, R.G. Lipid analyses: part II. In: Vergroesen, A.J.; Crawford, M., eds. The Role of Fats in Human Nutrition. London: Academic Press; 1989:p. 515–565.

Renaud, C.B.; Kaiser, K.L.E.; Comba, M.E. Historical versus recent levels of organochlorine contaminants in lamprey larvae of the St. Lawrence River basin, Quebec. Can. J. Fish. Aquat. Sci. 52:268–275; 1995a.

Renaud, C.B.; Kaiser, K.L.E.; Comba, M.E.; Metcalfe-Smith, J.L. Comparison between lamprey ammocoetes and bivalve molluscs as biomonitors of organochlorine contaminants. Can. J. Fish. Aquat. Sci. 52:276–282; 1995b.

Rezanka, T.; Dembitsky, V.M. Identification of unusual cyclopropane monounsaturated fatty acids from the deep-water lake invertebrate *Acanthogammarus grewingkii*. Comp. Biochem. Physiol. 109B:407–413; 1994.

Sargent, J.R.; Parkes, R.J.; Mueller-Harvey, I.; Henderson, R.J. Lipid biomarkers in marine ecology. In: Sleigh, M.A. Microbes in the Sea. Chichester, UK: Ellis Horwood Ltd.; 1987:p. 119–138.

Sasaki, G.C.; Capuzzo, J.M. Degradation of *Artemia* lipids under storage. Comp. Biochem. Physiol. 78B:525–531; 1984.

Shaikh, N.A. Extraction, purification, and analysis of lipids from animal tissues. In:

Fozzard, H.A. et al., eds. The Heart and Cardiovascular System. New York: Raven Press; 1986:p. 289–302.

Smith, R.E.H.; D'Souza, F.M.L. Macromolecular labeling patterns and inorganic nutrient limitation of a North Atlantic spring bloom. Mar. Ecol. Prog. Ser. 92:111-118; 1993.

Vanderploeg, H.A.; Gardner, W.S.; Parrish, C.C.; Liebig, J.L.; Cavaletto, J.F. Lipids and life-cycle strategy of a hypolimnetic copepod in Lake Michigan. Limnol. Oceanogr. 37:413–424; 1992.

Verreth, J.; Custers, G.; Melger, W. The metabolism of neutral and polar lipids in eleuthero-embryos and starving larvae of the African catfish *Clarias gariepinus*. J. Fish. Biol. 45:961–971; 1994.

Wainman, B.C.; Lean, D.R.S. Methodological concerns in measuring the lipid fraction of carbon fixation. Hydrobiologia 273:111–120; 1994.

Wainman, B.C.; Lean, D.R.S. Carbon fixation into lipid in small freshwater lakes. Limnol. Oceanogr. 37:956–965; 1992.

Wainman, B.C.; McQueen, D.J.; Lean, D.R.S. Seasonal trends in zooplankton lipid concentration and class in freshwater lakes. J. Plankton Res. 15:1319–1332; 1993.

Wood, R. Sample preparation, derivatization and analysis. In: Perkins, E.G., ed. Analysis of Fats, Oils and Lipoproteins. Champaign, IL: American Oil Chemists' Society; 1991:p. 236–269.

Yang, Z.; Parrish, C.C., Helleur, R.J. Automated gas chromatographic method for neutral lipid carbon number profiles in marine samples. J. Chromatogr. Sci. 34:556–568; 1996.

Yunker, M.B.; Macdonald, R.W.; Whitehouse, B.G. Phase associations and lipid distributions in the seasonally ice-covered Arctic estuary of the Mackenzie Shelf. Organic Geochem. 22:651–669; 1994.

2
Fatty Acids as Trophic and Chemical Markers in Freshwater Ecosystems

Guillermo E. Napolitano

2.1. Introduction

The elucidation of trophic relationships and the identification of sources and sinks of organic matter are important steps for understanding the dynamics of aquatic ecosystems (Pimm et al., 1991). The trophic relationships between aquatic organisms can be investigated in a number of ways, from the inspection of gut contents to the use of biochemical, immunological (Grisley and Boyle, 1985), and stable isotope analyses (Peterson and Fry, 1987). Some lipid species (fatty acids, fatty alcohols, hydrocarbons, and sterols) are limited to certain taxa, so if the lipid in question is metabolically stable (or retains its basic structure after consumption), it may be used to trace energy transfers through the food chain, thus helping to define predator–prey relationships.

Certain lipid markers can also indicate sources and sinks of discrete pools of reduced carbon in different compartments of the ecosystem, including the dissolved and particulate organic matter in the water column and sediments. Lipids and, in particular, fatty acids have been extensively used in marine and lacustrine biogeochemistry (Meyers and Ishiwatari, 1993; Saliot et al., 1991).

Methods for fatty acid extraction, quantification, and identification require substantial training and expertise (see Parrish, this volume), but they are often simpler than the methods required for the detailed analyses of other biomolecules. Furthermore, the study of lipids of aquatic organisms (in particular, marine fish oils) was a challenge and, at the same time, a catalyst for the rapid development and improvement of capillary gas chromatography (GC), the method of choice for the analysis of complex mixtures of fatty acids. Fatty acids in aquatic organisms typically have a chain length of between 12 and 24 carbon atoms and a variable number of ethylenic bonds. Fatty acids rarely occur in a free form in organisms; instead they are esterified to glycerol and other alcohols (see Parrish, this volume). Their content in different aquatic species varies greatly but normally accounts for about 2–15% of the dry weight (Lechevalier and Lechevalier, 1988).

Early studies of the composition and origin of lipids in marine phytoplankton (Chuecas and Riley, 1969; Ackman et al., 1968), zooplankton (Bottino, 1974), and fish (Ackman and Burgher, 1964) provided the necessary information on the occurrence and distribution of lipids in organisms and prompted similar studies in fresh water. Some early analyses, however, are not very reliable due to loss of polyunsaturated fatty acids (PUFA) and incomplete compound separations. This chapter examines the use of fatty acids as biochemical markers of trophic relation- ships between species and as indicators of sources and sinks of organic matter in freshwater ecosystems. I will only consider the occurrence of fatty acids in organ- isms and in the water column. For a discussion of fatty acids in sediments, readers are referred to the reviews by Bourbonniere and Meyers (1996), Meyers and Ishiwatari (1993), Parkes (1987), and Barnes and Barnes (1978).

2.2. Nomenclature

The shorthand nomenclature for fatty acids used here is of the form $18:2\omega6$, in which "18" designates the total number of carbon atoms, "2" the number of *cis* double bonds, and "$\omega6$" the position of the first double bond (the "6" being counted from the methyl end of the molecule). Thus, $18:2\omega6$ is equivalent to $\Delta9,12-18:2$ and $18:3\omega3$ is equivalent to $\Delta9,12,15-18:3$. Double bonds in a PUFA are separated by a -CH_2 group (methylene interrupted), unless otherwise stated. The prefixes *iso* and *anteiso* refer to a fatty acid with a single methyl branch located two or three carbons away from the terminal methyl group, respectively, whereas *br* refers to a methyl branch in an unspecified location of the molecule.

2.3. Characteristics of Fatty Acid Markers for Trophic Studies

The ideal fatty acid marker is uncommon in nature and yet can be quantitatively extracted from biological and environmental samples and analyzed with appropri- ate sensitivity. Because lipid metabolism and storage in animals are organ- specific, fatty acid markers should be extracted from individual tissues or body parts. For example, the fatty acids in triacylglycerols of adipose tissue of animals are particularly useful as trophic markers because they represent dietary fatty acids (Henderson and Sargent, 1981; Leger et al., 1981), whereas the liver ac- counts for most of the fatty acid synthesis by elongation and desaturation.

Another important aspect of fatty acid trophic markers is that the fatty acid composition of an animal represents the time-integrated dietary intake. The quantitative aspects of the integrative properties of dietary lipids are not well studied and may be species-specific. Bourdier and Amblard (1989) addressed this problem by performing feeding experiments with previously starved calanoid copepods (*Acanthodiaptomus denticornis*). Their results indicated that the rate of restoration of the lipid reserves depended on the algal species considered and that no noticeable recovery was observed before the second day of feeding. They also observed that the phospholipid fatty acid profile varied little between the starva- tion and feeding periods, but the fatty acids in storage lipids (e.g., $16:1\omega7$ and

18:1ω7) reflected those of the algal food after 20 days of feeding. Examinations of the gut content of an animal provide information of the last ingesta; lipid analyses integrate the processes of feeding, absorption, and deposition of energy reserves over a much longer period of time (see Arts, this volume).

2.4. Primary Sources and Trophic Transfer of Fatty Acids

2.4.1. Fatty Acid Composition of Algae and Cyanobacteria

Algae are at the base of the trophic ladder of aquatic ecosystems, providing energy and essential nutrients for primary consumers. The major acyl lipid classes in algae are phospholipids (e.g., phosphatidylcholine, phosphatidylethanolamine, phosphatidylserine, phosphatidylglycerol, phoasphatidylinositol), glycolipids (monogalactosyldiglyceride, digalactosyl glycerol, sulfolipids), triacylglycerols, sterol esters, and free fatty acids (Dembitsky and Rozentsvet, 1996; Douce et al., 1990; Pohl and Zurheide, 1982; Wood, 1974). Phospholipids are structural constituents of cellular membranes, whereas glycolipids are major components of the thylakoid membrane in chloroplasts (Douce et al., 1990). The triacylglycerols of algae are intracellular storage materials and can occur as clearly visible oil droplets (Vechtel et al., 1992; Wood, 1988). Sterol esters are normally minor lipid constituents in plants, and their cellular function is not clear (Petkov and Furnadzieva, 1993).

Fatty acids rarely occur in the nonesterified form (free fatty acids) at concentrations of more than a few percentages of the total lipids of intact or well-preserved plant cells (Parrish et al., 1991); large concentrations of free fatty acids are almost certainly related to enzymatic autolysis of lipids, primarily of phospholipids (Galliard, 1980). The recurrent reports on relatively high levels of free fatty acids in plant samples are likely to be the result of the activity of powerful lipases. Moreover, certain plant lipases can maintain their activity at very low temperatures and even in the presence of some of the organic solvents used in lipid extraction procedures (Parrish et al., 1991; Christie, 1982).

Analysis of fatty acids in natural populations of freshwater algae are not common and are limited to cyanobacteria (Kenyon, 1972; Kenyon et al., 1972) and several species of green algae and flagellates (including dinoflagellates and chrysophytes) (Cranwell et al., 1990, 1988). Although diatoms are among the major primary producers in freshwater ecosystems, their fatty acid compositions have not been adequately studied. The fatty acid composition of natural and laboratory algal populations can change dramatically as a function of environmental factors such as temperature, nutrient concentrations, and light (see Wainman et al., this volume), and these variations are the consequence of a number of interdependent factors affecting the cell cycle, lipid class composition, or membrane fluidity. Despite these adaptations to changing physiological and environmental conditions, there is a clear relationship between algal fatty acid composition and taxa status. Table 2.1 presents examples of the fatty acid com-

TABLE 2.1. Fatty acid composition (% of total fatty acids) of cultured and natural populations of freshwater algae.[a]

	I	II	III	IV	V	VI	VII	VIII	IX
14:0	2.1	3.6	1.4	0.5	3	5	6.9	2.8	4.8
16:0	18.5	17.3	12.7	39.3	16	10	28.8	23.0	10.3
16:1ω7	22.6	15.1	—	1.3	31	9	1.6	16.8	13.5
C16 PUFA	—	—	—	0.5	17	1	—	12.4	11.6
18:0	1.6	2.2	0.6	0.3	1	8	0.6	0.8	1.7
18:1ω9	1.9	3.5	11.6	12.9	2	16	29.0	2.6	1.6
18:1ω7	0.5	1.0	0.4	1.1	—	—	—	1.9	0.3
18:2ω6	11.4	4.3	12.9	10.7	1	26	0.5	4.5	0.5
18:3ω6	—	0.5	1.3	0.7	—	—	0.1	0.7	1.8
18:3ω3	24.6	23.2	22.2	12.2	1	5	0.2	20.6	2.8
18:4ω3	—	1.2	3.3	—	—	—	5.2	—	2.4
20:0	—	—	0.2	0.3	—	—	—	—	0.1
20:1ω9	—	—	0.3	0.6	—	—	—	—	2.1
20:4ω6	—	—	2.8	—	—	—	2.0	1.8	
20:5ω3	—	2.6	—	2.3	26	—	7.8	7.3	6.7
22:5ω3	—	—	—	—	—	—	0.1	—	tr
22:6ω3	—	1.8	—	0.02	—	—	11.0	0.3	10.9
Reference[b]	1	1	1	2	3	4	1	2	5

[a]Cyanophyta: I, *Oscillatoria* sp. (cultured); II, *Oscillatoria agardhii*. Chorophyta: III, *Scenedesmus quadricauda* (cultured); IV, *Cladophora* sp. Bacillariophyceae: V, *Navicula pelliculosa* (cultured). Chryptophyceae: VI, *Chryptomonas ovata* (cultured). Dinophyceae: VII, *Peridinium cinctum;* VIII, stream periphyton; IX, lake phytoplankton. tr, Trace.
[b]References: 1, Ahlgren et al., 1990; 2, Napolitano et al., 1994; 3, Kates and Volkani, 1966; 4, Beach et al., 1970; 5, Napolitano et al., 1995.

positions of the total lipids of cultures and natural populations of the major divisions and classes of freshwater algae including Cyanophyta (cyanobacteria), Chlorophyta (green), Bacillariophyceae (diatoms), Dinophyceae (dinoflagellates), and Chrysophyceae. Although the analyses of algal cultures are of limited use in ecological studies, this information can be used to demonstrate the biochemical potential and the chemical composition of algae under a set of environmental conditions. The fatty acid profiles shown in Table 2.1 are difficult to compare because they represent analyses performed on algae subjected to very different growing conditions and analyzed by different techniques over a period of many years. However, some generalizations on the occurrence of specific fatty acids or fatty acid groups are attempted in this section. Despite these limitations, it is apparent that the analyses of wild population of algae generally present a larger number of fatty acids and also present higher concentrations of some "atypical" fatty acids for a given taxon. For example, Table 2.1 shows that natural populations of *Oscillatoria* and *Cladophora* have conspicuous, or at least detectable, levels of some ω3 PUFA, such as 20:5ω3 and 22:6ω3, which are ordinarily absent in freshwater Cyanophyta and Chlorophyta, respectively. The occurrence of these "atypical" fatty acids in natural algal populations may be explained by the presence of detritus, animal remains, and a variable number of unidentified algae in the sample (including epiphytic species).

Saturated fatty acids (SAFA) 14:0, 16:0, and 18:0 are of little taxonomic value as biochemical markers because they are readily synthesized by most organisms and occur at various levels of concentrations in all algal groups (Table 2.1). Among the monounsaturated fatty acids (MUFA), 16:1ω7, 18:1ω7, and 18:1ω9 are potentially useful markers. Palmitoleic acid (16:1ω7) is the primary product of the Δ9 desaturase during the synthesis of unsaturated fatty acids in both plants and animals, explaining the wide distribution of this fatty acid in nature. Nevertheless, the distribution of 16:1ω7 among algae is far from homogeneous. Palmitoleic acid is a major, often prominent, constituent of the lipids of diatoms (Sicko-Goad, 1988; Wood, 1988, 1974) and some cyanobacteria (Ahlgren et al., 1992; Murata et al., 1992), sometimes exceeding concentrations of 30% of the total fatty acids. 16:1ω7 is much less abundant (>5% of the total fatty acids) in cultures and natural populations of green algae (Napolitano et al., 1994) and dinoflagellates (Ahlgren et al., 1992) (Table 2.1). Palmitoleic acid is a major fatty acid (15%) in the lipids of *Vaucheria* sp. (Xanthophyceae), an inhabitant of damp soils and stream banks (Nichols and Appleby, 1969).

The two major C18 MUFAs, 18:1ω7 (*cis*-vaccenic) and 18:1ω9 (oleic) acids, are potentially important biochemical markers. However, the lack of separation between these two fatty acids during earlier analyses performed on packed GC columns limits the use of the available information. Even today, when capillary columns provide complete separation of C18 MUFA, it is common to find reports with tables of fatty acid compositions with a single entry for "18:1."

Cis-vaccenic acid (18:1ω7) is a rather minor fatty acid in eucaryotic algae, normally accounting for 1–2% of the total fatty acids. Although lipids of marine cyanobacteria (e.g., *Synechococcus* sp.) may contain high proportions of 18:1ω7 (i.e., 21.1%) (Goodloe and Light, 1982), most recent analyses of freshwater cyanobacteria revealed a concentration of between 0.5–3.1% (Ahlgren et al., 1992, 1990). Murata et al. (1992) reported that 18:1ω7 plus 18:1ω9 made up 24% of the total fatty acids in the lipids of a filamentous strain of *Anabaena variabilis;* the authors, however, did not refer to the relative contribution of each isomer. Oleic acid (18:1ω9) is typically a major fatty acid in dinoflagellates and in some green algae (Ahlgren et al., 1992). Many analyses of natural algal populations frequently show high concentrations of 18:1ω9 in their lipids (Table 2.1). However, considering that 18:1ω9 may also be an important constituent of animal lipids and is relatively resistant to photodegradation (Armstrong et al., 1966), its presence in the particulate organic matter (POM) may also reflect the contribution of animal detritus.

Algae's PUFAs are important biochemical markers because they include some essential fatty acids that animals must assimilate from their diet. The interpretation of PUFA data reported in the literature must be exercised with caution, however, because identification is often based on GC retention data, without further derivatization or GC/mass spectrometry structural confirmation. C16

PUFA, including 16:2ω4, 16:2ω6, 16:3ω3, 16:3ω4, 16:4ω1, and 16:4ω3, can be abundant in the lipids of green algae and diatoms (Cranwell et al., 1990; Wood, 1974; Kates and Volcani, 1966). Linoleic acid (18:2ω6) is important in green algae (Napolitano et al., 1994; Nichols, 1965) and in a number of cyanobacteria, including species of *Spirulina, Nostoc, Synechococcus, Aphonocaspa, Oscillatoria, Lyngbia,* and *Anabaena* (Petkov and Furnadzieva, 1993; Kenyon, 1972; Holton et al., 1968; Nichols et al., 1968). Linoleic acid is a minor constituent of the lipids of diatoms and dinoflagellates. α-Linolenic acid (18:3ω3) is a major component in the lipids of most green algae and some cyanobacteria (Ahlgren et al., 1992) and also accounts for 40% of the total fatty acids of the freshwater heterotrophic Cryptophyceae (*Chylomonas paramecium*) (Beach et al., 1970). Two PUFA, 18:4ω3 and the unusual 18:5ω3, complete the series of potentially useful C18 PUFA markers. High proportions of 18:4ω3 can occur in certain strains of green and cyanobacteria. Nevertheless, this fatty acid is a major component of the lipids of most species of dinoflagellates examined so far (Ahlgren et al., 1992). The existence of 18:5ω3 was revealed less than two decades ago in the lipids of marine dinoflagellates (Joseph, 1977; Ackman et al., 1974), and its occurrence was later extended to other marine algal classes (Napolitano et al., 1988; Volkman et al., 1981). The distribution of 18:5ω3 in the freshwater environment seems to be much more restricted and unpredictable than in the ocean. Bourdier and Amblard (1988, 1987) reported 18:5ω3 among the major fatty acids in the lipids of phytoplankton during a bloom in Lake Pavin (up to 12% of the total fatty acids), when the dinoflagellate *Peridinium willei* dominated. However, fatty acids of a spring bloom dominated by the dinoflagellate *Peridinopsis penardii* showed that 18:5ω3 comprised only 0.01% of the total fatty acids (Napolitano et al., 1995).

C20 and C22 PUFA are typically absent in freshwater green algae (Pohl and Zurheide, 1982; Nichols, 1965) and cyanobacteria (Ahlgren et al., 1992; Murata et al., 1992; Ahlgren et al., 1990). Eicosapentaenoic acid (20:5ω3) is the prominent PUFA in virtually all species of diatoms so far studied (Napolitano et al., 1994; Sicko-Goad, 1988; Kates and Volcani, 1966). Docosahexaenoic acid (22:6ω3) is normally a minor fatty acid in diatoms (Napolitano et al., 1994) (Table 2.1) but can be a dominant constituent in the lipids of dinoflagellates (Napolitano et al., 1995) and some species of the small group of freshwater Chrysophyceae (Ahlgren et al., 1992).

Very-long-chain fatty acids (VLCF; carbon chain length between 24–30) are normally omitted in most analyses, assuming that there is nothing of importance beyond 24:1. VLCF may represent a substantial fraction of the total lipids in several green algae, including *Botryococcus braunii, Chlorella kessleri,* and *Euglena gracilis* (Rezanka, 1989), and they constitute a potential group of biochemical markers. Nevertheless, because terrestrial sources of VLCF may predominate in freshwater environments (see below), algal sources should be identified on the basis of characteristic fatty acid profiles and not by the presence of a few fatty acid components.

2.4.2. Fatty Acids as Trophic Markers of Algae and Cyanobacteria

Algae are the most important autochthonous source of dissolved and particulate organic matter in aquatic ecosystems. Table 2.2 summarizes the distribution of fatty acid markers in the major groups of freshwater algae. Two important conclusions can be drawn from the information presented in this table. First, no single fatty acid can be diagnostic of an algal group. Second, due to variations in the fatty acid compositions within phylogenetically close species and to the compounding effect of environmental factors, in most cases, fatty acid profiles can be used to distinguish algae at the class level only. Among the different algal groups, cyanobacteria are of particular biogeochemical importance because they thrive in environments that favor the preservation of organic matter, such as hypersaline and reducing waters (Parker et al., 1967). Together with certain bacteria, cyanobacteria represent a dominant portion of the flora in a number of extreme environments, such as hot springs and circumpolar and nutrient-depleted lakes. Fortunately, lipids from these two groups of procaryotes can be distinguished because cyanobacteria do not contain *iso-* and *anteiso*-fatty acids (Parker et al., 1967).

Several studies have focused on the importance of cyanobacteria and other microorganisms in the formation and trophodynamics of algal mats and similar structures in Antarctic lakes (Orcutt et al., 1986) and hot springs (Zeng et al., 1992a; Ward et al., 1985). Algal mats, such as those found in hot springs of the Yellowstone National Park (Wyoming), have a relatively complex trophic structure (Zeng et al., 1992a), in which the top layers are predominantly autotrophic (dominated by photosynthetic bacteria and cyanobacteria) and the bottom layers are heterotrophic. This transition in the trophic structure of the mat is reflected in its fatty acid composition. Lipids of the top layer of the microbial mat

TABLE 2.2. Potential fatty acid markers for freshwater algae.

	Cyanophyta	Chlorophyta	Bacillariophyceae	Cryptophyceae	Dinophyceae
10:0[a]	+				
16:1ω7	+		+		
C16 PUFA		+	+		
18:1ω9		+		+	+
18:1ω7	+				
18:2ω6	+	+		+	
18:3ω6	+				
18:3ω3	+	+		+	
18:4ω3					+
18:5ω3					+
20:5ω3			+		
22:6ω3					+

[a]Reported for a cultured marine species of *Trichodesmium* (Carpenter et al. 1997; Parker et al., 1967); freshwater congeners may be abundant but fatty acid data are not available.

are dominated by 16:0, 18:0, and 18:1, and their concentrations decrease with depth. This fatty acid profile coincides with that of *Synechococcus lividus,* a cyanobacteria isolated from the hot spring, which is thought to play an important role in the formation of the mat (Zeng et al., 1992a).

Fatty acid composition can also describe the stratification and the seasonal variations of different inputs of organic matter in lakes (Hama et al., 1992). The presence of large proportions of 22:6ω3 in the seston of Lake Vetchen indicated the dominance of dinoflagellates in the surface waters (Fredrickson et al., 1986). However, it was observed that the anoxic metalimnion of the lake supported a dramatically different community that included cyanobacteria (*Synechococcus-* like) and bacteria. The presence of these procaryotes was confirmed by a fatty acid profile enriched in 16:0, 16:1ω7, and particularly, 18:1ω7 and low concentrations of PUFA (Table 2.3).

Fatty acid compositions are also sensitive to the seasonal succession of phytoplankton species. For example, algal pigments and fatty acids in the particulate matter of Lake Kasumigaura varied seasonally, reflecting the taxonomic composition of the dominant phytoplankton species (Miyazaki, 1983). During the winter, when the diatoms (e.g., *Coscinodiscus* and *Cyclotella*) were abundant in the phytoplankton, the lipids in the POM contained high concentrations of 20:5ω3, a relatively high (~1.0) 16:1ω7/16:0 ratio and chlorophyll *c.* In spring and early summer, when green algae, (*Chlamydomonas*) dominated, the concentration of 18:3ω3 increased, the ratio 16:1ω7/16:0 decreased, and chlorophyll *b* increased.

The occurrence of 18:5ω3 in zooplankton and other grazers is especially interesting because this PUFA has a rather restricted distribution in the lipids of marine (Conte et al., 1994; Napolitano et al., 1988; Volkman et al., 1981; Joseph, 1977; Ackman et al., 1974) and freshwater algae (Napolitano et al., 1995; Ahlgren et al., 1992; Hama et al., 1992). Octadecapentaenoic acid (18:5ω3) has been undetected in many analyses of fatty acids of primary consumers, and there is little evidence for its effective transfer through the food chain. Laboratory experiments with marine bivalves have suggested that 18:5ω3 consumed by grazers is either used for energy or is rapidly elongated to C20 and C22 PUFA (Napolitano, unpublished

TABLE 2.3. Characteristic bacterial biomarkers.

Bacteria	Fatty acid marker	References
Gram-positive	*iso*15:0, *anteiso*15:0, 15:0, *anteiso* 17:0	Findlay and Dobbs, 1993b Komagata and Suzuki, 1987
Gram-negative	Cyclopropane, 2- and 3-hydroxy acids	Komagata and Suzuki, 1987
Sulfate-reducing	10Me16:0, *iso*17:0, *iso*17:1, cy17:0, 17:0, cy19:0; 16:1ω7; 18:1ω7	Findlay and Dobbs, 1993b Vainshtein et al., 1992
Acidophilic-thermophilic bacilli	ω-Cyclohexylundecanoic acid	Komagata and Suzuki, 1987
Methanotrophic bacteria	16:1ω6, Δ11-*trans*-16:1, 18:1ω8	
Thermophiles, mesophilic, psyhrophiles (e.g., *Clostridia*)	Cyclopropane 15:1	Chan et al., 1971

data). A report of the occurrence of 18:5ω3 in the lipids of the cladocera *Eurycercus lamellatus* and *Simocephalus vetulus* from a small pond (Desvilettes et al., 1994) was rather unexpected, and it may have originated in the lipids of *Gymnodinium* sp., a locally common phytoplankton species.

The fatty acid compositions and the relationship with the diet were studied in a large number of species of aquatic and semiaquatic insects from diverse habitats that use different feeding strategies (Hanson et al., 1985). Arachidonic acid (20:4ω6) and 20:5ω3 were among the major PUFA in most of the insects analyzed. The concentrations of 20:5ω3 in the aquatic insects were larger than those reported for terrestrial species (Dadd, 1983) and may represent an adaptation to aquatic life. The analysis of lipids from aquatic insects also indicates that the distribution of fatty acids markers of food sources differed predictably among functional feeding groups (Hanson et al., 1985). Filter-feeding species presented the highest concentrations of 20:5ω3 when compared with other feeding groups such as shredders, gatherers, and predators. The high concentration of 20:5ω3 in the filterers obviously reflects the relative importance of diatoms in their diet.

2.4.3. Fatty Acid Composition of Bacteria

A comprehensive review of the lipid and fatty acid composition of bacteria can be found in Ratledge and Wilkinson (1988). Bacterial fatty acids have been extensively studied, as evidenced by the large databases available on their lipid composition and chemotaxonomy (Boon et al., 1996; Lechevalier, 1982). Bacterial lipids and fatty acids have been used to assess biomass, taxonomic composition, and physiological status (White, 1988; White et al., 1979). Consequently, bacterial chemotaxonomy rapidly evolved as a very powerful tool with a wide range of applications in classic bacteriology and microbial ecology. These applications have created a demand for rapid and accurate methods for the identification of bacteria and the development of automated GC-computerized systems for the analysis of fatty acid composition of microorganisms (Haack et al., 1994; Landry, 1994; Stead et al., 1992; Moss, 1990). A detailed discussion of the occurrence and distribution of fatty acids in bacteria is beyond the scope of this chapter, so a summary of the principal fatty acids used as indicators of bacterial biomass will be presented.

Lipids of gram-positive bacteria are concentrated in the plasma membrane, whereas those of gram-negative bacteria have a greater complexity and consist of an arrangement of neutral lipids and lipoproteins, embedded in the polysacharides and proteins of the cytoplasm and the outer cellular envelope. Eubacteria can be distinguished from archaebacteria due to the presence of acid-stable ether lipids in the latest (Smith, 1988). There is a large structural variety in the fatty acids of bacteria; different types of fatty acids markers used in microbial ecology are presented in Table 2.3. Most bacterial fatty acids are either saturated or monounsaturated, including 16:0 and 16:1ω7. Oleic acid (18:1ω9) is present in bacteria, but *cis*-vaccenic acid (18:1ω7) is typically the dominant 18:1 isomer. The fatty acids that make bacteria unique, however, are the hydroxy, cyclopropane, odd, and branched-chain fatty acids (Lechevalier, 1982). Although 20:5ω3 and 22:6ω3 have been found in deep-sea bacteria (DeLong and Yayanos, 1986) and in the intestinal flora of certain fish (Ringo et al., 1992), PUFA and fatty acids with a chain length longer than 18 carbon atoms are typically absent in these procaryotes. These fatty acid structures contrast sharply with those of the eucaryotic algae, that typically contain large proportions of fatty acids with 20–22 atoms of carbons and three to six *cis* double bonds (Table 2.1).

2.4.4. Fatty Acids as Trophic Markers of Bacteria

Although there are a number of methods used for the qualitative and quantitative analysis of algal biomass (e.g., chlorophyll *a,* biovolume, dry mass), it is much more difficult to measure bacterial biomass in natural samples. A major obstacle to the determination of microbial biomass is that traditional enrichment and isolation techniques do not ensure a quantitative retrieval of natural bacterial populations (Findlay and Dobbs, 1993a,b). An additional problem in microbial ecology is that bacterial morphology is of little use in taxonomic determinations. Therefore, the use of specific biochemical markers acquires primary importance in the identification and quantification of bacteria (White, 1995; Tunlid and White, 1992). Microbial lipid markers have been broadly used in the biogeo-chemistry of marine (Saliot et al., 1991; Gillan and Johns, 1986) and lacustrine (Meyers and Ishiwatari, 1993) sediments, and more recently, these techniques have been applied to the microbiology of soils (Cavigelli et al., 1995). The number of studies of bacterial lipid markers in streams, rivers, and lake waters is much more limited (Findlay and Dobbs, 1993a).

Phospholipid fatty acids (PLFA) are considered an excellent surrogate for bacterial biomass and are an effective tool in bacterial taxonomy (White, 1995). A chemotaxonomic study using PLFAs of organism lipids in the euphotic zone in Ace Lake (Antarctic) presented clear evidence for a dramatic shift in the tax-onomic status of the dominant primary producers (Volkman et al., 1988). PLFA signatures in Ace Lake indicated that although eucaryotic phytoplankton domi-nated the surface waters (as evidenced by the high concentrations of 16:4ω3, 18:2ω6, 18:3ω3, and 18:4ω3), photosynthetic procaryotes (leading to high con-centrations of 14:0, 15:0, 16:0, 18:0, 16:1ω7, and 18:1ω9) dominated in deeper waters. These fatty acid profiles, together with the pigment analysis, suggested that the deepest layers of the euphotic zone of Ace Lake were populated by photosynthetic bacteria belonging to the genus *Chlorobium* (Kenyon, 1972). The results of a similar study performed in the same lake a few years later confirmed the clear differences in the microbial communities at each depth, although the presence of a different bacterial community was indicated (Mancuso et al., 1990). This later study corroborated that lipids in the POM of the upper layers of the euphotic zone were dominated by PUFA-producing microeucaryotes, but the lower layers presented high concentrations of 10-methyl-16:0 and *iso*17:1ω7, characteristic of the sulfur-reducing bacteria *Desulfobacter* and *Desulfovibrio,* respectively. A fatty acid ratio defined as (Σ *iso*15:0 + Σ *anteiso*15:0)/16:0 was also used to estimate the proportion of bacteria present in the POM at different depths of Ace Lake and indicated a near-tenfold increase in bacterial biomass from surface to the depth of 23 m (Mancuso et al., 1990). Comparable results were obtained in a study of lipid markers in POM of a coastal brackish pond near Cape Cod (Wakeham and Canuel, 1990). This study showed that fatty acids of photoeucaryotes (e.g., C16 PUFA, 18:3ω3, and 18:4ω3) were abundant in the oxic surface waters, whereas bacterial fatty acids (18:1ω7 and *anteiso*15:0) dominated the deepest anoxic zone and the water–sediment interface. These differences between surface and deeper waters were accentuated during the stratification of the pond in late summer, when anaerobic processes increased, favoring the production of bacterial fatty acids (especially 18:1ω7) in the metalimnion and hypolimnion.

Animal lipids may present large concentrations of PUFA that are derived from autotrophic sources such as 18:3ω3, 20:5ω3, and 22:6ω3 but may also contain minor amounts of branched and odd-numbered saturated fatty acids of bacterial origin. For example, the proportions of these branched and odd-numbered fatty acids in the triacylglycerols of zooplankton are good indicators of the relative importance of algae and bacterial biomass as alternative sources of food for filter feeders (Desvilettes et al., 1994). Considering that all animals normally host a population of microbes of one type or another in their digestive systems, the occurrence of bacterial fatty acids in their lipids should not be considered rare. Small quantities of bacterial fatty acid markers in animal fats and oils likely represent the degree of detail and the quality of a particular analysis. For example, the lipids of the mummichog fish (*Fundulus heteroclitus*) contain small proportions of cyclopropanoid (*cis*-9,10-methylenehexadecanoic and *cis*-9,10-methyl-eneoctadecanoic acids) and *iso, anteiso,* and branched fatty acids characteristic of bacteria (Cosper et al., 1984). Results of field and laboratory experiments indicated that the bacterial fatty acids found in the tissues of *F. heteroclitus* originated from both dietary sources and from the metabolism of commensal bacteria in the fish intestine (Cosper et al., 1984).

It has been reported that the lipids of beavers contained up to 3% of the total fatty acids as *trans*-11–18:1 and a conjugated diunsaturated *cis*-9, *trans*-11–18:2 (Käkelä and Hyvärinen, 1996; Käkelä et al., 1996). These fatty acids are not common constituents of the lipids of most vertebrates or of the beaver's diet (Scholz and Boon, 1993), but they are normally found in the adipose tissue of ruminants. The presence of substantial amounts of these bacterial fatty acids in the lipids of beavers (particularly in the digestive tract) suggested that they originated from microbes that hydrogenate dietary 18:2ω6 in the beaver's intestine (Christie, 1981).

2.4.5. Fatty Acid Markers from Allochthonous Sources

Freshwater environments contain organic matter derived from a number of autochthonous and allochthonous sources, including algae, vascular plants, and decaying plant and animal materials. Lipids from these different sources can have characteristic fatty acid signatures, which allows an assessment of their contribution to the organic matter pool and to higher levels of the aquatic food web. Tables 2.4 and 2.5 present examples of ecological studies that used fatty acid markers to trace trophic relationships and sources of organic matter, respectively.

Studies have shown that lipids from aquatic and terrestrial plants have different and diagnostic fatty acids signatures and can therefore be used to characterize the predominant energy source in aquatic systems. Although the dissolved and particulate lipids in the organic matter of lakes and other large bodies of water show typical "aquatic fatty acid" profiles, lipids of terrestrial origin gradually become more important with the proximity to land (Meyers et al., 1984). This transition from aquatic to terrestrial fatty acids is often reflected by the relative contribution of medium-chain (~C16) and longer-chain (~C26) fatty acids of phytoplankton and terrestrial plants, respectively. Accordingly, the C16/C26 fatty acid ratio in the lipids of the POM of lakes increases with distance from land, reaching maximum values close to the bottom of the euphotic zone and farther from potential sources of terrigenous materials (Meyers et al., 1984).

The classification of the different inputs of organic carbon to aquatic ecosystem can be complicated by the loss and transformation of the more labile components due to physicochemical (Armstrong et al., 1966) and biological processes (Meyers et al., 1984; Rhead et al., 1971). During these processes, PUFAs (e.g., $18:3\omega3$, $18:4\omega3$, $18:5\omega3$, $20:5\omega3$, $22:6\omega3$) are preferentially degraded, favoring the preservation of the less unsaturated and longer fatty acids of terrestrial plant origin (i.e., C20–C32 SAFAs and MUFAs) and ultimately altering the original fatty acid profile (Meyers et al., 1984). Other changes in the fatty acid composition of the organic matter of aquatic systems may be the result of atmospheric transport of particles and aerosols from distant sources. During precipitation, rainwater scavenges gases and aerosols that contain a vast array of organic materials of both biogenic (Kawamura and Kaplan, 1983) and anthropogenic origins. Fatty acids are an important fraction of the organic matter in rainwater, where they can exceed the concentrations of ubiquitous hydrocarbons (n-alkanes) by a factor of 10 (Kawamura and Kaplan, 1983). Fatty acids in rainwater of a coastal area predominantly consisted of straight-chain compounds of C12, C14, C16, and C18 of algal origins and minor amounts of C20–C30 compounds from terrestrial plants (Kawamura and Kaplan, 1983).

2.4.6. Fatty Acids as Trophic Markers in Vertebrates

Comparative studies of fatty acid compositions have clearly shown that the concentrations of $18:2\omega6$ and $20:4\omega6$ in the lipids of freshwater animals are higher than in their marine counterparts (Linko et al., 1992; Muje et al., 1989; Gonzalez-Baro and Pollero, 1988; Bell et al., 1986; Kaitaranta and Linko, 1984). These differences are related to mechanisms for the regulation of membrane fluidity and the diet (see Landrum and Fisher, this volume). Ackman and Takeuchi (1986) observed that in wild salmon (*Salmo salar*), the content of $20:4\omega6$ was six times higher than in hatchery fish, and they suggested that this deficit may have been in

TABLE 2.4. Fatty acids as biochemical markers of trophic relationships in freshwater ecosystems.

Lower trophic level	Upper trophic level	Fatty acid marker	Reference
Bacterioplankton	*Cryptomonas ovata, C. marsonii*	15:0, 17:0, *iso*17:0, *anteiso*17:0	Rieman, 1985; Desvilettes et al., 1994
Symbiotic cyanobacteria	*Azolla caroliniana* (aquatic fern)	12:1, 14:1ω5	Caudales et al., 1992
Symbiotic bacteria	*Castor canadensis* (beaver)	Δ11-*trans*18:1, *cis*-9, *trans*-11−18:2	Käkelä et al., 1996
Intestinal bacteria	*Fundulus heteroclitus* (Cyprinodont)	Cyclopropanoid 17:0, 19:0	Cosper et al., 1984
Diatoms	Aquatic insects (Diptera and Trichoptera)	20:5ω3	Hanson et al., 1985
Periphytic diatoms	Stoneroller minnows (*Campostoma anomalum*)	20:5ω3	Napolitano et al., 1996
Periphytic green algae	Stoneroller minnows (*Campostoma anomalum*)	18:3ω3	Napolitano et al., 1996
Heterozostera tasmanica (seagrass)	Zooplankton (isopodes) and sea garfish (*Hyporhamphus melanochir*)	18:1ω9, 18:2ω6, 18:3ω3, 16:3ω3, Δ3-*trans*-16:1	Nichols et al., 1986
Angiosperm detritus	*Macrobrachioum boorelli* (decapod)	18:1ω9, 18:2ω6	Gonzalez-Baro et al., 1988
Aquatic insects	Atlantic salmon (*Salmo salar*)	20:4ω6	Ackman and Takeuchi, 1986; Bell et al., 1994
Freshwater fish	Harbour seal (*Phoca vitulina*), ringed seal (*Phoca hispida*)	18:2ω6, 18:3ω3, 20:4ω6	Smith et al., 1996; Käkelä et al., 1993, 1995
Freshwater fish	River dolphin (*Inia geoffrensis*)	18:2ω6	Ackman et al., 1971

TABLE 2.5. Fatty acids as biochemical markers of sources of organic matter in freshwater ecosystems.

Source	Reservoir	Fatty acid marker	Reference
Bacterial assemblages	POC in Antarctic lake	ΣBranched fatty acids/16:0 ratio	Mancuso et al., 1990
Gram-negative bacteria	Lake sediments	Hydroxy fatty acids	Fukushima et al., 1992
Aerobic bacteria	Submerged biofilms	18:1ω7	Scholz and Boon, 1993
Sulfur-oxidizing bacteria	Lake seston	18:1ω7	Fredrickson et al., 1986
Sulfate-reducing bacteria	Wetland sediments, lake seston	iso17:0, br17:0, 17:1ω6	Boon et al., 1996; Fredrickson et al., 1986
Methanotrophic bacteria	Wetland sediments	16:1ω6, trans-16:1ω5, 18:1ω8	Boon et al., 1996
Anaerobic photosynthetic bacteria	Brackish pond POM	16:1ω7, 18:1ω7	Wakeham and Canuel, 1990
Thermus aquaticus, Bacillus stearothermophilus	Hot springs	iso17:0, iso15:0	Dobson et al., 1988; Ray et al., 1971
Cyanobacteria	Lake seston	18:1ω7	Fredrickson et al., 1986
Ceratium sp. (dinoflagellate)	Lake seston	22:6ω3	Fredrickson et al., 1986
Phytoplankton	Lake POM	C16 and C20 PUFA	Miyazaki et al., 1986; Mermoud et al., 1981
Green algae	Stream periphyton	18:3ω3	Napolitano et al., 1994; Steinman et al., 1988; McIntire et al., 1969
Diatoms	Stream periphyton	20:5ω3	Napolitano et al., 1994; Steinman et al., 1988; McIntire et al., 1969
Diatoms	Lake phytoplankton	16:1ω7/18:1ω9 ratio	Bourdier and Amblard, 1988
Prymnesiophytes	Brackish pond POM	18:4ω3, 18:5ω3, 20:5ω3, 22:6ω3	Wakeham and Canuel, 1990
Green algae		18:1ω9, 18:2, 18:3, 16:4	Wakeham and Canuel, 1990
	Antarctic lake POM	16:4, 18:2, 18:3, 18:4	Volkman et al., 1988
Chlorobium sp.	Antarctic lake POM	18:1ω9, 18:0, 14:0, 15:0	Volkman et al., 1988; Kenyon, 1972
Chlamydomonas sp.	Lake phytoplankton	18:1ω3	Miyazaki, 1983
Coscinodiscus sp. and Cyclotella sp.	Lake phytoplankton	20:5ω3	Miyazaki, 1983
Vascular plant detritus	Wetland sediments, lake POM	Long-chain fatty acids (C20–C32), low C16/C26 ratio	Boon et al., 1996; Kawamura et al., 1987; Meyers et al., 1984

part responsible for a common fin disease. It is believed that the high proportions of 20:4ω6 in wild salmon originates in the lipids of aquatic insects, which constitute an important portion of their natural food (Ackman and Takeuchi, 1986). The analysis of several species of freshwater invertebrates that are natural food of salmon indicated, however, that not only insects but also the Gammaridae and Oligochaeta can be a source of 18:2ω6 and 20:4ω6 for fish (Bell et al., 1994).

Growing environmental and conservationist concerns have stimulated the use of alternative methods applied to the ecology of aquatic and semiaquatic mammals. Many of these investigations have used fatty acid markers to determine food resources for populations of aquatic mammals (for a comparative discussion, see also Landrum and Fisher, this volume). A comparison of the fatty acids in the blubber of a population of harbor seals, *Phoca vitulina* from the Ungava Peninsula (a land-locked environment in northern Quebec) and their marine counterpart, showed that the inland animals have a fatty acid profile enriched in components of freshwater origin, such as 18:2ω6, 18:3ω3, and 20:4ω6; the blubber of the marine populations of *P. vitulina* contained higher concentrations of 20:1ω9 and 22:1ω11 (Smith et al., 1996). These MUFAs are characteristic components of the wax esters of marine calanoid copepods, which are eaten by zooplanktivorus fish and then are transferred to higher trophic levels of the food chain (Clarke et al., 1987, Sargent, 1976).

The contribution of marine versus freshwater fatty acids food sources can be assessed by its ω3/ω6 fatty acid ratio in the predator tissues. A typical ω3/ω6 ratio for freshwater fish ranges from 0.5 to 3.8, whereas for marine fish the ratio varies from 4.7 to 14.4 (Henderson and Tocher, 1987). The ω3/ω6 fatty acid ratio in the blubber of "marine" and "freshwater" seals are >6.3 and 2.0, respectively (Smith et al., 1996). The fatty acids in the blubber of the river dolphin, *Inia geoffrensis* (an entirely freshwater species), have a ω3/ω6 ratio of 0.27 (Ackman et al., 1971). Comparable results on individual fatty acid components and ω3/ω6 fatty acid ratios have been obtained to distinguish marine from lake populations of the ringed seal, *Phoca hispida,* from Finnish waters (Käkelä, 1996; Käkelä et al., 1995; Käkelä et al., 1993).

Endogenous and dietary fatty acids play an important role in maintaining the optimal degree of membrane fluidity in the peripheral tissues of small homothermous animals exposed to near-zero water temperatures. The anatomical distribution and the fatty acid composition of adipose tissue of aquatic and semiaquatic rodents from high-latitude environments present a valuable opportunity to study the importance of dietary fatty acids in cold adaptation. Beavers (*Castor canadensis* and *C. fiber*) and the muskrats (*Onidatra zibethicus*) accumulate large amounts of dietary 18:3ω3 (up to 18% and 20% of the total fatty acids, respectively) in peripheral adipose tissues of the tail and feet (Käkelä and Hyvärinen, 1996). The concentrations of 18:3ω3 in the core tissues such as the mesenteries, the perirenal tissues, and the liver are much lower (<1% of the total fatty acids). Linolenic acid (18:3ω3) is an essential fatty acid for many animals, and beavers obtain it from green plants, bark, and wood (Peng, 1992; Dunlop-Johns et al., 1991). The lipids of cold water animals also contain substantial amounts of C20 and C22 ω3 PUFA, arising from dietary sources or derived via elongation and desaturation of 18:3ω3. A significant dietary effect on the fatty acid composition of beavers and muskrats is that the liver of beavers, which are exclusively herbivorous, contained very low concentrations (~0.2% of the total fatty acids) of 22:6ω3, a typical component of the lipids of fish and aquatic invertebrates. By contrast, the liver of the muskrats, which not only feed on plants but also prey on fish and mussels, contained high proportions of 22:6ω3 (6.9% of the total fatty acids) (Käkelä and Hyvärinen, 1996).

2.5. Research Needs

Future research in the area of fatty acids markers in freshwater environments should expand the database on primary sources of lipids, extend their use to unexplored trophic relationships, and optimize the use of statistical treatments on new and existing data. An examination of the available information of the fatty acid composition of freshwater algae reveals an uneven coverage of the different algal taxa and, in particular, a lack of adequate studies on diatoms, often the major primary producer of planktonic and benthic communities.

Fatty acids markers should be used to evaluate the transference of nutrients and metabolites in unconventional trophic linkages, such as in the case of organisms involved in symbiotic relationships. There are only a few studies on the translocation of photosynthetic products from symbiotic algae to the host in marine (see Bishop et al., 1976, and references therein) and freshwater species (Caudales et al., 1992). Because the species involved in symbiotic relationships are often from philogenetically unrelated taxa, it would be possible to detect symbiosis by the study of fatty acid signatures in lipid extracts of the host.

Studies should also address the quantitative aspects of the trophic transference of fatty acids and the assessments of organic matter pools in aquatic ecosystems. The bulk of the detailed quantitative information on fatty acid composition of aquatic lipids is presented as percentages rather than actual pool sizes. This type of data precludes the possibility of evaluating net transfer of biomass and energy from one trophic level to the next, limiting the use of biochemical markers to a qualitative evaluation of trophic relationships. Further work should integrate the use of complementary information derived from other lipid and nonlipid materials. The information from different sources, in particular from the large number of fatty acids identified in each matrix analyzed, should be integrated by means of statistical analyses to produce unambiguous patterns or chemical marker signatures.

2.6. Conclusions

The information presented in this chapter illustrates that the analysis of the fatty acid composition of organisms and the identification of potential biochemical markers is a valuable method for studying trophodynamics and biogeochemical cycles in aquatic ecosystems. In particular, fatty acid markers can be successfully used to differentiate between bacterial and algal biomass or between aquatic and terrestrial primary production. This information can then be used to elucidate the contribution of the different carbon sources to higher trophic levels of the aquatic food web.

Although there is no single fatty acid showing all the features of the ideal biochemical marker, information on the distribution and trophic transference of fatty acids with unusual structures and uncommon unsaturation patters can result in a valuable contribution to ecological studies. Among the different fatty acids groups, the ω3-PUFAs (e.g., 18:4ω3, 18:5ω3, 20:5ω3, and 22:6ω3) are the most important fatty acid markers of algal biomass. The usefulness of these ω3-PUFAs as biochemical markers is derived from their virtual absence in most bacteria and terrestrial plants, their uneven distribution in algae, and their retention and limited synthesis by animals.

Due to the intrinsic and environmental variability of the fatty acid composition of closely related species of primary producers and other organisms at the bottom of the food web, carbon sources are reflected more closely by fatty acid groups or profiles (fatty acid signatures) rather than by single chemical compounds. Furthermore, the power of the biochemical marker approach can be enhanced by the contribution of studies on the occurrence and distribution of other lipid and nonlipid compounds, such as biogenic hydrocarbons (Nevenzel, 1989), sterols (Patterson, 1991), pigments (Conte et al., 1994; Brown and Jeffrey, 1992), and the use of stable isotope analyses (Peterson and Fry, 1987).

Fatty acid data are sometimes analyzed to classify the samples into groups according to certain similarities (e.g., sources of organic carbon) (Boon et al., 1996). In a second type of approach, the presence of individual markers is used to infer the presence and abundance of a certain taxon. Regardless of the nature of the approach, a salient feature of the fatty acid method is that, with little or no modifications, it can be applied to a wide range of matrices and levels of organization, from organic matter aggregates and bacteria to mammals.

Acknowledgments.   Writing of this chapter was supported in part by the appointment of G.E.N. to the Oak Ridge National Laboratory (ORNL) Research Associates Program administered jointly by ORNL and by the Oak Ridge Institute for Science and Education. ORNL is managed by Lockheed Martin Energy Research Corp. for the U.S. Department of Energy under contract DE-ACO5-96OR22464. H. Heras (Universidad de La Plata, Argentina) and S. M. Adams (ORNL) provided constructive comments on a previous version of this work. This chapter is Environmental Sciences Division publication 4672.

References

Ackman, R.G.; Burgher, R.D. Cod liver oil: component fatty acids as determined by gas-liquid chromatography. J. Fish. Res. Bd. Can. 21:319–326; 1964.

Ackman, R.G.; Takeuchi, T. Comparison of fatty acids and lipids of smolting hatchery-fed and wild Atlantic salmon *Salmo salar.* Lipids 21:117–120; 1986.

Ackman, R.G.; Manzer, A.; Joseph, J.D. Tentative identification of an unusual naturally-occurring polyenoic fatty acid by calculation from precision open-tubular GLC and structural element retention data. Chromatographia 7:107–114; 1974.

Ackman, R.G.; Eaton, C.A.; Litchfield, C. Composition of wax esters, triglycerides and diacyl glyceryl ethers in the jaw and blubber fats of the Amazon River dolphin (*Inia geoffrensis*). Lipids 6:69–77; 1971.

Ackman, R.G.; Tocher, C.S.; McLachlan, J. Marine phytoplankter fatty acids. J. Fish. Res. Bd. Can. 25:1603–1620; 1968.

Ahlgren, G.; Gustafsson, I-B.; Boberg, M. Fatty acid content and chemical composition of freshwater microalgae. J. Phycol. 28:37–50; 1992.

Ahlgren, G.; Lundstedt, L.; Brett, M.; Forsberg, C. Lipid composition and food quality of some freshwater phytoplankton for cladoceran zooplankters. J. Plankton Res. 12:809–818; 1990.

Armstrong, F.A.J.; Williams, P.M.; Strickland, J.D.H. Photo-oxidation of organic matter in sea water by ultraviolet radiation, analytical and other applications. Nature 211:481–483; 1966.

Barnes, M.A.; Barnes, W.C. Organic compounds in lake sediments. In: Lerman, A., ed. Lakes. Chemistry, Geology, Physics. New York: Springer-Verlag; 1978:p. 127–152.

Beach, D.H.; Harrington, G.W.; Holtz, G.G. The polyunsaturated fatty acids of marine and freshwater Cryptomonads. J. Protozool. 17:501–510; 1970.

Bell, J.G.; Guioni, C.; Sargent, J.R. Fatty acid composition of 10 freshwater invertebrates which are natural food organisms of Atlantic salmon parr (*Salmo salar*): a comparison with commercial diets. Aquaculture 128:301–313; 1994.

Bell, M.V.; Henderson, R.J.; Sargent, J.R. The role of polyunsaturated fatty acids in fish. Comp. Biochem. Physiol. 83B:711–719; 1986.

Bishop, D.G.; Bain, J.N.; Downton, J.S. Ultrastructure and lipid composition of zoox-anthellae from *Tridacna maxima.* Aust. J. Plant Physiol. 3:33–40; 1976.

Boon, P.I.; Virtue, P.; Nichols, P.D. Microbial consortia in wetland sediments: a biomarker analysis of the effects of hydrological regime, vegetation and season on benthic microbes. Mar. Freshwat. Res. 47:27–41; 1996.

Bottino, N.R. The fatty acids of Antarctic phytoplankton and euphausids. Fatty acid exchange among trophic levels of the Ross Sea. Mar. Biol. 27:197–204; 1974.

Bourbonniere, R.A.; Meyers, P.A. Sedimentary geolipid records of historical changes in the watersheds and productivities of Lakes Ontario and Erie. Limnol. Oceanogr. 41:352–359; 1996.

Bourdier, G.A.; Amblard, C.A. Variabilités verticaux et temporelles des acides gras d'un phytoplancton lacustre au cours d'un cycle nycthemeral. Hydrobiologia 157:57–68; 1988.

Bourdier, G.A.; Amblard, C.A. Evolution de la composition en acides gras d'un phytoplancton lacustre (Lac Pavin, France). Int. Rev. Ges. Hydrobiol. 72:81–95; 1987.

Bourdier, G.A.; Amblard, C.A. Lipids in *Acanthodiaptomus denticornis* during starvation and fed on three different algae. J. Plankton Res. 11:1201–1212; 1989.

Brown, M.R.; Jeffrey, S.W. Biochemical composition of microalgae from the green algal classes Chlorophyceae and Prasinophyceae. 1. Amino acids, sugars and pigments. J. Exp. Mar. Biol. Ecol. 161:91–113; 1992.

Carpenter, E.J.; Harvey, R.H., Fry, B.; Capone, D.G. Biogeochemical tracers of the marine cyanobacterium *Trichodesmium.* Deep-Sea Res. 44:27–38; 1997.

Caudales, R.; Moreau, R.A.; Wells J. M. Cellular lipid and fatty acid compositions of cyanobionts from *Azolla caroliniana*. Symbiosis 14:191–200; 1992.

Cavigelli, M.A.; Robertson, G.P.; Klug, M.J. Fatty acid methyl ester (FAME) profiles as measures of soil microbial community structure. Plant Soil 170:99–113; 1995.

Chan, M.; Himes, R.H.; Akagi, J.M. Fatty acid composition of thermophilic, mesophilic and psychrophilic clostridia. J. Bacteriol. 106:876–881; 1971.

Christie, W.W. Lipid Analysis. Oxford: Pergamon Press; 1982.

Christie, W.W. The composition, structure and function of lipids in tissues of ruminant animals. In: Christie, W.W., ed. Lipid Metabolism in Ruminant Animals. Oxford: Pergamon Press; 1981:p. 95–91.

Chuecas, L.; Riley, J.P. Component fatty acids of the total lipids of some marine phytoplankton. J. Mar. Biol. Assn. U.K. 49:97–116; 1969.

Clarke, A.; Lesley, J.H.; Hopkins, C.C.E. Lipid in an Antarctic food chain: *Calanus, Bolinopsis, Beroe*. Sarsia 72:41–48; 1987.

Conte, M.H.; Thompson, A.; Eglinton, G. Primary production of lipid biomarker compounds by *Emiliania huxleyi*. Results from an experimental mesocosm study in fjords of Southwestern Norway. Sarsia 79:319–331; 1994.

Cosper, C.I; Vining, L.C.; Ackman, R.G. Sources of cyclopropanoid fatty acids in the mummichog *Fundulus heteroclitus*. Mar. Biol. 78:139–146; 1984.

Cranwell, P.A.; Jaworski, G.H.M.; Bickley, H.M. Hydrocarbons, sterols, esters and fatty acids in six freshwater chlorophytes. Phytochemistry 29:145–151; 1990.

Cranwell, P.A.; Creghton, M.E.; Jaworski, G.H.M. Lipids of four species of freshwater chrysophytes. Phytochemistry 27:1053–1059; 1988.

Dadd, R.H. Essential fatty acids: insects and vertebrate compared. In: Mittler, T.E.; Dadd, R.H., eds. Metabolic Aspects of Lipids Nutrition in Insects. Boulder, CO: Westview Press; 1983:p. 107–147.

DeLong, E.F.; Yayanos, A.A. Biochemical function and ecological significance of novel bacterial lipids in deep sea procaryotes. Appl. Environ. Microbiol. 51:730–737; 1986.

Dembitsky, V.M.; Rozentsvet, O.A. Distribution of polar lipids in some marine, brackish and freshwater green macrophytes. Phytochemistry 41:483–488; 1996.

Desvilettes, C.; Bourdier, G.; Breton, J.C.; Combrouze, P. Fatty acids as organic markers for the study of trophic relationships in littoral cladoceran communities of a pond. J. Plankton Res. 16:643–659; 1994.

Dobson, G.; Ward, D.M., Robinson, N.; Eglinton, G. Biogeochemistry of hot spring environments: extractable lipids of cyanobacterial mats. Chem. Geol. 68:155–179; 1988.

Douce, R.; Joyard, J.; Block, M.A.; Dorne, A-J.; Harwood, J.L.; Bowyer, J.R. Glycolipid analyses and synthesis in plastids. In: Harwood, J.L.; Bowyer, J.R., eds. Methods in Plant Biochemistry. vol. 4. 1990:p. 471–503.

Dunlop-Jones, N.; Jialing, H.; Allen, L.H. An analysis of the acetone extractives of the wood and bark from fresh trembling aspen:implications for deresination and pitch control. J. Pulp Paper Sci. 17:J60–J66; 1991.

Findlay, R.H.; Dobbs, F.C. Quantitative description of microbial communities using lipid analysis. In: Kemp, P.F.; Sherr, B.F.; Sherr, E.B.; Cole, J.J., eds. Aquatic Microbial Ecology. Boca Raton, FL: Lewis Publisher; 1993a:p. 271–284.

Findlay, R.H.; Dobbs, F.C. Analysis of microbial lipids to determine biomass and detect the response of sedimentary microorganisms to disturbance. In: Kemp, P.F.; Sherr, B.F.; Sherr, E.B.; Cole, J.J., eds. Aquatic Microbial Ecology. Boca Raton, FL: Lewis Publisher; 1993b:p. 347–358.

Fredrickson, H.L.; Cappenberg, T.E.; Leeuw, J.W. Polar lipid ester-linked fatty acids composition of Lake Vechten seston: an ecological application of lipid analysis. FEMS Microbiol. Ecol. 38:381–396; 1986.

Fukushima, K; Kondo, H.; Sakata, S. Geochemistry of hydroxy acids in sediments. 1. Some freshwater and brackish water lakes in Japan. Organic Geochem. 18:913–922; 1992.

Galliard, T. Degradation of acyl lipids: hydrolytic and oxidative enzymes. In: Stumpf, P.K., ed. The Biochemistry of Plants. Lipids: Structure and Function. New York: Academic Press; 1980:p. 85–119.

Gonzalez-Baro, M.; Pollero, R.J. Lipid characterization and distribution among tissues of the freshwater crustacean *Macrobrachium borellii* during an annual cycle. Comp. Biochem Physiol. 91B:711–715; 1988.

Gillan, F.T.; Johns, R.B. Chemical markers for marine bacteria: fatty acids and pigments. In: Johns, R.B., ed. Biological Markers in the Sedimentary Record. Methods in Geochemistry and Geophysics, vol. 8. Amsterdam: Elsevier; 1986:p. 291–309.

Goodloe, R.S.; Light, R.J. Structure and composition of hydrocarbons and fatty acids from a marine blue-green *Synechococcus* sp. Biochim. Biophys. Acta 710:485–492; 1982.

Grisley, M.S., Boyle, P.R. A new application of serological techniques to gut content analysis. J. Exp. Mar. Biol. Ecol. 90:1–9; 1985.

Haack, S.K.; Garchow, H.; Odelson, D.A.; Forney, L.J.; Klug, M.J. Accuracy, reproducibility, and interpretation of fatty acid methyl ester profiles of model bacterial communities. Appl. Environ. Microbiol. 60:2483–2493; 1994.

Hama, T.; Matsunaga, K.; Handa, N.; Takahashi, M. Fatty acid composition in photosynthetic products of natural phytoplankton population in Lake Biwa, Japan. J. Plankton Res. 14:1055–1065; 1992.

Hanson, B.J.; Cummings, K.W.; Cargill, A.S.; Lowry, R.R. Lipid content, fatty acid composition, and the effect of diet on fats of aquatic insects. Comp. Biochem. Physiol. 80B:257–276; 1985.

Henderson, R.J.; Sargent, J.R. Lipid biosynthesis in rainbow trout, *Salmo gairdnerii,* fed diets of differing lipid content. Comp. Biochem. Physiol. 69C:31–37; 1981.

Henderson, R.J.; Tocher, R.D. The lipid composition and biochemistry of freshwater fish. Prog. Lipid Res. 26:281–347; 1987.

Holton, R.W.; Blecker, H.H.; Stevens, T.S. Fatty acids in blue-green algae, possible relationships to phylogenetic position. Science 160:545–547; 1968.

Joseph, J.D. Identification of 3,6,9,12,15-octadecapentaenoic acid in laboratory-cultured photosynthetic dinoflagellates. Lipids 10:395–403; 1977.

Kaitaranta, J.K.; Linko, R.R. Fatty acid in the roe lipids of common food fishes. Comp. Biochem. Physiol. 79B:331–334; 1984.

Käkelä, R. Fatty acid compositions in subspecies of ringed seal (*Phoca hispida*) and several semiaquatic mammals: site-specific and dietary differences. Ph.D. thesis, University of Joensuu, Finland; 1996.

Käkelä, R.; Hyvärinen, H. Fatty acids in extremity tissues of Finnish beavers (*Castor canadensis* and *Castor fiber*) and muskrats (*Ondatra zibethicus*). Comp. Biochem. Physiol. 113B:113–124; 1996.

Käkelä, R.; Hyvärinen, H.; Vainiotalo, P. Unusual fatty acids in the depot fat of the Canadian Beaver (*Castor canadensis*). Comp. Biochem. Physiol. 113B:625–629; 1996.

Käkelä, R.; Ackman, R.G.; Hyvärinen, H. Very long chain polyunsaturated fatty acids in the blubber of ringed seals (*Phoca hispida* sp.) from Lake Saimaa, Lake Ladoga, the Baltic Sea, and Spitsbergen. Lipids 30:725–731; 1995.

Käkelä, R.; Hyvärinen, H.; Vainiotalo, P. Fatty acid composition in liver and blubber of the Saimaa ringed seal (*Phoca hispida saimensis*) compared with that of the ringed seal (*Phoca hispida botnica*) and grey seal (*Halichoerus grypus*) from the Baltic. Comp. Biochem. Physiol. 105B:553–565; 1993.

Kates, M.; Volcani, B.E. Lipid compositions of diatoms. Biochim. Biophys. Acta 116:264–278; 1966.

Kawamura, K.A.; Kaplan, I.R. Organic compounds in the rainwater of Los Angeles. Environ. Sci. Technol. 17:497–501; 1983.

Kawamura, K.; Ishiwatari, R; Ogura, K. Early diagenesis of organic matter in the water column and sediments: microbial degradation and resynthesis of lipids in Lake Haruna. Org. Geochem. 11:251–264; 1987.

Kenyon, C.N. Fatty acid composition of unicellular strains of blue-green algae. J. Bacteriol. 109:827–834; 1972.

Kenyon, C.N.; Rippka, R.; Stainier, R.Y. Fatty acid composition and physiological properties of some filamentous blue-green algae. Arch. Microbiol. 83:216–236; 1972.

Komagata, K.; Suzuki, K-I. Lipids and cell wall analysis in bacterial systematics. Methods Microbiol. 19:161–207; 1987.

Landry, W.L. Identification of *Vibrio vulnificus* by cellular fatty acid composition using the Hewlett-Packard 5898A microbial identification system: collaborative study. J. AOAC Int. 77:1492–1499; 1994.

Lechevalier, H.; Lechevalier, M.P. Chemotaxonomic use of lipids—an overview. In: Ratledge, C.; Wilkinson, S.G., eds. Microbial Lipids. New York: Academic Press; 1988:p. 869–902.

Lechevalier, M.P. Lipids in bacterial taxonomy. In: Laskin, A.I.; Lechevalier, H.A., eds. CRC Handbook of Microbiology. Boca Raton, FL: CRC Press, Inc.; 1982:p. 435–541.

Leger, C.; Fremont, L.; Boudon, M. Fatty acid composition of the lipids in the trout—I. Influence of dietary fatty acids on the triglyceride fatty acid desaturation in serum, adipose tissue, liver, white and red muscle. Comp. Biochem. Physiol. 69B:99–105; 1981.

Linko, R.R.; Rajasilta, M.; Hiltunen, R. Comparison of lipids and fatty acids composition in vendace (*Coregonus albula* L.) and available plankton feed. Comp. Biochem. Physiol. 103A:205–212; 1992.

McIntire, C.D.; Tinsley, I.J. Lowry, R.R. Fatty acids in lotic periphyton: another measure of community structure. J. Phycol. 5:26–32; 1969.

Mancuso, C.A.; Franzmann, P.D.; Burton, H.R.; Nichols, P.D. Microbial community structure and biomass estimate of a methanogenic Antarctic lake ecosystem as determined by phospholipid analyses. Microb. Ecol. 19:73–95; 1990.

Mermoud, F.; Clerc, C.; Guelacar, F.O.; Buchs, A. Free fatty acids and sterols in the plankton of Lake Leman. Arch. Sci. Genève 34:367–374; 1981.

Meyers, P.A.; Ishiwatari, R. Lacustrine organic geochemistry—an overview if indicators of organic matter sources and diagenesis in lake sediments. Organic Geochem. 20:867–900; 1993.

Meyers, P.A.; Leenheer, M.J.; Eadie, B.J.; Maule, S.J. Organic geochemistry of suspended and settling particulate matter in Lake Michigan. Geochim. Cosmochim. Acta 48:443–452; 1984.

Miyazaki, T. Compositional changes of fatty acids in particulate matter and water temperature, and their implications to the seasonal succession of phytoplankton in a hypereutrophic lake, Lake Kasumigaura, Japan. Arch. Hydrobiol. 99:1–14; 1983.

Miyazaki, T.; Irie, J.; Ogawa, T.; Ichimura, S.E. Fatty acids in lipids from particulate organic matter in a eutrophic lake, Lake Nakanuma, Japan. Int. Rev. Ges. Hydrobiol. 71:101–113; 1986.

Moss, C.W. The use of cellular fatty acids for identification of microorganisms. In: Fox, A.; Morgan, S.L.; Larson, L.; Odham, G., eds. Analytical Microbiology Methods. Chromatography and Mass Spectrometry. New York: Plenum Press; 1990:p. 59–69.

Muje, P.; Agren, J.J.; Lindqvist, O.V.; Hanninen, O. Fatty acid composition of Vendace (*Coregonus albula* L.) muscle and its plankton fed. Comp. Biochem. Physiol. 92B:75–79; 1989.

Murata, N.; Wada, H.; Gombos, Z. Modes of fatty-acid desaturation in cyanobacteria. Plant Cell Physiol. 33:933–941; 1992.

Napolitano, G.E.; Shantha, N.C.; Hill, W.R.; Luttrell, A.E. Lipid and fatty acid compositions of stream periphyton and stoneroller minnows (*Campostoma anomalum*): trophic and environmental implications. Arch. Hydrobiol. 137:211–225; 1996.

Napolitano, G.E.; Heras, H.; Stewart, A.J. Fatty acid composition of freshwater phytoplankton during a red tide event. Biochem. Syst. Ecol. 23:65–69; 1995.

Napolitano, G.E. The relationship of lipids with light and chlorophyll measurements in freshwater algae and periphyton. J. Phycol. 30:943–950; 1994.

Napolitano, G.E.; Hill, W.R.; Guckert, J.B.; Stewart, A.J.; Nold, S.C.; White, D.C. Changes in periphyton fatty acid composition in chlorine polluted streams. J. North Am. Benthol. Soc. 13:237–249; 1994.

Napolitano, G.E.; Ratnayake, W.N.M.; Ackman, R.G. All-*cis*-3,6,9,12,15-octadecapentaenoic acid: a problem of resolution in the GC analysis of marine fatty acids. Phytochemistry 27:1751–1755; 1988.

Nevenzel, J. Biogenic hydrocarbons in marine organisms. In: Ackman, R.G., ed. Marine Biogenic Lipids, Fats and Oils. Boca Raton, FL: CRC Press; 1989:p. 3–71.

Nichols, B.W. Light-induced changes in the lipids of *Chlorella vulgaris.* Biochim. Biophys. Acta 106:274–279; 1965.

Nichols, B.W.; Appleby, R.S. The distribution and synthesis of arachidonic acid in algae. Phytochemistry 8:1907–1915; 1969.

Nichols, B.W.; Wood, B.J.B.; James, A.T. The occurrence and biosynthesis of gamma-linolenic acid in a blue-green algae *Spirulina platensis.* Lipids 3:46–50; 1968.

Nichols, P.D.; Klump, D.W.; Johns, R.B. Lipid components and utilization in consumers of a seagrass community: an indicator of carbon source. Comp. Biochem. Physiol. 83B:103–113; 1986.

Orcutt, D.M.; Parker, B.C.; Lusby, W.R. Lipids of blue-green algal mats (modern stromatolites) from Antarctic Oasis Lakes. J. Phycol. 22:523–530; 1986.

Patterson, G.W. Sterols of algae. In: Patterson, G.W.; Nes, W.D., eds. Physiology and Biochemistry of Sterols. Champaign, IL: American Oil Chemists' Society; 1991:p. 118–157.

Parker, P.L.; Van Baalen, C.; Maurer, L. Fatty acids of eleven species of blue-green algae: geochemical significance. Science 155:707–708; 1967.

Parkes, R.J. Analysis of microbial communities within sediments using biomarkers. In: Fletcher, M.; Gray, T.R.G.; Jones, J.G., eds. Ecology of Microbial Communities. London: Cambridge University Press; 1987:p. 147–177.

Parrish, C.C.; DeFreitas, A.S.W.; Bodennec, G.; MacPherson, E.J.; Ackman, R.G. Lipid composition of the toxic marine diatom, *Nitzschia pungens.* Phytochemistry 30:113–116; 1991.

Peng, A.C. Fatty acids in vegetables and vegetable products. In: Chow, C.K., ed. Fatty Acids in Foods and Their Health Implications. New York: Marcel Decker; 1992:p. 185–236.

Peterson, B.J.; Fry, B. Stable isotopes in ecosystem studies. Annu. Rev. Ecol. Syst. 18:293–320; 1987.

Petkov, G.D.; Furnadzieva, S.T. Non-polar lipids of some microalgae. Arch. Hydrobiol. 96:79–84; 1993.

Pimm, S.L.; Lawton, J.H.; Cohen, J.E. Food web patterns and their consequences. Nature 350:669–674; 1991.

Pohl, P.; Zurheide, F. Fat production in freshwater and marine algae. In: Marine Algae in Pharmaceutical Science. New York: Walter de Gruyter; 1982:p. 64–80.

Ratledge, C.; Wilkinson, S.G. Microbial Lipids. London, Academic Press; 1988.

Ray, P.H.; White, D.C.; Brock, T.D. Effect of temperature on the fatty acid composition of *Thermus aquaticus*. J. Bacteriol. 106:25–30; 1971.

Rezanka, T. Very long chain fatty acids from the animal and plant kingdoms. Prog. Lipid Res. 28:147–187; 1989.

Rhead, M.M.; Eglinton, G.; Draffan, G.H.; England, P.J. Conversion of oleic acid to saturated fatty acids in seven estuary sediments. Nature 232:327–330; 1971.

Rieman, B. Potential importance of fish predation and zooplankton grazing on natural populations of freshwater bacteria. Appl. Environ. Microbiol. 50:187–193; 1985.

Ringo, E.; Jostensen, J.P.; Olsen, R.E. Production of eicosapentaenoic acid by freshwater *Vibrio*. Lipids 27:564–566; 1992.

Saliot, A.; Laureillard, J.; Scribe, P.; Sicre, M.A. Evolutionary trends in the lipid biomarker approach for investigating the biogeochemistry of organic matter in the marine environment. Mar. Chem. 36:233–248; 1991.

Sargent, J.R. The structure, metabolism and function of lipids in marine organisms. In: Malins, D.; Sargent, J.R., eds. Biochemical and Biophysical Perspectives in Marine Biology. New York: Academic Press; 1976:p. 149–212.

Scholz, O.; Boon, P.I. Biofilms on submerged River Red gum (*Eucalyptus camaldulensis* Dehnh, Myrtaceae) wood in billabongs: an analysis of bacterial assemblages using phospholipid profiles. Hydrobiologia 259:169–178; 1993.

Sicko-Goad, L.; Simmons, M.L.; Lazinsky, D.; Hall, J. Effect of light cycle on diatom fatty acid composition and quantitative morphology. J. Phycol. 24:1–7; 1988.

Smith, P.F. Archaebacteria and other specialized bacteria. In: Microbial Lipids. London: Academic Press; 1988:p. 489–547.

Smith, R.J.; Hobson, K.A.; Koopman, H.N.; Lavigne, D.M. Distinguishing between populations of fresh- and salt-water harbour seals (*Phoca vitulina*) using stable-isotope ratios and fatty acid profiles. Can. J. Fish. Aquat. Sci. 53:272–279; 1996.

Stead, D.E.; Sellwood, J.E.; Wilson, J.; Viney, I. Evaluation of a commercial microbial identification system based on fatty acid profiles for rapid, accurate identification of plant pathogenic bacteria. J. Appl. Bacteriol. 72:315–321; 1992.

Steinman, A.D.; McIntire, C.D.; Lowry, R.R. Effects of irradiance and age on the chemical constituents of algal assemblages in laboratory streams. Arch. Hydrobiol. 114:45–61; 1988.

Tunlid, A.; White, D.C. Biochemical analysis of biomass, community structure, nutritional status, and metabolic activity of microbial communities in soil. In: Stotzky, G.; Bollag J-M., eds. Soil Biochemistry. New York. Marcel Decker; 1992:p. 229–262.

Vainshtein, M.; Hippe, H.; Kroppenstedt, J. Cellular fatty acid composition of *Desul-*

fovibrio species and its use in classification of sulfate-reducing bacteria. Syst. Appl. Microbiol. 15:554–566; 1992.

Vechtel, B.; Eichenberger, W.; Ruppel, H.G. Lipid bodies in *Eremosphaera viridis* De Bary (Chlorophyceae). Plant Cell Physiol. 33:41–48; 1992.

Volkman, J.K.; Burton, H.R.; Everitt, D.A.; Allen, D.I. Pigment and lipid compositions of algal and bacterial communities in Ace Lake, Vestfold Hills, Antarctica. Hydrobiologia 165:41–57; 1988.

Volkman, J.K.; Smith, D.J.; Eglinton, G.; Forsberg, T.E.; Corner, D.S.E. Sterol and fatty acid composition of four marine Haptophycean algae. J. Mar. Biol. Assn. U. K. 61:509–517; 1981.

Wakeham, S.G.; Canuel, E.A. Fatty acids and sterols of particulate matter in a brackish and seasonally anoxic coastal salt pond. Adv. Organic Geochem. 16:703–713; 1990.

Ward, D.M.; Brassell, S.C.; Eglinton, G. Archaebacterial lipids in hot-spring microbial mats. Nature 318:656–659; 1985.

White, D.C. Chemical ecology: possible linkage between macro- and microbial ecology. Oikos 74:177–184; 1995.

White, D.C. Validation of quantitative analysis for microbial biomass, community structure, and metabolic activity. Arch. Hydrobiol. Beih. Ergebn. Limnol. 31:1–18; 1988.

White, D.C.; Davis, W.M.; Nickes, J.S.; King, J.D.; Bobbie, R.J. Determination of the sedimentary microbial biomass by extractable lipid phosphate. Oecology 40:51–62; 1979.

Wood, B.J.B. Lipids of algae and protozoa. In: Ratledge, G.; Wilkinson, S.G., eds. Microbial Lipids. New York: Academic Press; 1988:p. 807–865.

Wood, B.J.B. Fatty acids and saponifiable lipids. In: Stewart, W.D.P., ed. Biochemistry and Physiology of Algae. Berkeley: University of California Press; 1974:p. 236–265.

Zeng, Y.B.; Ward, D.M.; Brassell, C.; Eglinton, G. Biogeochemistry of hot spring environments. 2. Lipid compositions of Yellowstone (Wyoming, U.S.A.) cyanobacterial and *Chloroflexus* mats. Chem. Geol. 95:327–345; 1992a.

Zeng, Y.B.; Ward, D.M.; Brassell, S.C.; Eglinton, G. Biogeochemistry of hot spring environments. 3. Apolar and polar lipid in the biologically active layers of a cyanobacterial mat. Chem. Geol. 95:347–360; 1992b.

3

Irradiance and Lipid Production in Natural Algal Populations

Bruce C. Wainman, Ralph E.H. Smith, Hakumat Rai, and
John A. Furgal

3.1. Introduction

Lipids are important to aquatic ecosystems, as essential dietary components for animals (including some of economic importance in both wild and cultured food production) (Olsen, this volume), as vectors for movement of hydrophobic materials (including many important contaminants), and as the proximate agents of toxicity in a variety of organisms (Landrum and Fisher, this volume). Microalgae, including phytoplankton and attached forms such as ice algae and periphyton, are major producers of aquatic lipids. A substantial body of measurements of lipid synthesis by natural populations of microalgae has developed, thanks largely to the relative ease with which ^{14}C and simple chemical extraction protocols can be applied to measure the intracellular allocation of recent photosynthate (Morris, 1981; Morris et al., 1974). In practice, the term *photosynthate* here refers to carbon incorporated (and therefore labeled) within the span of typical primary production experiments (usually 4–24 hours). Such measurements have revealed substantial variation in the synthesis and relative allocation of photosynthate to lipids, which may be related to environmental and taxonomic factors (Madariaga, 1992; Wainman and Lean, 1992).

In culture, deficiency of nitrogen, phosphorus, or silicon is an effective stimulus to increased lipid synthesis in many species of microalgae (Lombardi and Wangersky, 1991; Parrish and Wangersky, 1987; Taguchi et al., 1987; Shifrin and Chisholm, 1981). There is evidence from the field for significant increases in lipid synthesis in situations of depleted nutrient supplies (Smith and D'Souza, 1993; Palmisano et al., 1988; Parrish, 1987). Possibly more important than the influence of nutrients, however, is that of light. The enzyme acetyl CoA carboxylase catalyzes the first committed step of fatty acid synthesis and is both strongly light regulated and rate-limiting to fatty acid synthesis in the chloroplasts of higher plants (Post-Beitenmiller et al., 1992; Gurr and Harwood, 1991; Harwood, 1988). De novo fatty acid synthesis in plants also depends largely on NADPH generated in the light reactions of photosynthesis. It is then no surprise that the lipid synthesis rate and the relative allocation of photosynthate to lipid both tend to increase with incubation irradiance (Wainman and Lean, 1992; Rivkin, 1989; Cuhel and Lean, 1987a; Smith et al., 1987; Li and Platt, 1982). Over the whole photic zone, with irradiance ranging from saturating to strongly limiting in most cases, the influence of light is likely to be far larger than that of any variations in nutrient supply.

It is nonetheless commonly observed that the changes in relative allocation to lipids in response to varying incubation irradiance are small compared with those observed in protein and polysaccharide components. Perhaps, partly for this reason, most attempts to determine the environmental controls on the synthesis of lipids by natural communities of microalgae have emphasized the patterns observed under saturating incubation irradiance (Wainman and Lean, 1996, 1992; Cuhel and Lean, 1987a; Smith et al., 1987; Lancelot and Mathot, 1985a). It is important to realize, however, that the relative allocation to lipids (typically 10–30% of total photosynthate) is generally smaller than that to either protein or polysaccharide (typically 30–60%). Even modest photosynthetically active radiation (PAR)-dependent changes in the proportion of total photosynthate directed to lipid therefore translate into large relative changes in lipid synthesis, often comparable with those observed for protein or polysaccharide (Wainman and Lean, 1992; Cuhel and Lean, 1987a; Smith et al., 1987; Li and Platt, 1982). In addition, the irradiance required to saturate lipid synthesis in a variety of freshwater phytoplankton communities is systematically higher than that for total carbon incorporation (Wainman and Lean, 1992; Priscu et al., 1987), indicating that observations made only at saturating light intensities are insufficient for accurate measurement of integrated lipid synthesis throughout the photic zone.

Although photosynthesis can proceed only during the illuminated phase of the diel cycle, algae can reallocate photosynthate to support continued synthesis of essential cellular components in darkness. Dark synthesis is well documented for protein, which can sometimes be synthesized, overnight, at rates rivaling those observed during the light phase (Cuhel and Lean, 1987a,b; Lancelot and Mathot, 1985a; but cf. Fernandez et al., 1992). If an accurate determination of lipid synthesis and the environmental factors that control it is desired, then the possibility of changes during the dark phase must be considered.

The high degree of light regulation of fatty acid synthesis in higher plants would suggest that lipid synthesis should be minimal in darkness. Catabolism of storage lipids (principally triacylglycerols) in higher plants, however, generally involves a high degree of interorganelle cooperation and transportation, and extensive catabolism is associated mainly with specialized situations such as germination of oil seeds (Gurr and Harwood, 1991) and not with short-term energy storage on the diel time scale. We may then anticipate relatively little synthesis or catabolism of lipids overnight, and measurements of photosynthate allocation in natural microalgae often support this expectation (Cuhel and Lean, 1987a; Lancelot and Mathot, 1985a). Lipid synthesis in plants can, however, be supported in darkness by using reductant generated from the pentose phosphate pathway (Gurr and Harwood, 1991). Catabolism is difficult to measure in isolation from the other factors that can effect losses of photosynthate from natural microalgal populations (e.g., excretion and predation), but there is evidence that overnight lipid catabolism can be extensive in some populations and may at times be the principal source of carbon for protein synthesis in the dark (Priscu et al., 1987). This seemingly inconsistent physiological behavior in different algal communities may reflect control by environmental factors that are currently only poorly understood.

Our purpose here is to improve our understanding of lipid synthesis in natural populations of microalgae by explicitly examining the implications of the varying irradiance environment in the photic zone and the overnight metabolic and reallocation activities of the algae. We deal specifically with data for temperate freshwater phytoplankton, which have been much less studied in regard to lipid synthesis and photosynthate allocation than their marine counterparts. We try to show that lipid synthesis when integrated over the photic zone and the diel cycle is not the same as that observed only during the light phase under saturating irradiance but that it is usefully predictable from light-phase measurements alone.

3.2. Metabolism and Reallocation

3.2.1. Lipids in Relation to Other Macromolecular Classes

The majority of studies reporting measurements of lipid synthesis by microalgae have concentrated on daytime rates, but a considerable number have also reported on the metabolism and reallocation of recently fixed lipid carbon as revealed by overnight changes in labeling patterns (Marañón et al., 1995; Fernandez et al., 1994, 1992; Smith and D'Souza, 1993; Hawes, 1990; Glover and Smith, 1988; Cuhel and Lean, 1987a; Lancelot and Mathot, 1985a; McConville et al., 1985). A widely accepted view, seemingly consistent with the biochemical fundamentals as we know them, is that overnight changes in phytoplankton mainly involve a continuation of protein synthesis at the expense of polysaccharides (including low-molecular-weight metabolites [LMW]), with comparatively minor changes in total lipids (Cuhel and Lean, 1987a; Lancelot and Mathot, 1985a,b). Study of overnight changes has focused largely on the patterns observed when the preceding incubation light intensity was sufficient to saturate total photosynthesis, although several authors have noted that energy supply during the light period can influence subsequent metabolism and reallocation significantly in the context of nighttime protein synthesis (Fernandez et al., 1992; Cuhel and Lean, 1987a,b).

A closer examination of the literature gives cause to question whether reallocation involving lipid is always small in extent or consistent in direction. Of the reallocation studies published to date, few show more than one example of a diel pattern and rarely is it obvious that diel patterns were systematically and regularly measured. Where more comprehensive results are reported, they can reveal considerable variations. For example, Glover and Smith (1988) found that the share of total labeled photosynthate in marine *Synechococcus* sp. at the end of the dark period decreased by 10% at some stations but increased by >50% at others, compared with the share at the end of the preceding light period. The estimated concentration of newly synthesized lipid varied similarly. Smith and D'Souza

(1993) found that the overnight change in the share of new photosynthate in lipid was small for North Atlantic phytoplankton but that the concentration of new (labeled) lipid at dusk versus dawn indicated overnight changes ranging from a 50% loss to a 25% gain. There is as yet little indication of what controls the size and direction of overnight changes in lipid, although light, nutrients, temperature, and taxonomic affiliation of the dominant phytoplankton all appear to influence daytime lipid synthesis (Fernandez et al., 1994; Smith and D'Souza, 1993; Madariaga, 1992; Wainman and Lean, 1992; Rivkin, 1989).

The first systematic examination of the overnight reallocation of lipids in fresh waters was recently carried out in one of the Laurentian Great Lakes. This study provides an opportunity to evaluate the variability of overnight reallocation involving lipids following incubation under irradiance ranging from saturating to strongly limiting. Besides total lipid synthesis, the synthesis of three major lipid classes was also determined. The neutral lipids are primarily associated with storage and are composed largely of triacylglycerol, the glycolipids comprise mainly the pigments and sugar-containing lipids associated with the plastid, and the phospholipids comprise mainly the lipids associated with nonplastid membranes (Gurr and Harwood, 1991; Parrish 1987). Together, the three classes normally comprise the great majority of algal lipids (Parrish, 1987). Given the differences in function and biosynthetic pathways that exist among the three classes, they may be expected to respond differently to environmental factors such as light and nutrients (Smith and D'Souza, 1993; Parrish and Wangersky, 1987) and may help explain the variability of total lipid, which is actually a collection of functionally diverse macromolecules. Information on lipid class synthesis and composition for natural microalgae, especially in fresh water, is very scarce to date.

Oligotrophic Georgian Bay in Lake Huron (Munawar, 1988) has been the subject of studies to determine the influence of varying light and nutrient status on productivity and biosynthetic patterns of freshwater phytoplankton (Furgal and Smith, 1997; Furgal, 1995; Smith and Maly, 1993; Maly, 1992). The results published to date have concerned only events during the light phase of incubations, but incubations were routinely performed throughout a light–dark cycle, and here we report on the diel patterns observed.

Full details of sampling and incubation protocols can be found in Furgal (1995) and Furgal and Smith (1997), and only the essential details are given here. Whole water samples were taken before dawn from a depth of 5 m at approximately 3-week intervals from June 10 to November 4, 1993. The samples were subsequently incubated under cool-white fluorescent light for 6 h, followed by a 12-h dark period. This rather aberrant diel cycle was dictated by logistics but nonetheless provided a long enough light period to permit extensive labeling of the algae and thus the chance to track overnight reallocation. PAR was supplied at 30, 60, 150, and 600 μmol photons $\cdot$ m^{-2} $\cdot$ s^{-1} and at in situ temperatures. Total photosynthesis and photosynthate allocation were determined by ^{14}C incorporation. The extraction protocol for photosynthate allocation was the same variation of Li and Platt's (1982) method used by Smith and D'Souza (1993). In brief, the first step was an extraction in 2:1 chloroform:methanol, followed by separation of the alcohol fraction (containing LMW) and the chloroform fraction (containing lipids). The insoluble residue from the first extraction step was then extracted in hot 5% trichloroacetic acid to solubilize polysaccharides, leaving proteins as the final insoluble residue. The total lipid extract was subsequently separated into three major lipid classes, neutral, glyco-, and phospholipids, by silica gel chromatography (Smith and D'Souza, 1993).

Environmental variables monitored through the sampling period included soluble reactive silicate (SRSi) and phosphate, representing the two nutrients most likely to be limiting for phytoplankton growth in Georgian Bay (Furgal and Smith, 1997; Munawar, 1988). Phosphorus limitation was additionally assessed by the phosphorus deficiency index (PDI) (Lean and Pick, 1981), which uses the ratio of P-saturated P uptake rate to light-saturated C uptake rate to characterize P deficiency. A PDI <10 indicates severe deficiency and a PDI <50 indicates moderate deficiency; higher values indicate slight or no deficiency. The light adaptation parameter I_k, which estimates the irradiance at which light saturation of total photosynthesis sets in (Kirk, 1983), was calculated from the observed photosynthesis-irradiance responses to characterize the photoadaptive state of the phytoplankton.

Nutrient availability and deficiency, and photoadaptive state, varied strongly among sampling dates (Table 3.1). SRSi was depleted early in the stratified period, whereas the PDI showed a period of extreme deficiency in the early part of the stratified season (PDI <10) but only slight or no deficiency on other dates (PDI >50). I_k values were comparatively low before and after stratification (June 10 and November 4) and higher at other times, consistent with a relatively shade-adapted community when the water column was deeply mixed in spring and fall. The samples from Georgian Bay therefore represented phytoplankton of widely varying physiological status.

TABLE 3.1. Seasonal variation of physicochemical conditions and chlorophyll a (Chl a) biomass observed in surface samples from Colpoys Bay (Georgian Bay) in 1993.[a]

Date	Temperature (°C)	Daylength (h)	SRSi (μmol $\cdot$ L^{-1})	SRP (μg $\cdot$ L^{-1})	PDI	I_k	Chl a (μg $\cdot$ L^{-1})
June 10	4.5	15.4	2.2	3.1	79	70	0.78
July 1	10.5	15.4	2.0	2.4	8	150	1.73
July 22	17.5	14.9	9.5	1.9	1	402	0.51
Aug. 10	18.0	14.1	12.5	1.5	107	205	0.92
Sept. 2	16.8	13.1	10.4	2.3	63	170	0.20
Sept. 23	12.6	12.2	6.6	0.2	nd[b]	173	0.19
Oct. 14	8.4	11.1	9.8	0.4	120	nd[b]	0.32
Nov. 4	4.9	10.1	14.2	2.6	198	65	0.77

[a]SRSi and SRP are soluble reactive silica and phosphorus, respectively, I_k is the photoadaptation parameter (μmol photons $\cdot$ m^{-2} $\cdot$ s^{-1}), and PDI is the dimensionless phosphorus deficiency index.
[b]nd, No data.

3.2.2. *Diel Versus Light-Phase Allocation and Synthesis*

Photosynthetic light responses and allocation patterns at the end of the light phase varied through the observation period in Georgian Bay, and a full analysis of the relationship of such responses to environmental variations is given elsewhere (Furgal et al., 1998; Furgal and Smith, 1997). The seasonal average allocation

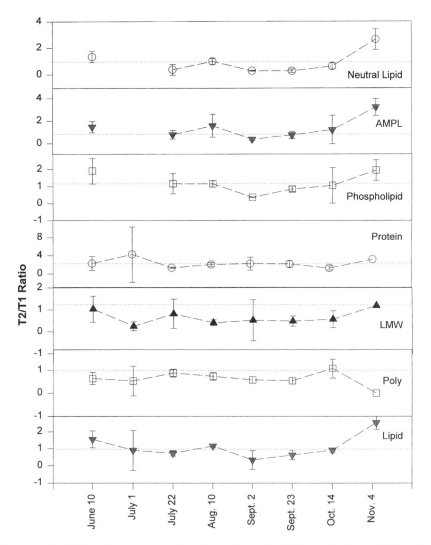

FIGURE 3.1. Ratio of percentage allocation at the end of the dark phase (T2) to the percentage allocation at the end of the preceding light phase (T1) for macromolecular components of Georgian Bay phytoplankton. The horizontal dotted lines denote a ratio of unity. AMPL, acetone mobile polar lipids (mostly glycolipid and pigments); LMW, low-molecular-weight material.

pattern (and SE, $n = 8$ dates) after 6 hours of incubation under saturating irradiance (150–600 μmol photons · m⁻² · s⁻¹) was 25% (10%) to protein, 13% (9%) to LMW, 46% (16%) to polysaccharide, and 16% (11%) to lipid, with both average and SE expressed as percentage of total photosynthate. Of interest here is the observation that allocation patterns also changed during the dark incubations. To illustrate this, we have calculated the T2/T1 ratio, where T1 denotes the percentage of total photosynthate in a given class at the end of the light phase and T2 is the corresponding value at the end of the dark period.

For phytoplankton previously exposed to saturating irradiance, the T2/T1 ratios for protein, LMW, and polysaccharide were generally consistent with a tendency of phytoplankton to continue protein synthesis overnight, at the expense of polysaccharide and LMW reserves (Fig. 3.1). The general pattern corresponded to that documented for several other freshwater and marine locations. Although not highly obvious from Figure 3.1, the extent of overnight reallocation to protein was quite variable among dates.

The lipids displayed T2/T1 ratios that were sometimes significantly greater than unity and sometimes significantly less (t-test, $P <.05$). There was no significant correlation between the T2 and T1 values for percentage of photosynthate in lipids. However, the concentration of newly synthesized, labeled, lipid at the end of the dark period was highly correlated with the value at the end of the light period according to linear regression analysis (Fig. 3.2), with 86% of the variation in the T2 value explained by the T1 value (Table 3.2). This is a type of part-whole

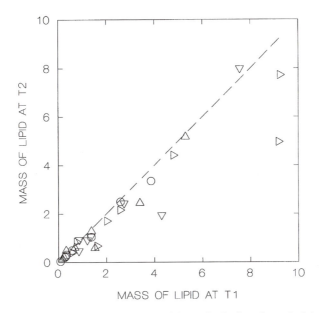

FIGURE 3.2. Concentration of new, labeled, lipid (μg C · L⁻¹) at the end of the dark phase (T2) versus the end of the light period (T1), with dashed line showing 1:1 correspondence. PAR during light period was 600 (▷), 150 (▽), 60 (△), or 30 (○) μmol photons · m⁻² · s⁻¹.

TABLE 3.2. Regression results for the model (mass at T2 = constant + slope × mass at T1), where T1 is the end of the light phase (6 hours) and T2 is the end of the ensuing dark phase (18 hours).[a]

Class	Irradiance	Slope ± 95% C.I.	n	r^2	P^b
Total lipid	30	0.90 ± 0.06	8	0.99	<.001
	60	0.91 ± 0.16	7	0.94	<.001
	150	0.99 ± 0.16	8	0.88	<.001
	600	0.69 ± 0.19	8	0.89	<.001
	All	0.79 ± 0.11	31	0.86	<.001
Protein	30	1.21 ± 0.12	8	0.98	<.001
	60	1.16 ± 0.10	7	0.99	<.001
	150	1.03 ± 0.03	7	0.99	<.001
	600	1.11 ± 0.20	8	0.99	<.001
	All	1.04 ± 0.05	30	0.98	<.001
LMW	30	0.62 ± 0.36	8	0.68	.020
	60	0.27 ± 0.16	7	0.68	.022
	150	0.19 ± 0.02	7	0.42	.114
	600	0.45 ± 0.12	8	0.91	<.001
	All	0.40 ± 0.06	30	0.86	<.001
Polysaccharide	30	0.14 ± 0.32	8	0.14	.417
	60	0.30 ± 0.18	7	0.69	.021
	150	0.33 ± 0.22	7	0.68	.020
	600	0.22 ± 0.12	8	0.77	.012
	All	0.21 ± 0.06	30	0.62	<.001

[a]Irradiance (μmol photons $m^{-2} \cdot s^{-1}$) is for the light phase of the incubation.
[b]P, overall probability level for the regression model.

correlation, because the T2 value has to depend at least partly on the T1 value, but it does show that overnight reallocation is not large enough to prevent useful predictions of diel lipid synthesis from measurements over the light phase only. The slope of the relationship was significantly less than unity (0.79). Regressions for each irradiance level separately also indicated both high degrees of correlation and slopes equal to or less than unity (although not always significantly, as judged by the confidence intervals for the slopes; Table 3.2). The apparent loss of new lipid overnight was maximal at the highest and lowest irradiances.

Is the high predictability of diel synthesis from the light-phase synthesis a feature of lipids only, or might it apply to other photosynthate classes? The mass of new, labeled, protein at T2 was highly predictable from the mass at T1 (Table 3.2) and indicated a very modest overnight net synthesis of protein from recently acquired photosynthate. The T2 masses of LMW and polysaccharide were less predictable than for lipid or protein (Table 3.2), apparently reflecting a substantial variability in respiratory and other losses overnight.

3.2.3. Budgets for Overnight Activity

Budgets based on models for overnight reallocation in some marine phytoplankton communities have been elegantly elucidated by Lancelot and Mathot

(1985a,b). For diatom-dominated communities in Belgian coastal waters, there was, on average, a relatively large reallocation from polysaccharide and LMW to protein, and a small loss from lipid, during the night (Lancelot and Mathot, 1985a). We used the eight sampling dates from Georgian Bay to calculate the average overnight metabolism and reallocation rates for our freshwater community over a substantial part of the main productive season. Unlike Lancelot and Mathot (1985a,b), we were able to examine the patterns not only for phytoplankton previously exposed to saturating irradiance but also those exposed to limiting irradiance. We assumed that the loss of labeled carbon from the particulate phase represented primarily respiratory and excretory losses from the phytoplankton, which we termed *carbon loss rate.* Consistent with Lancelot and Mathot (1985a,b), we also defined a quantity termed *metabolism rate,* which is the sum of the carbon loss term and nighttime protein synthesis.

Both the loss and the metabolism rates increased with increasing previous incubation irradiance (Fig. 3.3A). The difference between the two rates was relatively small, reflecting a relatively small conversion of metabolized carbon into new protein. Even at saturating irradiance, only a small fraction of the mobilized carbon was reincorporated into protein, suggesting that energy supply rate alone does not account for the low levels of nighttime protein synthesis. Nutrient limitation also seems an unlikely explanation, considering that SRSi and the PDI suggested little or no shortage of Si or P on most sampling dates (Table 3.1).

The rate of overnight protein synthesis increased with previous incubation irradiance (Fig. 3.3B), up to the second-highest irradiance, at which photosynthesis was usually close to light saturation (Furgal and Smith, 1997; and cf. I_k values in Table 3.1). A similar dependence of overnight protein synthesis on previous irradiance has been reported for Lake Ontario phytoplankton (Cuhel and Lean, 1987a,b), although the overnight protein synthesis rates reported for Lake Ontario were far larger than the current results for Georgian Bay. The major source of carbon for protein synthesis was polysaccharide, and polysaccharide loss rates increased with previous incubation irradiance (Fig. 3.3B). On average, LMW (not shown) and lipids (Fig. 3.3C) were also lost overnight, and the loss rate increased with irradiance. Polysaccharide contributed 45–75% of the total overnight metabolism, the proportion varying in a nonsystematic way among irradiance levels. LMW contributed 18–23% whereas lipid contributed a rather variable share (7–35%, averaging 19%), with considerable but nonsystematic differences among irradiance levels.

As the confidence intervals in Fig. 3.3 show, there was considerable variation around the mean seasonal value for overnight rates of gain or loss from each photosynthate class. Regression analysis showed that for lipid, the variation was best explained by the rate of lipid synthesis over the preceding light phase. The percentage explained variation ranged from 72% to 97% among irradiance levels and the slopes from -0.345 to -0.172, with no significant differences among irradiance levels. Thus, the rate of lipid loss (a negative rate of carbon exchange) in the dark increased as the rate of lipid synthesis in the light increased. Factors

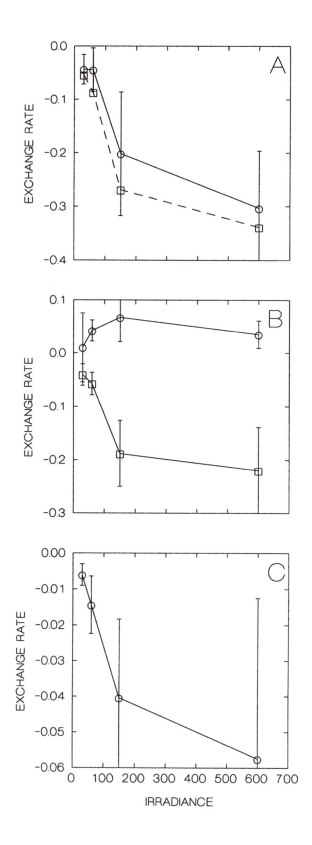

that promote lipid synthesis during the day therefore appear to stimulate greater lipid consumption at night. Night synthesis rates for protein and loss rates for polysaccharide and LMW also tended to be significantly and positively correlated with day rates, with r^2 values of 0.83–0.97. Only protein and polysaccharide at the highest irradiance levels gave poor correlations (r^2 of 0.21 and 0.54, respectively). No single environmental factor (e.g., PDI or I_k) gave better prediction of overnight synthesis or loss rates than did the day rate of synthesis for any of the photosynthate classes.

We used multiple regression to ask whether physicochemical variables, together with day rates, might give better explanations of the night rates. For example, we found that daylength and day rate together could provide reasonable predictions of overnight protein synthesis at the highest irradiance level ($r^2 = 0.78$; $P = .02$). In general, however, any gains in predictive power were minimal. It was particularly surprising that nutrients, temperature, and daylength had little apparent influence on the night rates because all are considered important influences on the physiology of microalgae generally and on photosynthate allocation specifically. Taxonomic succession may have confounded the situation (Fernandez et al., 1992; Feuillade et al., 1992; Madariaga, 1992; Rivkin, 1989), but we lack species counts with which to test this idea.

In summary, the diel reallocation patterns observed in Georgian Bay largely confirmed generalizations that have been made on the basis of apparently more limited data sets (Cuhel and Lean, 1987a; Lancelot and Mathot, 1985a). In particular, the results showed that lipids are subject to a net loss overnight but to a small extent compared with polysaccharides. The loss is predictable from the day rates so that production over the 24-hour cycle can, on average, be estimated as 87% (average T2/T1 slope among irradiance levels, Table 3.2) of that over the light phase only. This pattern holds true for Georgian Bay phytoplankton, ranging widely in nutrient status and photoadaptive state and after exposure to irradiance ranging from strongly limiting to saturating.

3.2.4. Reallocation Among Lipid Classes

All three major lipid classes (neutral, glyco-, and phospholipids) contain fatty acids that are synthesized primarily in the plastid but then modified by reactions outside the plastid. In the case of glycolipids of higher plants, about half of the fatty acids never leave the plastid during the synthetic process. The other fatty acids of glycolipids, and all those contained in neutral and phospholipids, are thought to be exported to the cytoplasm as C16 or C18 molecules and then

---

FIGURE 3.3. Relationship of PAR during the light period (μmol photons $\cdot$ m^{-2} $\cdot$ s^{-1}) to (A) overnight respiration ($\bigcirc$) and metabolism ($\square$) rates, (B) carbon exchange rates (positive denotes net increase or synthesis; negative denotes net loss or catabolism) for protein ($\circ$) and polysaccharide ($\square$), and (C) carbon exchange rates for total lipids. Respiration, metabolism, and exchange rates in C $\cdot$ Chl a^{-1} $\cdot$ h^{-1}, showing means ± 2 SE bars.

modified by reactions associated with the endoplasmic reticulum (Gurr and Harwood, 1991; Somerville and Browse, 1991). The transportation of fatty acids between plastid and other cell compartments and the assembly of the finished lipids are not immediately dependent on the strongly light-regulated activity of acetyl CoA carboxylase and can occur in the dark (Gurr and Harwood, 1991).

Figure 3.1 shows that for Colpoys Bay the share of total photosynthate in neutral lipid tends to decrease more overnight than does the share in phospholipids or glycolipids, suggesting some reallocation from neutral lipids to the other classes. As for total lipid, however, all three classes could gain significantly in a share of photosynthate overnight, as seen on the first and last sampling dates. The mass of lipid in each class at the end of the dark period (T2) was significantly and positively related to the mass at the end of the previous light phase (T1) when regression analysis was applied to the results for all irradiance levels combined. There was considerable scatter, however, with r^2 values of only 0.53, 0.51, and 0.40 for neutral, phospho-, and glycolipid, respectively. Within irradiance levels, there was generally no significant (i.e., $P > .05$) relationship between mass of lipid at T2 and at T1.

The first and last sampling dates suffered from poor replication of the lipid class synthesis measurements, including some large values for overnight lipid class synthesis at some irradiance levels. For the intervening dates, all under thermally stratified conditions, a somewhat more consistent pattern emerged when mean values for overnight change in lipid mass were calculated (Table 3.3). Loss of neutral lipids was consistently greater than for the other two lipid classes, and the loss of neutral lipids increased with irradiance. The overnight change in total lipids tended to be dominated by the changes in glyco- and phospholipids, which together made up the majority of the total lipids (cf. allocation among classes at dusk; Table 3.3).

Our results therefore indicate that the neutral lipids are more prone to catabolism at night than other classes but give little evidence of overnight synthesis of any of the major lipid classes. Whether the same would be true for phytoplankton communities that appear more active in overnight synthesis of macromolecules (at least protein; Cuhel and Lean, 1987a,b; Lancelot and Mathot, 1985a) would be an interesting subject of investigation. At least in the limited setting of the strat-

TABLE 3.3. Mean ratios for the mass of lipid (μgC·L^{-1}) at dawn (T2) versus dusk (T1) for neutral lipids (NL), glycolipids (GL), phospholipids (PL), and total lipids (TL).[a]

Irradiance	T2/T1 NL	T2/T1 GL	T2/T1 PL	T2/T1 TL	% NL	% GL	% PL
30	0.88 ± 0.40	1.17 ± 0.70	1.07 ± 0.57	0.86 ± 0.40	2.1 ± 1.3	2.0 ± 0.9	4.9 ± 2.0
60	0.63 ± 0.31	0.78 ± 0.53	0.88 ± 0.48	0.85 ± 0.29	2.7 ± 1.2	2.5 ± 0.6	6.1 ± 2.6
150	0.58 ± 0.28	0.75 ± 0.39	0.86 ± 0.41	0.85 ± 0.38	4.9 ± 2.2	3.8 ± 1.8	6.5 ± 3.0
600	0.26 ± 0.16	0.60 ± 0.32	0.77 ± 0.32	0.63 ± 0.27	7.5 ± 1.0	3.8 ± 0.4	7.2 ± 1.2
All	0.63 ± 0.45	0.86 ± 0.42	0.88 ± 0.37	0.85 ± 0.42	4.1 ± 2.5	3.0 ± 1.2	6.2 ± 2.4

[a]All as mean ±2 SE, in phytoplankton from Georgian Bay incubated at various irradiances (μmol photons · m^{-2} · s^{-1}), together with the mean percentage of total photosynthate (±2 SE) in each lipid class at dusk.

fied season in Georgian Bay, however, it seems valid to assume that the synthesis of glycolipids, phospholipids, and total lipids over the light phase is closely related to and only 10–15% larger than the net synthesis over the full diel cycle.

3.3. Irradiance and Lipid Synthesis

3.3.1. Photosynthetic Parameters

There are few published data on the photosynthetic parameters associated with any of the major biomolecular classes and lipid in particular. Most of the studies reporting on lipid synthesis by microalgae have concentrated on light-saturated lipid production (i.e., L_m), the lipid analogue of P_m (total photosynthesis at light saturation), or the lipid fraction of carbon fixation at light saturation (L_m/P_m). However, much photosynthesis (and presumably, lipid synthesis) occurs at less than light saturation and should properly be approached through analysis of the complete light response of the phytoplankton.

Although many models for photosynthetic light response have been developed over the years, most of them characterize the response by use of the light-saturated rate plus a parameter that characterizes the slope of the response at low irradiance. This parameter is commonly termed α (Platt et al., 1980) and is a measure of the efficiency of light use. I_k ($= P_m/\alpha$), the light adaptation parameter, is also often used as an index of the photoadaptive state of the phytoplankton. The lipid analogs of these parameters, α-lipid and I_k-lipid, have been used previously to explain the production of lipid (Wainman and Lean, 1992). A final photosynthetic parameter, β, the slope of the light-inhibited portion of the P-I curve, may also be required if photoinhibition is appreciable but was not needed in the data analyzed here. All parameters reported in this chapter are chlorophyll-specific values, in keeping with general usage.

One of the greatest problems in looking at photosynthetic lipid production is the paucity of complete sets of lipid production versus irradiance data. For this study, three sets of data, two from Canada and one from Germany, comprising 49 separate dates on five different bodies of fresh water, have been collected. The first data set arises from 24 experiments on Anstruther (44° 45′ N, 78° 12′ W), Bay (45° 01′ N, 77° 52′ W), and Jack (44° 42′ N, 78° 02′ W) Lakes in 1988 and has been partially reported in Wainman and Lean (1992). The second data set comes from eight experiments on Colpoys Bay of Lake Huron (44° 50′ N, 81° 2′ W) in 1993 (Furgal et al., 1998; Furgal and Smith, 1997; Furgal and Smith, unpublished data). The final data come from 16 experiments on the Schöhsee (54° 15′ N) in Germany carried out in 1995 (Rai, unpublished data).

All lipid extractions were part of sequential extractions of lipid, protein, carbohydrate, and LMW. Wainman and Lean (1992) used the sequential extraction method detailed in Cuhel and Lean (1987). Furgal and Smith (1997; and unpublished data) and Rai (unpublished data) used the Li and Platt method as modified by Smith and D'Souza (1993). There is little functional difference between these methods for lipid extraction except that the combined ethanol, ethanol:diethyl ether lipid solvent system used by Cuhel and Lean (1987a) and the chloroform:methanol solvent used by Li and Platt (1982) yield crude lipid extracts and likely overestimate lipid synthesis levels, perhaps by as much as 15% (Wainman and Lean, 1994).

Areal carbon fixation and fixation of carbon into lipid were estimated by substituting calculated photosynthetic parameters (see above) into the modified Talling equation as suggested by Kirk (1983, Equation 11.5, p. 287). PAR was estimated using the "YPHOTO" program of the "primary production model" of Fee (1990).

3.3.2. Light Saturation Parameter, I_k

The light saturation parameter, I_k, is normally highest when insolation is high and lowest when light levels are low. The temperate lakes in this study experience

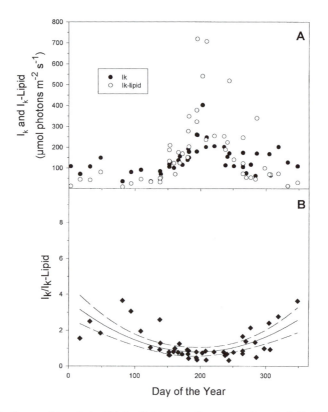

FIGURE 3.4. Seasonal trends in (A) the light saturation parameters I_k, I_k-lipid (μmol photons $\cdot$ m^{-2} $\cdot$ s^{-1}) and (B) the light saturation ratio, I_k/I_k-lipid. The solid line in B is a second-order linear regression of the data, and the broken lines are the 95% confidence intervals.

large seasonal alterations in insolation and, as expected, I_k is highest in the summer when light levels are high and lowest in the winter (Fig. 3.4A). Interestingly, I_k-lipid is remarkably high in the summer, and thus lipid production is light saturated at levels 50–200% higher than overall photosynthesis in midsummer (Fig. 3.4A). This result is particularly clear in Figure 3.4B, in which the light saturation ratio (I_k/I_k-lipid) indicates that overall photosynthesis is light-saturated at two to four times that of lipid synthesis in the winter but the reverse in summer. The biological significance of this is that the lipid synthesis can take advantage of increased light availability more than other biosynthetic activities, which suggests that in the upper reaches of the epilimnion lipid synthesis is relatively great, particularly in the summer.

Although we made the assumption that light was the most important factor driving the relationship between I_k and I_k-lipid, the light saturation ratio is actually related more closely to water temperature ($r^2 = 0.46$; Figure 3.5) than to daylength ($r^2 = 0.31$; Fig. 3.6). The relationship between temperature and the light saturation parameters is not simply due to the correlation between daylength and temperature. Although it is often assumed that water temperature is closely related to daylength, this is usually *not* the case over the year. In our data, the relationship with daylength only explains 35% of the variance in water temperature. It is likely that both water temperature and daylength, although somewhat autocorrelated, are significant factors behind changes in light saturation parameters. If water temperature is also important, then the lipid synthesis machinery of the summer species of phytoplankton is particularly well adapted to high insolation and warm water. The most obvious implication of this is that in the summer the epilimnetic algae, by virtue of existing in a brighter and warmer environment, are able to synthesize relatively greater amounts of lipid than their hypolimnetic counterparts.

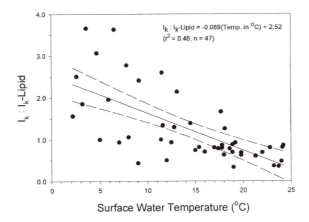

FIGURE 3.5. Relationship between the light saturation ratio (I_k/I_k-lipid) and water temperature (°C). The solid line is a first-order linear regression of the data, and the broken lines are the 95% confidence intervals.

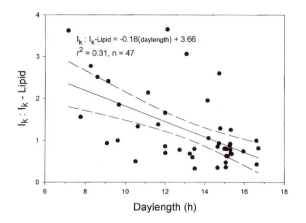

FIGURE 3.6. Relationship between the light saturation ratio (I_k/I_k-lipid) and daylength. The solid line is a first-order linear regression of the data, and the broken lines are the 95% confidence intervals.

3.3.3. Production Efficiency Parameter, α

The efficiency with which algae use light to fix carbon is summarized by α, the slope of the initial linear portion of the production versus light intensity curve. A high α indicates that a relatively small increase in light is required to fix additional carbon (i.e., the photosynthetic apparatus is efficiently using light to fix carbon). During periods of high insolation, it is unnecessary, and metabolically costly, to maintain a high efficiency of light harvest, which may explain why summer tends to be a time of low α. In our data, the seasonal trends in α and α-lipid do *not* appear to be clearly related to daylength. Some of the highest values for α are in

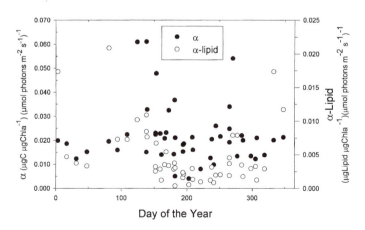

FIGURE 3.7. Seasonal values for α (μg C $\cdot$ μg Chl a^{-1}) $\cdot$ (μmol photons $\cdot$ m^{-2} $\cdot$ s^{-1})$^{-1}$ and α-lipid (μg lipid $\cdot$ μg Chl a^{-1}) $\cdot$ (μmol photons $\cdot$ m^{-2} $\cdot$ s^{-1})$^{-1}$.

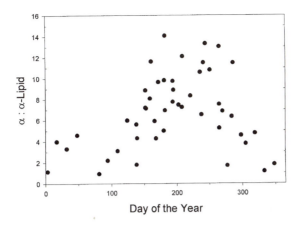

FIGURE 3.8. Seasonal trend in the ratio of α to α-lipid (the α-ratio) that expresses the relative efficiency of conversion of photons to particulate carbon and particulate lipid.

the midsummer when insolation is highest, and many late fall and winter α values are as high as some values in the summer (Fig. 3.7).

For this chapter, the ratio of the total photosynthetic efficiency to the efficiency of lipid synthesis (α/α-lipid; Fig. 3.8) is used instead of the absolute values to test hypotheses about column-integrated lipid synthesis. The α-ratio has the benefit of eliminating the highly dimensioned α in favor of a dimensionless ratio. This is a benefit because chlorophyll a measurements are notoriously variable between laboratories, and the data for this chapter come from four different laboratories using different methods on a number of different years. The α-ratio indicates how lipid synthesis is changing relative to overall synthesis and shows that midsummer

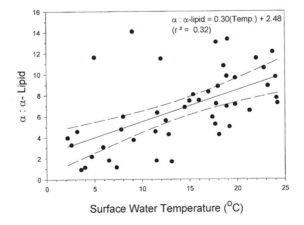

FIGURE 3.9. Relationship between the ratio of α to α-lipid (the α-ratio) and the water temperature (°C). The solid line is a first-order linear regression of the data, and the broken lines are the 95% confidence intervals.

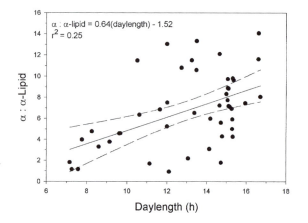

$$\alpha : \alpha\text{-lipid} = 0.64(\text{daylength}) - 1.52$$
$$r^2 = 0.25$$

FIGURE 3.10. Relationship between the ratio of α to α-lipid (the α-ratio) and daylength. The solid line is a first-order linear regression of the data, and the broken lines are the 95% confidence intervals.

lipid production requires relatively large amounts of light. The epilimnion in midsummer is characterized by high light levels and high water temperatures. The α-ratio is weakly, but significantly, related to water temperature (Fig. 3.9; $r^2 = 0.32$) and daylength (Fig. 3.10; $r^2 = 0.25$). These results suggest that the high temperatures and long days of summer are associated with a diminished efficiency of lipid synthesis compared with other cell constituents. Summer also brings other interrelated changes, such as diminished nutrient availability. The phytoplankton community at this time consists of those species that can exploit these physical and chemical features, and phytoplankton species vary in their capacity for lipid production. We do not have the data to test whether the taxonomic composition of the phytoplankton is directly related to lipid synthesizing ability, although in the limited data from Wainman and Lean (1992), there is no relationship (Wainman et al., 1993).

Nitrate levels appear to be inversely related ($r^2 = 0.33$) to the increase in the α-ratio, but the nitrate data are skewed toward the detection limit of nitrate due to the extremely low levels of nitrate nitrogen in the water of meso- to oligotrophic lakes. This is also true for phosphorus, measured as total phosphorus; a relationship appears to exist but phosphorus data are skewed.

3.3.4. Areal Lipid Production

The lipid percent of areal production averaged $16.8 \pm 6.4\%$ over the entire data set, with a maximum of 39.3% and a minimum of 6.1% (Fig. 3.11). There was no clear seasonal trend in areal lipid synthesis despite wide fluctuations in the pattern of lipid production. The mean lipid percent of carbon fixation ($L_m/P_m * 100$) was $16.5 \pm 5.4\%$, with a maximum of 31.1% and a minimum of 6.2% (Fig. 3.12), with

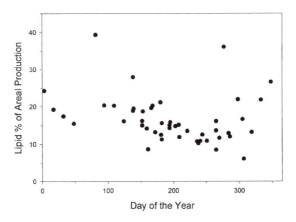

FIGURE 3.11. Seasonal values for areal lipid production divided by total areal production × 100 (the lipid percent of areal production).

highest values in midsummer and lowest in late fall and winter. Although the means and ranges were similar and the relationship between the two sets of number was significant ($P <.001$), the r^2 was only 0.30 (Fig. 3.13). When the two points with large leverage (indicated by arrows in Fig. 3.14) were removed from the data, the r^2 fell to 0.14 ($P = .015$). The lipid percent of areal production was weakly related to water temperature ($r^2 = 0.13$; $P = .013$) but not to daylength ($r^2 = 0.05$; $P = .143$). The lipid percent of areal production was closely related ($r^2 = 0.62$; $P <.001$) to the α-ratio, such that when α-lipid was close to α, areal production of lipid as a percentage of total carbon fixation was greatest (Fig. 3.14).

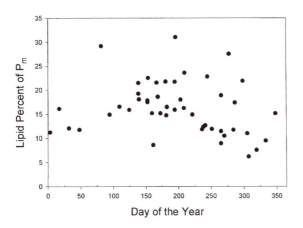

FIGURE 3.12. Seasonal values for the lipid percent of P_m.

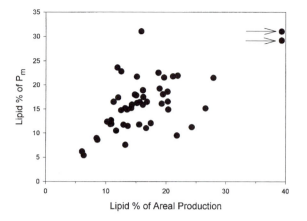

FIGURE 3.13. Relationship between the lipid percent of areal production and the lipid percent of P_m. Points marked by arrows have very large leverage in the regression of the two variables (see text).

The difference between the lipid percent of areal production and the lipid percent of P_m (hereafter termed *the areal bias*) changed over the year (Fig. 3.15). The areal bias is an expression of the difference between the lipid percent of production at light saturation and the lipid percent of production expressed on an areal basis at any one time. The lipid percent of P_m was greater than the lipid percent of areal production in summer (i.e., negative bias), whereas the reverse was true in winter. The areal bias was related to the changes in the photosynthetic parameters I_k and α. For example, the areal bias varied with I_k-lipid (Fig.

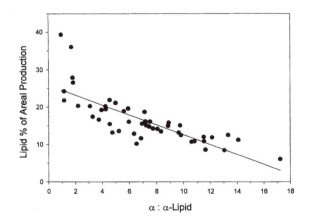

FIGURE 3.14. Relationship between the lipid percent of areal production and the ratio of α to α-lipid (the α-ratio). The solid line is for the regression, lipid percent of areal production = 25.9 − 1.32 * α-ratio ($r^2 = 0.62$, $n = 48$).

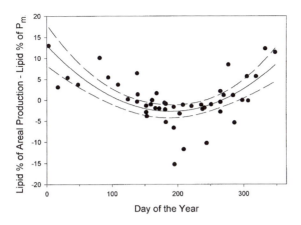

FIGURE 3.15. Seasonal trend in the difference between the lipid percent of areal production and the lipid percent of P_m (the areal bias). The solid line is a second-order linear regression of the data, and the broken lines are the 95% confidence intervals.

3.16), with an r^2 of 0.70 (P <.001). Values for I_k-lipid of about 100 μmol photons · m^{-2} · s^{-1} marked the boundary between positive and negative areal bias. The areal bias was also related to the α-ratio ($r^2 = 0.55$; P <.001) and α-lipid ($r^2 = 0.49$; P <.001). The consequence of this relationship is that high α-lipid values lead to a lipid percent of areal production that is greater than the lipid percent of P_m (positive bias). A hyperbolic relationship appeared to exist between the areal bias and the light saturation ratio such that bias was negative when I_k and

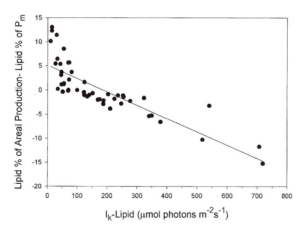

FIGURE 3.16. Relationship between the difference between the lipid percent of areal production and the lipid percent of P_m (the areal bias) and I_k-lipid (μmol photons · m^{-2} · s^{-1}). The solid line is for the linear regression, areal-P_m lipid difference = 5.1 − 0.028 * I_k-lipid (μmol photons · m^{-2} · s^{-1}), ($r^2 = 0.70$, $n = 48$).

I_k-lipid were of comparable magnitude, whereas bias was positive when I_k was much larger than I_k-lipid.

3.3.5. Implications of the Irradiance Response

This section of the chapter has been chiefly concerned with the photosynthetic parameters α, I_k, their lipid counterparts α-lipid and I_k-lipid, and the relationship of these parameters to the percentage of photosynthate allocated to lipids at either light saturation or on an areal basis. The two main seasonal changes in the lipid production versus light intensity curve were that (1) the light saturation of lipid production (I_k-lipid) occurred at higher levels than overall production in summer but less than overall production in winter (Fig. 3.4) and (2) the efficiency of lipid production (α-lipid) decreased relative to the overall efficiency of carbon fixation (α) in summer (Fig. 3.8).

The seasonal change in α/α-lipid (α-ratio) and I_k/I_k-lipid (light saturation ratio) were, not surprisingly, related to the light environment. The light saturation ratio was lowest when days were longest (Fig. 3.6) and waters were warmest (Fig. 3.5), whereas the α-ratio was greatest during those conditions (Figs. 3.9 and 3.10). Low values for α and high values for I_k are indicators of high-light adaptation (Platt et al., 1982), so lipid synthesis in midsummer was more high-light adapted than was the overall biosynthesis. The winter situation has the opposite profile of photosynthetic parameters, indicating that the lipid synthesis is shade-adapted relative to total biosynthesis. The consequence is that the share of total biosynthesis directed to lipid is maximal in the upper photic zone during summer and lower in the photic zone during winter.

Seasonal changes in lipid production compared with overall production at light saturation are best illustrated by the lipid percent of P_m (Fig. 3.12). In summer, when the lipid synthesizing apparatus is more light adapted than the overall photosynthetic apparatus, the lipid percent of production is about 15–30% of P_m. In winter, the lipid production is 5–20% of P_m. If lipid production is related to lipid content and it is where it has been measured (Taguchi et al., 1987), then the lipid content of epilimnetic algae in summer may be relatively high when lipid synthesis is light saturated (300–800 μmol photons · m^{-2} · s^{-1} in the summer) but decreases with depth. However, this inference depends also on the rates of vertical motion associated with mixing, and situations in which the phytoplankton are static in the vertical light gradient may be unusual. If mixing strength is high, then no vertical differentiation of lipid content would be expected.

The lipid percent of areal production presents a somewhat different view of algal lipid production. There is no clear seasonal trend in the lipid percent of areal production (Fig. 3.11), and it is remarkably stable over the year, with a mean of $16.8 \pm 6.4\%$, compared to protein or carbohydrate fraction of carbon fixation (Cuhel and Lean, 1987a, Table 4). This seasonal invariability suggests that, usually, lipid is constitutive and not susceptible to "luxury storage." This would be consistent with measurements that suggest that the lipid pool of phytoplankton is dominated by lipids associated with structural functions unless the cells are placed under rather severe stress such as strong nutrient limitation (Parrish and Wangersky, 1987). There are times when lipid synthesis is much higher (30–40% of total production), and this may well be luxury storage, perhaps in response to nutrient limitation.

The reason for the relative stability in the lipid percent of areal synthesis across seasons is that the increase in α-lipid is concurrent with a decrease in I_k-lipid so overall synthesis of lipid is maintained. The variability that was observed in lipid production was related to changes in the photosynthetic parameters, particularly α. When α-lipid and α are similar, then the lipid percent of areal production is high. Because this parity between α's occurs most often in the late fall and winter and they are farthest apart in summer (Fig. 3.8), one would expect that areal lipid production would be highest in winter and lowest in summer; but that is not so (Fig. 3.11) because these are times when P_m and I_k are low. The real highs in areal lipid production appeared when the α-ratio and the light saturation ratio are high, which is a rare situation because they are usually inversely related. The seasonality in the lipid percent of P_m is not, unfortunately, a good predictor of the areal production of lipid relative to total production (Fig. 3.13). Predicted areal production from P_m has a strong bias related to season. Perhaps it should not be surprising that P_m, which reflects only part of the photoadaptive strategy of the algae, is not indicative of areal production. Nonetheless, most work to date on the environmental control of lipid synthesis by phytoplankton has emphasized observations at P_m and may need to be re-evaluated in the light of the present findings.

3.4. Conclusions

A plausible interpretation of the present observations is that phytoplankton adapt the photosynthetic parameters governing lipid production to maintain the column-integrated proportion of biosynthesis directed to lipids within a fairly narrow range and, thus, a balanced cell composition. This would be consistent with the idea that the total lipid pool is, primarily, a constitutive component and that mixing is, on average, strong enough that the plants respond mainly to the average effective light climate. Focusing on behavior in selected portions of the vertical light gradient, such as the light-saturated zone, gives only a partial and biased view of phytoplankton activity. Future research could profitably focus on the average light climate as the single variable most likely to give good predictions of column-integrated lipid production.

3.5. Research Directions

A number of studies could be undertaken to further understand the control of algal lipid production.

1. The lake light climate as well as the hypsographical data could be measured so that algal lipid budgets for whole lakes could be calculated.

2. The photosynthetic parameters for lipid synthesis for individual species need to be worked out so that the contributions of individual species to the lipid production of the phytoplankton community can be understood.
3. The conditions/species that lead to large episodic increases in lipid production require further study, particularly in fresh water.
4. The photosynthetic parameters associated with the production of the other biomolecules could be calculated so that the biochemical profile of the whole lake could be studied.

References

Cuhel, R.L.; Lean, D.R.S. Influence of light intensity, light quality, temperature and daylength on the uptake and assimilation of carbon dioxide and sulfate by lake plankton. Can. J. Fish. Aquat. Sci. 44:2118–2132; 1987a.

Cuhel, R.L.; Lean, D.R.S. Protein synthesis by lake plankton measured using *in situ* carbon dioxide and sulfur assimilation. Can. J. Fish. Aquat. Sci. 44:2102–2117; 1987b.

Fee, E.J. Computer programs for calculating *in situ* phytoplankton photosynthesis. Can. Tech. Rep. Fish. Aquat. Sci. 1740; 1990.

Fernandez, E.; Balch, W.M.; Marañón, E.; Holligan, P.M. High rates of lipid biosynthesis in cultured, mesocosm and coastal populations of the coccolithophore *Emiliania huxleyi*. Mar. Ecol. Prog. Ser. 114:13–22; 1994.

Fernandez, E.; Serret, P.; Madariaga, I.; Harbour, D.S.; Davies, A.G. Photosynthetic carbon metabolism and biochemical composition of spring phytoplankton assemblages enclosed in microcosms: the diatom-*Phaeocystis* sp. succession. Mar. Ecol. Prog. Ser. 90:89–102; 1992.

Feuillade, M.; Feuillade, J.; Pelletier, J-P. Photosynthate partitioning in phytoplankton dominated by the cyanobacterium *Oscillatoria rubescens*. Arch. Hydrobiol. 125:441–461; 1992.

Furgal, J.A. Effects of major environmental variables and ultraviolet light on photosynthetic carbon assimilation and allocation in a natural assemblage of Great Lakes phytoplankton. M.Sc. thesis, University of Waterloo, Ontario, Canada; 1995.

Furgal, J.A.; Smith, R.E.H. Ultraviolet radiation and photosynthesis by phytoplankton of varying light and nutrient status. Can. J. Fish. Aquat. Sci. 54:1659–1667; 1997.

Furgal, J.A.; Taylor, W.D.; Smith, R.E.H. Environmental control of photosynthate allocation in the phytoplankton of Colpoys Bay (Lake Huron). Can. J. Fish. Aquat. Sci. 55:726–736; 1998.

Glover, H.E.; Smith, A.E. Diel patterns of carbon incorporation into biochemical constituents of *Synechococcus* spp. and larger algae in the Northwest Atlantic Ocean. Mar. Biol. 97:259–267; 1988.

Gurr, M.I.; Harwood, J.L. Lipid Biochemistry, 4th ed. New York: Chapman & Hall; 1991.

Harwood, J.L. Fatty acid metabolism. Annu. Rev. Plant Physiol. 39:101–138; 1988.

Hawes, I. Photosynthate partitioning in Antarctic freshwater phytoplankton: *in situ* incubations. Freshwat. Biol. 24:193–200; 1990.

Holmes, R.W. The Secchi disk in turbid coastal water. Limnol. Oceanogr. 25:688–694; 1970.

Kirk, J. Light and Photosynthesis in Aquatic Systems. Cambridge: Cambridge University Press; 1983.

Lancelot, C.; Mathot, S. Biochemical fractionation of primary production by phytoplankton in Belgian coastal waters during short- and long-term incubations with ^{14}C-bicarbonate. I. Mixed diatom population. Mar. Biol. 86:219–226; 1985a.

Lancelot, C.; Mathot, S. Biochemical fractionation of primary production by phytoplankton in Belgian coastal waters during short- and long-term incubations with ^{14}C-bicarbonate. II. *Phaeocystis pouchetti* colonial population. Mar. Biol. 86:227–232; 1985b.

Lean, D.R.S.; Pick, F.R. Photosynthetic response of lake phytoplankton to nutrient enrichment: a test for nutrient limitation. Limnol. Oceanogr. 26:1001–1019; 1981.

Li, W.K.W.; Platt, T. Distribution of carbon among photosynthetic end-products in phytoplankton of the eastern Canadian Arctic. J. Phycol. 18:466–471; 1982.

Lombardi, A.T.; Wangersky, P.J. Influence of phosphorus and silicon on lipid class production by the marine diatom *Chaetoceros gracilis* grown in turbidostat cage cultures. Mar. Ecol. Prog. Ser. 77:39–47; 1991.

Madariaga, I. Interspecific differences in the photosynthetic carbon metabolism of marine phytoplankton. Mar. Biol. 114:509–515; 1992.

Maly, J.A. Vertical distribution and flux of phytoplankton and other suspended particles in a nearshore zone of western Georgian Bay (Colpoys Bay). M.Sc. thesis, University of Waterloo, Ontario, Canada; 1992.

McConville, M.J.; Mitchell, C.; Wetherbee, R. Patterns of carbon assimilation in a microalgal community from annual sea ice, east Antarctica. Polar Biol. 4:135–141; 1985.

Marañón, E.; Fernandez, E.; Anadon, R. Patterns of macromolecular synthesis by natural phytoplankton assemblages under changing upwelling regimes. J. Exp. Mar. Biol. Ecol. 188:1–28; 1995.

Morris, I. Photosynthesis products, physiological state and phytoplankton growth. In: Platt, T., ed. Physiological Bases of Phytoplankton Ecology. Can. Bull. Fish. Aquat. Sci. 210:83–102; 1981.

Morris, I.; Glover, H.E.; Yentsch, C.S. Products of photosynthesis by marine phytoplankton: the effect of environmental factors on the relative rates of protein synthesis. Mar. Biol. 27:1–9; 1974.

Munawar, M. Limnology and Fisheries of Georgian Bay and the North Channel Ecosystems. Boston: Kluwer Academic; 1988.

Palmisano, A.C.; Lizotte, M.P.; Smith, G.A.; Nichols, P.D.; White, D.C.; Sullivan, C.W. Changes in photosynthetic carbon assimilation in Antarctic sea-ice diatoms during spring bloom: variations in synthesis of lipid classes. J. Exp. Mar. Biol. Ecol. 116:1–13; 1988.

Parrish, C.C. Time series of particulate and dissolved lipid classes during spring phytoplankton blooms in Bedford Basin, a marine inlet. Mar. Ecol. Prog. Ser. 35:129–139; 1987.

Parrish, C.C.; Wangersky, P.J. Particulate and dissolved lipid classes in cultures of *Phaeodactylum tricornutum* grown in cage culture turbidostats with a range of nitrogen supply rates. Mar. Ecol. Prog. Ser. 35:119–128; 1987.

Platt, T.; Harrison, W.G.; Irwin, B.; Horne, E.P.; Gallegos, C.L. Photosynthesis and photoadaptation of marine phytoplankton in the Arctic. Deep-Sea Res. 29:1159–1170; 1982.

Post-Beitenmiller, D.; Roughan, G.; Ohlrogge, J.B. Regulation of plant fatty acid biosynthesis. Analysis of acyl-coenzyme A and acyl-acyl carrier protein substrate pools in spinach and pea chloroplasts. Plant Physiol. 100:923–930; 1992.

Priscu, J.C.; Priscu, L.R.; Vincent, W.F.; Howard-Williams, C. Photosynthate distribution by microplankton in permanently ice-covered Antarctic desert lakes. Limnol. Oceanogr. 32:260–270; 1987.

Rivkin, R.B. Influence of irradiance and spectral quality on the carbon metabolism of phytoplankton. I. Photosynthesis, chemical composition and growth. Mar. Ecol. Prog. Ser. 55:291–304; 1989.

Shifrin, N.S.; Chisholm, S.W. Phytoplankton lipids: interspecific differences and effects of nitrate, silicate and light-dark cycles. J. Phycol. 17:374–384; 1981.

Smith, R.E.H.; D'Souza, F.M.L. Macromolecular labelling patterns and inorganic nutrient limitation of a North Atlantic spring bloom. Mar. Ecol. Prog. Ser. 92:111–118; 1993.

Smith, R.E.H.; Maly, J.A. Photosynthetic carbon assimilation and allocation by surface and deep phytoplankton in Colpoys Bay (western Georgian Bay). Can. J. Fish. Aquat. Sci. 50:2235–2244; 1993.

Smith, R.E.H.; Clement, P.; Cota, G.F.; Li, W.K.W. Intracellular photosynthate allocation and the control of Arctic marine ice algal production. J. Phycol. 23:124–132; 1987.

Somerville, C.; Browse, J. Plant lipids: metabolism, mutants and membranes. Science 252:80–87; 1991.

Taguchi, S.; Hirata, J.A.; Laws, E.A. Silicate deficiency and lipid synthesis of marine diatoms. J. Phycol. 23:260–267; 1987.

Wainman, B.C.; Lean, D.R.S. A comparison of photosynthate allocation in lakes. J. Great Lakes Res. 22:803–809; 1996.

Wainman, B.C.; Lean, D.R.S. Methodological concerns in measuring the lipid fraction of carbon fixation. Hydrobiologia 273:111–120; 1994.

Wainman, B.C.; Lean, D.R.S. Carbon fixation into lipid in small freshwater lakes. Limnol. Oceanogr. 37:956–965; 1992.

Wainman, B.C.; Pick, F.R.; Hamilton, P.; Lean, D.R.S. Lipid production and phytoplankton species composition in small lakes. Arch. Hydrobiol. 128:197–207; 1993.

4

Lipids in Freshwater Zooplankton: Selected Ecological and Physiological Aspects

Michael T. Arts

4.1. Introduction

The high-energy density of lipids relative to proteins or carbohydrates coupled with the small body size of most planktonic invertebrates makes lipids the energy storage biomolecules of choice for freshwater and marine zooplankters. Zooplankton lipids often comprise a significant portion of their dry weight, with estimates ranging as high as 60–65% (Arts et al., 1993; Cavaletto et al., 1989). Furthermore, lipid concentrations are often reported as a percentage of dry weight, but the dry weight of zooplankton includes a significant contribution of less digestible chitin. Therefore, the contribution of lipid to the digestible portion of zooplankton is substantially higher than suggested from estimates of zooplankton lipid calculated simply as a function of total dry weight. The most variable component of zooplankton lipid is usually the energy reserve component. This class of lipid is dominated by the triacylglycerols (TAG) and, to a lesser extent in fresh waters, the wax esters. Most of the variability in individual lipid content reflects changes in TAG levels because this class of lipids most closely tracks seasonal changes in factors such as food supply and temperature, which are known to influence the deposition or loss of lipids (Arts et al., 1993). Phospholipids are a more refractory lipid class because they are an important constituent of membrane lipid.

Zooplankton lipid is primarily dietary in origin, and current research based on ^3H_2O and ^{14}C-acetate incorporation rates (Goulden and Place, 1990) suggests that very little de novo synthesis of lipid occurs. Thus, the amount and type of lipid contained within zooplankton tissues are often a good reflection of their recent feeding success and selectivity. Various other external factors such as temperature and salinity and internal factors such as reproductive condition will also influence zooplankton lipid content. Dynamic interactions between these factors produce the temporal patterns in the zooplankton lipid content, and these patterns profoundly affect the flow of energy within the ecosystems of lakes and wetlands. The role of zooplankton as gatherers and repositories of lipid and their intermediate position in the trophic chain make them a critical link between the lower and higher trophic levels.

There is currently a proliferation of interest in the role of essential fatty acids (EFAs) in animal health (Hall, 1996) and in human health (Layne et al., 1996; Osterud et al., 1995; Renaud et al., 1995). Recently, researchers have begun to examine the role of EFAs in structuring zooplankton (Hagen et al., 1995; Müller-Navarra, 1995a; Norsker and Stottrup, 1994) and fish (Bell and Sargent, 1996; Schwalme, 1994) populations and communities. Much research has also focused on algal EFA content because variations in EFA concentration and composition at this trophic level largely determine the amount and type of EFAs available for the rest of the food chain. The concept has been exploited by the mariculture industry, in which intense investigations centered around manipulating EFAs in feed stock are well underway to guarantee EFA availability for a variety of cultured organisms (Reis et al., 1996; and see Olsen, this volume).

The role of lipids as solvents for lipophilic contaminants is another reason to examine seasonal patterns of lipids in zooplankton. The sensitivity of lipids to factors that affect the feeding success of zooplankton makes them a useful stress detector, but only if the other conditions known to affect lipid levels are also carefully monitored and controlled. Thus, with the proper controls and ecological insights, lipids can be an excellent indicator of stress, particularly chronic stress (see Désy, 1996; Himbeault, 1995; Dauble et al., 1985).

My primary goal in this chapter is to illustrate selected aspects of lipid studies involving, primarily, zooplankton and to convince the reader of the usefulness of incorporating some measure of lipid content/composition in future studies. I also provide suggestions for what I believe to be important new research avenues in the field of lipids and highlight areas of research that are undergoing rapid expansion. Although the emphasis of this chapter is on freshwater zooplankton, lipid research in marine systems is included. Additional information on the ecophysiology of lipids in marine crustacean zooplankton can be found in Arts (1997). Although concepts presented here derive from work on zooplankton, reference is also made to other groups of organisms.

4.2. Usefulness of Areal Energy Reserve Estimates

Lipid levels in invertebrates are usually expressed as lipid content (i.e., the amount of lipid per animal) or as lipid concentration (i.e., the amount of lipid per unit mass of tissue). Lipid concentration has been variously measured as lipid/dry weight, lipid/ash-free dry weight, lipid/lipid-free dry weight, and lipid/wet weight, with no clear consensus as to which measure is most appropriate. Lipid/lipid-free dry weight may be preferred because this measure suffers less from problems of autocorrelation. Depending on the application, one expression may be preferred over the other, although, in principle, if the weights of the organisms are provided, conversion from one expression to the other is straightforward. Expression of lipid as lipid content is useful when one desires to characterize absolute differences in lipid, whereas the use of lipid concentration is more suited to providing information on the relative change in lipid over the season or in response to stress. An additional way to express lipid levels is as a function of surface area (Arts et al., 1992). In this approach, lipid contents are determined from stratified samples collected from the pelagic regions of lakes or wetlands. The measured lipid contents (μg per animal) are then multiplied by the areal density of the organisms and the results expressed as areal lipid (total lipid) or areal energy reserves (TAG plus wax esters, if they are present). Expression of lipids in the form of areal energy reserve may offer some unique insights into community structure and ecosystem functioning.

In saline Redberry Lake in central Saskatchewan, for example, three species of zooplankton (*Diaptomus sicilis, D. nevadensis,* and *Daphnia pulex*) are dominant (Arts et al., 1993). Expression of total lipid contents on an individual versus an areal basis provides a different representation of the seasonal lipid patterns in this lake (Fig. 4.1). Areal lipid patterns are a function of the density of organisms as well as the individual lipid contents so that population responses are readily evident. For example, in terms of individual lipid, lipid levels of *D. nevadensis* would appear to dwarf the contribution of the other two species in winter and spring (Fig. 4.1). However, by virtue of the relatively high densities of *D. sicilis,* it is clear that this species' contribution to overall zooplankton lipid levels on a community basis is much higher. *D. nevadensis,* because of constraints imposed by the size of its feeding basket, prefers copepodites and smaller *Daphnia* over adult copepods as prey items (Arts et al., 1993). Copepodites of *D. sicilis* are typically most abundant in Redberry Lake from July to September, and this is clearly reflected in the sharp increase in areal lipids of *D. nevadensis* observed during these months (Fig. 4.1). During this period, areal lipids increase fivefold, whereas individual lipids increase from a low of $\approx$60 μg per animal to a high of $\approx$115 μg per animal. Thus, the areal representation of lipids more clearly accentuates the strong seasonal dependence of *D. nevadensis* on appropriately sized prey items (Fig. 4.1).

4.3. Time Course of Lipid Deposition/Loss

The flux of energy reserve lipids is a highly dynamic process and is influenced by a variety of factors (see above). To measure the effects of these factors on lipid energy reserves, some a priori estimate of the time required for an observable change to be detected in the organism(s) under study is required. The time course of lipid deposition and/or loss is an important issue for several reasons.

First, detailed data on phytoplankton community structure made synchronously with measurements of zooplankton lipid reserves provide information about the suitability of a particular algal species for herbivorous zooplankton (Arts et al., 1993, 1992). This is not a trivial association because most zooplankton–phytoplankton interactions cannot be modeled in the laboratory. This is partially because many of the algal species (and zooplankton) found in nature have either not been cultured or are very difficult, expensive, or time-consuming to culture. In addition, it is not usually possible in the laboratory to recreate the species assem-

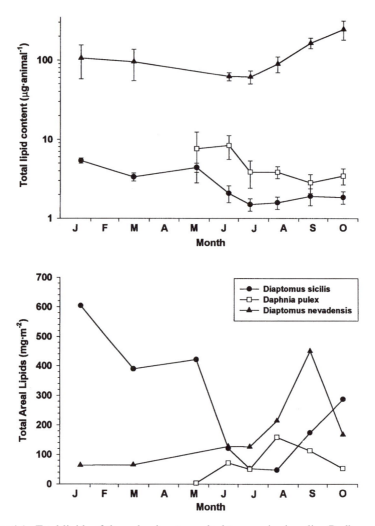

FIGURE 4.1. Total lipids of three dominant zooplankton species in saline Redberry Lake, Saskatchewan, from January to October 1994 expressed on an individual basis (top panel) and on an areal basis (lower panel). Data are for adult animals only. Note the log scale in the upper panel. Error bars are 95% confidence intervals.

blages and physiological conditions of phytoplankton as they exist in the lake/ wetland. Therefore, correlative field studies of the type mentioned here are often more realistic despite their more circumstantial nature when compared with laboratory feeding studies. Field studies of this type also provide information on the rate of zooplankton lipid deposition/loss under a variety of special conditions (e.g., during algal blooms, during periods of unusually cold or warm water temperatures).

Second, a successful demonstration of cause and effect between a stressor and the lipid energy reserves of the study organism depends on a knowledge of the normal time course of lipid deposition/loss. This information is necessary to apply proper statistical and logistical experimental design to chronic or acute eco-toxicological experiments in which the goal is to further our understanding of the effects of anthropogenic stressors on lipid energetics.

Third, during food web manipulation experiments there is often a time lag between the applied perturbation and the observed effect (Carpenter and Kitchell, 1992). Because lipids usually comprise the greatest fraction of energy-storing biomolecules in freshwater zooplankton, information on their rates of accrual and decline as they pass among trophic levels can lead to a better understanding of food web dynamics. This is particularly true if one examines the trophic level farthest removed from the perturbation. Thus, better estimates of the time required for lipids to pass between trophic levels should enhance our ability to predict the duration of time lags in food web studies.

Despite the importance of lipid energy reserves, their flux rates are seldom measured. Gut fluorescence analyses provide information on recent feeding success in terms of pigment intake but do not provide explicit information on rates of lipid acquisition (Ohman, 1988). Detailed ingestion studies (Pond et al., 1995; Bradshaw et al., 1990) provide information on ingestion and assimilation efficiency and are an important component in the determination of the time course and biochemical specifics of lipid acquisition in zooplankton.

Rapid transitions in small-scale local climatic events are potentially useful in providing information on rates of lipid energy reserve deposition or loss. This is because algal lipid reserves are known to be directly influenced by light and temperature in the light saturated zone (Wainman et al., this volume); lipid levels generally increase with increasing light and temperature. For example, over the course of a few days during a recent Group for Aquatic Productivity (GAP) workshop, photosynthetically active radiation (PAR) was reduced to 50% below that of a typical sunny day, air temperatures plummeted, and strong winds cooled the epilimnetic surface waters by nearly 2°C (Fig. 4.2). Concomitant with these physical events, concentrations of TAG in the phytoplankton declined, followed by a corresponding TAG of adult male copepods (*Diaptomus sicilis*). Males were chosen for these analyses because they obviate somewhat the need to be concerned with reproductive condition because, with females, greater care must be taken to get representative samples of gravid versus nongravid individuals. After only 2 d, there was an early indication of a decline in TAG levels in the copepods in response to the decline in algal TAG, and certainly, after 4 d, the response was clearly evident.

Bourdier and Amblard (1989) investigated the effect of starvation and diet variability on the lipid composition and content of the calanoid copepod *Acantho-diaptomus denticornis* in controlled laboratory experiments. They demonstrated that it took ~20 d following starvation to restore all the TAG of the copepods and <20% of the wax esters. Further, they showed that the rate of lipid restoration following starvation was dependent on the species of algae fed to the copepods.

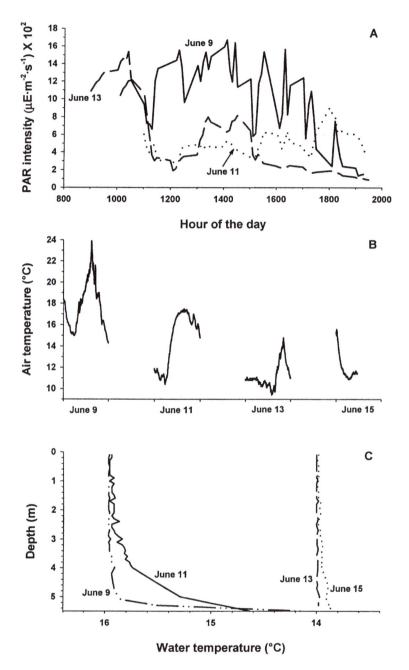

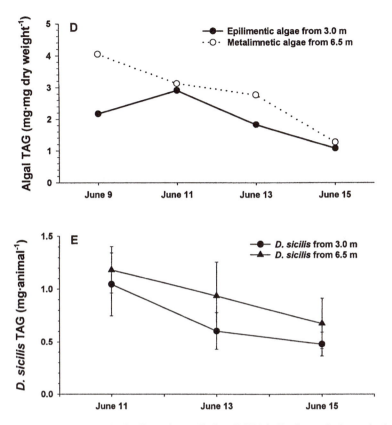

FIGURE 4.2. (A) Photosynthetically active radiation (PAR) in Redberry Lake at the lake's surface on 3 separate days collected with a Li-Cor quantum sensor (2π geometry) attached to LI-1000 data logger. (B) Air temperature above the surface of Redberry Lake. (C) Water temperature in the epilimnion of Redberry Lake. (D) Energy reserve lipids (TAG, tri-acylglycerols) of epi- and meta-limnetic algal communities in Redberry Lake on four separate dates during June 1993. (E) Energy reserve lipids (TAG) of ♂ calanoid copepods (*Diaptomus sicilis*) in Redberry Lake collected from two depths and on three separate dates. Error bars are 95% confidence intervals.

This result shows that recovery from food deficiency depends on the algal species present during the restoration period and that it is important for zooplankton to synchronize their peak population densities with those brief, key periods during the year (in temperate lakes) when highly digestible lipid-rich algae such as flagellates and diatoms dominate. Because lipid energy reserves are so critical for reproduction and, in some species, overwintering success, this result also serves to highlight the importance of these key periods of sensitivity when natural and anthropogenic perturbations could potentially have the greatest effect on zoo-plankton populations and communities.

4.4. Lipids as Indices of Stress

Zooplankton are often indirectly affected by stressors (e.g., toxicants) that affect algae. These effects can be of a more acute nature, as when the abundance of edible algae declines, or more chronic and subtle, as when the FA composition of algae is affected by the presence of the stressor (Laura et al., 1996; Mooney et al., 1995; Guckert et al., 1992). Zooplankton are also directly susceptible to a wide variety of stressors. In either situation, a strong case can be made that lipid levels and/or FA composition can and will respond in several ways to presence of stressors (see below). Stressors, in this regard, are not limited to anthropo-genically derived stressors; natural stressors such as starvation or exposure to suboptimal food, salinity, nutrient deficiency, and/or exposure to cyanobacterial toxins may also influence lipid levels and FA composition. It is therefore critical to realize, particularly in field studies, that observed changes in lipid concentra-tions or FA compositions will integrate the effects of both anthropogenic and natural stressors. This integration will also include a genetic component because an organism's response to stressors and its ability to secure food, reproduce, and so on will also influenced by the phenotypic expression of its genes. Finally, lipids will play a prominent role in sequestering nonpolar contaminants (Arts et al., 1996, 1995), and this will have a direct effect on bioaccumulation and depuration rates (see Landrum and Fisher, this volume).

4.4.1. Ratio of Storage to Membrane Lipids

Because storage lipids (i.e., TAG and wax esters) are the most variable of all the lipid classes and the most responsive to factors that affect the organisms' ability to obtain and process food, exposure to contaminants often results in a reduction in an organism's storage lipid content. For example, Capuzzo et al. (1984) demons-trated a decline in storage lipids in American lobster when they were exposed to petroleum hydrocarbons. It is often useful to standardize the observed changes in storage lipids against the less variable and primarily "structural" lipid classes (phospholipids and sterols) as proposed by Fraser (1989). Using this approach, Guckert et al., (1992) showed that periphyton exposed to municipal effluent had lower endogenous energy reserves (i.e., a decreased storage to membrane lipid ratio) than unexposed periphyton. Similarly, Himbeault (1995) showed that pe-riphyton collected on either rocks or artificial substrates had a reduced TAG-to-phospholipid ratio immediately downstream of a municipal water treatment plant. Although this approach has not yet been widely applied in studies involving zooplankton, the study of Himbeault (1995) clearly demonstrated that the TAG-to-phospholipid ratio in the mayfly *Baetis tricaudatus* was reduced when the animals were exposed to water containing 1 and 10% municipal effluent.

4.4.2. Maternal Lipid Investment

Zooplankton allocate lipids to their eggs to provide their offspring with fuel for growth and development and to offset the potential of starvation during the early life stages. In cladocerans, these energy reserves are highly visible within the eggs carried in the transparent brood chamber. The eggs progress through five developmental stages (Threlkeld, 1979). In the first developmental stage, the eggs contain a single, readily measurable, spherical lipid droplet. In freshwater Cladocera, these lipids have been characterized as neutral lipids belonging to the TAG family (Goulden and Henry, 1984). The product of egg number (clutch size) and mean lipid droplet volume in these eggs represents the maternal lipid investment (MLI) for that brood (Arts and Sprules, 1988). It has been demonstrated (for the cladoceran *Holopedium gibberum*) that most of the variation in MLI among females is due to between-lake differences and secondarily to differences between females within a lake (Arts and Sprules, 1988).

MLI is a useful stress indicator for three reasons. First, reproduction is often a sensitive endpoint in toxicology. Second, MLI relates to something with real biological meaning. Because cladocerans provide no parental care for their offspring beyond the egg stage, maternal investment in these organisms is represented exclusively by the investment they make in their eggs (MLI) and the protection the brood chamber affords the developing embryo. Finally, because MLI is the product of two interdependent parameters (i.e., clutch size and mean lipid volume), it is likely to be more sensitive than either parameter on its own. Thus, the use of MLI in standard toxicity testing is more likely to result in statistically significant differences (i.e., greater sensitivity) between the control and experimental treatments.

Désy (1996) demonstrated in 15-day chronic laboratory experiments that *Daphnia pulex* exposed to the herbicide triallate ($C_{10}H_{16}Cl_3NOS$) at 125 $\mu g \cdot L^{-1}$ had significantly lower MLI as compared with control *Daphnia*. Furthermore, greater differences between the control and the experimental treatment were observed for MLI than for either clutch size of mean egg volume when these data were corrected for the effects of body length (Désy, 1996, Table 6.4). MLI decreased from 5.00 ± 0.53 SE to 4.12 ± 0.67, representing a decrease of 17.6%, whereas clutch size decreased only slightly (and not significantly) from 10.7 ± 0.8 to 10.3 ± 1.1 (a decrease of 3.7%) and mean egg volume decreased 11.9% from 4.13 ± 0.15 to 3.64 ± 0.19.

4.4.3. Visible Lipid Energy Stores

Storage lipids are associated with pigments that make them highly visible in freshwater and marine zooplankton. This visibility allows for inexpensive microscopical techniques to measure lipid stores. Tessier and Goulden (1982) first proposed that a visible lipid index for *Daphnia* would be useful in assessing food limitation and evaluating stress in cladoceran populations. Using Tessier and Goulden's index, Holm and Shapiro (1984) provided empirical evidence that the cyanobacteria *Aphanizomenon flos-aquae* was a nutritionally inadequate food source for *Daphnia*. Dauble *et al.* (1985) extended the value of this index by commenting on its use as a stress indicator for bioaccumulation studies with *Daphnia magna*. Arts and Evans (1991) developed a similar but more quantitative optical-digital method to measure visible TAG and wax ester energy reserves in calanoid copepods. They proposed that the distribution of storage lipids within the body segments would be useful as a predictive and quantitative tool for monitoring and assessing the status of copepod populations with respect to their ability to resist starvation or, alternatively, as an index of their recent feeding success. Finally, in a study on the effects of acid stress on zooplankton communities, Locke and Sprules (1993) stained zooplankton with a lipophilic dye (Sudan IV) designed to accentuate the lipid stores in preserved samples (Bjorkman and Shapiro, 1986). They demonstrated, using in situ enclosure experiments, that even though some species of acid-tolerant species (e.g., *Sida crystallina*) persist in an enclosure following acidification, their lipid energy reserves, as measured by their lipid-ovary index scores, actually decreased. Interestingly, acid-tolerant species that appeared to benefit from the reduced competition and increased food supplies such as the cyclopoid copepods (e.g., *Diacyclops thomasi*) also contained increased levels of storage lipids (Locke and Sprules, 1993).

4.4.4. Fatty Acid Composition and Abundance

Examination of FA composition and abundance (FA profiles) in zooplankton in relation to stress has not received the attention it deserves. Several examples in algae (Mooney et al., 1995; Sicko-Goad et al., 1989) and periphyton (Napolitano et al., 1994) illustrate how FA profiles change as a function of stress. Because zooplankton obtain their FAs from the diet, there is good reason to believe that zooplankton FA profiles will also be affected by the presence of stressors. Because FAs are fundamental components of membranes, the effects of stressors on FA composition and the subsequent health of zooplankton merit further attention.

4.5. Ultraviolet Radiation and Zooplankton Lipids

Significant reductions of stratospheric ozone have been observed over the past decade, and there is concern that ultraviolet B (UV-B) radiation (300–320 nm) reaching the earth's surface will also increase (Madronich, 1995). UV-B radiation can penetrate to ecologically significant depths in natural waters, thus the observed increases in surface UV-B radiation have the potential to adversely affect phytoplankton productivity, especially in shallow lakes, wetlands, and rivers. There is new evidence to suggest that effects of UV-B radiation in freshwater systems may be exacerbated by anthropogenic acidification through associated processes that reduce the amount of dissolved organic carbon in lakes (Schindler et al., 1996; Yan et al., 1996).

The effects of UV radiation (UVR) on zooplankton have, to date, focused on acute exposures resulting in direct mortality or reductions in fecundity rather than indirect influences such as food chain effects (Williamson et al., 1994; Dey et al., 1988). Although such direct effects have been supported, studies that document the indirect effects of UV-B radiation on lipid metabolism in zooplankton are just starting to emerge (Hessen et al., 1997). In addition, although zooplankton are capable of a wide variety of adaptive responses to minimize their exposure to UV-B (e.g., active swimming, vertical migration), studies on the effects of UV-B radiation on specific life stages (eggs, larvae, juveniles) are still in their infancy (Siebeck and Böhm, 1994).

Zooplankton obtain lipids largely from their diet; therefore, indirect effects of UV-B on the algal, cyanobacterial, and bacterial food of zooplankton are perhaps the most likely mechanism whereby UV-B radiation may influence zooplankton lipids. Although numerous studies have demonstrated overall reductions in gross photosynthetic rates (reviewed in Häder, 1996), few studies have examined the effects of UV-B radiation on carbon allocation in freshwater algae (but see Arts and Rai, 1997). Furthermore, in two different studies, exposure to UV-B radiation (at fluxes below that required to inhibit photosynthetic carbon assimilation) differentially altered the FA composition by reducing the concentrations of poly-unsaturated fatty acids (PUFA) of marine algae (Goes et al., 1994; Wang and Chai, 1994).

In larvae of marine and, presumably, freshwater fish, there is a large demand for long-chain PUFA during early development. Fish must obtain these EFAs through their diet (see Olsen, this volume). Deficits in EFAs, in particular 22:6ω3, have been shown to affect the number of rods in the photoreceptor population of the eyes of herring (Bell and Dick, 1993) and to hamper the feeding of herring under low light intensities (Bell et al., 1995). Thus, any reductions in algal PUFA as a result of exposure to elevated levels of UVR could have profound consequences for larval feeding success. In addition, it is tempting to speculate that visual acuity in the zooplankton will similarly be affected by the concentration of PUFA in their diet. The compound eye of *Daphnia,* for example, is important in the maintenance of position in the water column (vertical migration) and for predator avoidance.

In the above contexts, more studies are urgently required on the effects of UVR on carbon allocation in algae, including the effects of UVR on FA profiles as well as how these changes in carbon allocation patterns and FA profiles influence both zooplankton and larval fish survivorship, growth, and fecundity.

4.6. Research Needs and Suggested Future Directions

Many of the authors in this book have commented on the apparent lack of information on lipid dynamics in freshwater compared with marine ecosystems. Although this certainly has been true, it is also apparent that freshwater lipid researchers have made significant strides in recent years. The advent of microgravimetric extraction and quantification (Gardner et al., 1985; Cavalleto and Gardner, this volume), the Iatroscan TLC-FID analyzer, and advances in gas chromatography techniques (Parrish, this volume) have greatly facilitated studies on total lipid, lipid classes, and FAs, respectively, of zooplankton, other invertebrates, and algae. These types of studies have allowed researchers to begin to gain an understanding of the role of various algal and cyanobacterial species as factors affecting zooplankton nutrition, survival, and reproduction. Still, many unanswered questions remain.

4.6.1. Geographical Disparities

Most of the work that has been done on lipid dynamics of freshwater zooplankton comes from the midlatitudes and, most commonly, from the Northern Hemisphere. Studies on high-altitude or tropical zooplankton populations (Wyngaard et al., 1994) are comparatively rare. Also notably absent are studies on lipid energetics, FA composition from some of the world's ancient lakes such as the African rift lakes and Lake Baikal (however, see Morris, 1983). These lakes with their endemic zooplankton species (e.g., the calanoid copepod *Epischura baikalensis*) and lipid-rich phytoplankton species (e.g., *Botryococcus braunii*) may well provide unique insights into novel manifestations of lipid dynamics and strategies in situations in which the evolutionary relationships among organisms have been maintained in relative isolation for millennia. Equally important is the opportunity to document present "baseline" lipid contents for a variety of species now subject to increasing anthropogenic stress in these systems.

4.6.2. Physicochemical Disparities

The total volume of the world's inland saline lakes is only slightly less (~83%) than that of freshwater lakes (Hammer, 1986). The definition of saline is still debated, but the one that curries the most favor is that saline lakes are lakes where the salt concentration exceeds 3.0 mg $\cdot$ L^{-1} (Hammer, 1986). Applying this definition to all inland lakes means that there are a substantial number of "saline" waters that presently support "freshwater" zooplankton such as *Daphnia*. One such lake is saline (total dissolved solids = 21 mg $\cdot$ L^{-1}) Redberry Lake in central Saskatchewan, which has viable populations of *D. pulex* and *Diaptomus sicilis* (Arts et al., 1993). Thus, many saline lakes (some of them quite large) support viable daphnid and copepod populations. The lipid chemistry and ecology of zooplankton inhabiting these inland saline lakes are poorly understood. For example, it is unknown whether nutrient limitation or EFA limitation (or food quality in the broadest sense) or predation is the dominant mechanism limiting *Daphnia,* copepods, or rotifers. The food webs and nutrient regimes in these systems often deviate substantially from the more familiar points of reference (i.e., the freshwater lakes). Food chain length and complexity become increasingly truncated along salinity gradients producing more "simplified" communities (i.e., shorter and few links). Because of this inherent simplicity, these communities provide useful natural laboratories for the examination of the effects of perturbation, food quality, temperature, and so on on lipid energy transfers and fatty acid composition and abundance. In more simplified communities, it is potentially easier to ascribe "cause and effect" to observations made following natural or artificial perturbations because the links are fewer, and therefore transfers of biomolecules from one trophic level to the next are likely to be more direct.

4.6.3. Essential Fatty Acids

Recently, much interest has been engendered in the so-called EFAs, members of the (ω3) and (ω6) series of FAs. For example, the studies of Müller-Navarra (1995a,b) have suggested that *Daphnia* growth rates may be well correlated with the sestonic content of eicosapentaenoic acid (20:5ω3). It is well known that fish and, in fact, all vertebrate species studied to date have an absolute requirement for certain of these PUFAs (Sargent et al., 1995). In addition, there are mammalian health implications because inadequate representation of these compounds in the diet has been implicated by numerous clinical studies in a wide variety of pathological conditions or added "health-risk potentialities" (Craig-Schmidt et al., 1996; Osterud et al., 1995; van Houwelingen et al., 1995).

One of the first steps necessary to demonstrate whether EFAs play a major role limiting the production of freshwater organisms is to document where and when these compounds exist in the lake or wetland. This requires that careful inventories of the FA composition of both primary producers (Ahlgren et al., 1990) and higher-order consumers (Bell et al., 1994; Hanson et al., 1985) be conducted in a variety of systems. This is a formidable task both because of the large numbers of species and because of the seasonal variability in FA composition. Changes in FA composition and biomass will also accompany changes in the different life stages of an organism, and these changes will be at least partially explained by shifts in diet as the animal matures. Relevant questions in this context might be, How does the FA and, in particular, the EFA composition change in the same species across trophic gradients or between freshwater and saline lakes at similar trophic levels? At what times of the year are sestonic EFA concentrations greatest, and which organisms appear most capable to capitalize on these nutritionally important particles? Are there important "bottlenecks" in the production of EFAs in the phytoplankton? How are the FAs of freshwater zooplankton modified by fish following dietary lipid transfer (Desvilettes, 1994)?

Fish have an absolute EFA requirement of roughly 0.5–1.0% of the diet (Sargent et al., 1995). Similar sorts of estimates for freshwater zooplankton are not yet available; however, the zooplankton may well be highly responsive to the presence of food sources containing relatively high concentrations of EFAs (but see Goulden et al., this volume) or at least have evolved effective strategies to store and conserve EFAs. If the appearance of EFAs in the seston follows more or less predictable seasonal patterns, then peaks in feeding or population density should occur when EFAs are most abundant or obtainable. By extension, if the procurement of EFAs is a strong driving force in the evolution of vertebrate communities, then fish might be expected to show some preference for zooplankton species rich in EFA.

4.6.4. Effects of Temperature Changes

Altering FA composition to affect changes in membrane fluidity is one of the principal methods that organisms use to adapt and acclimate to changes in temperature (Gurr and Harwood, 1991). Thus, membrane FA composition plays a significant role in overwintering survivorship and/or in the ability of organisms to colonize particular environments. A large body of literature, particularly in fish and mammals, relates to the effects of changes in membrane FAs on membrane fluidity (Gurr and Harwood, 1991). Information on algae (Nishida and Murata, 1996; Tatsuzawa et al., 1996) and zooplankton (Farkas et al., 1984, 1981; Farkas, 1979), particularly from freshwater species, is comparatively rare. Recent predictions concerning the effects of climate change suggest that increases in air temperature over a 20-year period (1970–1990) have resulted in increases in the length of the ice-free season, mean and maximum water temperature, and the heat contents of boreal forest lakes (Schindler et al., 1990). Given these trends in temperature, there is a need to better quantify the relationships among membrane FA composition, membrane fluidity, overwintering success, and colonization ability of organisms.

4.6.5. Diapause

Diapause is a potent phenomenon structuring pelagic zooplankton communities (Dahms, 1995; Hairston et al., 1995; Santer and Lampert, 1995). Resting stages function to replenish populations following periods when conditions for the active stages become uninhabitable. Diapausing eggs can be centuries old and still be viable (Hairston et al., 1995). Given that EFAs are so vital to the health of mature animals, the FA composition and in particular EFA concentration in the diet may have a significant effect on the viability of resting eggs (see Jónasdóttir and Kiorboe, 1996). In a similar manner, the survivorship and growth of newly hatched *Daphnia* from ephippia may well be influenced by the concentrations and composition of FAs in the resting stages. Because lipids play such a crucial role both in the maintenance of cell membrane functionality and as metabolic fuel for newly hatched animals, FA composition likely has an important role to play in the viability of resting stages.

4.6.6. Lipids as Allelopathic Compounds and Chemical Feeding Deterrents

It is now known that algae produce compounds that inhibit the growth of other algae or sometimes themselves (auto-growth inhibition). This phenomenon is called allelopathy, and there are examples from both the marine (Kakisawa et al., 1988) and freshwater (Yamada et al., 1993) literature. Allelopathic compounds are typically unsaturated FAs (Suzuki et al., 1996; Yamada et al., 1993; Kakisawa et al., 1988) or other compounds that fit the broad definition of lipid. Some of the compounds produced by algae also deter the feeding of zooplankton (Shaw et al., 1995a,b) or other invertebrates such as gastropods (Sawai et al., 1994). As with the allelopathic compounds, chemical deterrents are often lipidic in nature (Shaw et al., 1995a,b; Sawai et al., 1994).

By suppressing the growth of other algal species, allelopathic compounds can indirectly affect zooplankton community structure and production rates. By controlling the length of time a bloom remains dominant (through direct suppression of zooplankton feeding), chemical deterrents may be able to affect phytoplankton community structure while simultaneously exerting direct control on zooplankton production rates. Because of these direct and indirect effects on zooplankton community structure and production, allelopathic and chemical deterrent compounds have the potential to affect the rate of energy transfers in pelagic ecosystems. Research into the identification and characterization of these compounds has progressed faster for marine than for freshwater systems because of the discernable threat to and possible usefulness of these compounds to the mariculture industry. For example, allelopathic compounds show great promise in terms of reducing the abundance of "undesirable" or "nuisance" species in pond- or tank-rearing systems designed to grow algal species for the production of economically important compounds. Also, in terms of harvesting wild-species with desirable chemical constituents (i.e., "biomolecular farming"), application of allelopathic compounds at key periods may be a useful way of maximizing yields from natural systems. Finally, allelopathic chemicals may, in the future, serve an important role as antifouling compounds. More research into the chemical structure and role of these allelopathic compounds in freshwater ecosystems is clearly required before the detailed mechanisms and generality of their use in controlling algal and zooplankton production and community composition can be assessed.

4.7. Conclusions

Lipids, particularly energy reserve lipids, are sensitive to environmental perturbation, both natural and anthropogenic. This feature makes lipids a useful marker of stress. The ratio of storage to membrane lipids, maternal lipid investment, and the tracking of visible lipid energy stores are potentially useful indices of stress in zooplankton. Changes in FA profiles as an index of stress in zooplankton is an area that holds great promise but it is also one in which more research and validation are urgently required.

The various expressions of energy reserve lipids (lipid content and concentration, areal energy reserves, and the visible lipid indices) provide unique perspectives on the physiological status of zooplankton and the relative contribution each species makes to the standing pools of storage lipid energy in freshwater ecosystems. By monitoring changes in these standing stocks of storage lipids, one can gain insights into the time course of lipid deposition and loss rates in organisms both within and between trophic levels.

The presence of UVR is a fact of existence for pelagic organisms; however, recent evidence suggesting that UV-B radiation is increasing has raised concerns. Rates of photosynthesis are clearly depressed in many algae and some marine algae have demonstrated altered FA profiles following exposure to realistic fluxes of UV-B. Perhaps more significantly, these changes in FAs include reductions in some of the EFAs. The implications of these reductions in EFAs following exposure to UV-B radiation in terms of survival and production rates of freshwater zooplankton should prove an interesting challenge.

Many critical and fascinating questions regarding the ecophysiology and chemistry of lipids in freshwater ecosystems remain unanswered. I have, in this chapter, tried to highlight but a few of these. The role of EFAs in altering zooplankton community structure, standing biomass, production rates, and perhaps the diapause process merits further study. Lipids are key players in the phenomenon of allelopathy and also act as chemical deterrents to foil zooplankton feeding. Further elucidation of the structure of these activity suppressing compounds and their role in structuring pelagic communities will undoubtedly be an emerging area of great importance both from the academic and commercial perspective.

Acknowledgments. I am grateful to Environment Canada, through the National Water Research Institute, for giving me the opportunity to write this chapter. Chris Jones (George Mason University) provided me with the 1993 PAR data from Redberry Lake. I thank Eric von Elert (Limnological Institute, University of Constance) for helpful discussion on the role of allelopathic and chemical deterrent compounds in marine and freshwater ecosystems. David R. S. Lean (University of Ottawa, Ontario), Barbara Santer (Max Planck Institut für Limnology, Plön, Germany), and Bruce C. Wainman (McMaster University, Ontario) provided valuable comments on an earlier draft of this chapter.

References

Ahlgren, G.; Lundstedt, L.; Brett, M.T.; Forsberg, C. Lipid composition and food quality of some freshwater phytoplankton for cladoceran zooplankters. J. Plankton Res. 12:809–818; 1990.

Arts, M.T. Ecophysiology of lipids in pelagic crustacean zooplankton communities. In: Cooksey, K.E., ed. Molecular Approaches to the Study of the Ocean. London: Chapman & Hall; 1997:p. 329–341.

Arts, M.T.; Evans, M.S. Optical-digital measurements of energy reserves in calanoid copepods: Intersegmental distribution and seasonal patterns. Limnol. Oceanogr. 36:289–298; 1991.

Arts, M.T.; Rai, H. Effects of ultraviolet radiation on the production of lipid, carbohydrate, and protein in three freshwater algal species. Freshwat. Biol. 38:597–610; 1997.

Arts, M.T.; Sprules, W.G. Evidence for indirect effects of fish predation on maternal lipid investment in *Holopedium gibberum*. Can. J. Fish. Aquat. Sci. 45:2147–2155; 1988.

Arts, M.T.; Headley, J.V.; Peru, K.M. Persistence of herbicide residues in *Gammarus lacustris* (Crustacea: Amphipoda) in prairie wetlands. Environ. Toxicol. Chem. 15:481–488; 1996.

Arts, M.T.; Ferguson, M.E.; Glozier, N.E.; Robarts, R.D.; Donald, D.B. Spatial and temporal variability in lipid dynamics of common amphipods: assessing the potential for uptake of lipophilic contaminants. Ecotoxicology 4:91–113; 1995.

Arts, M.T.; Robarts, R.D.; Evans, M.S. Energy reserve lipids of zooplanktonic crustaceans from an oligotrophic saline lake in relation to food resources and temperature. Can. J. Fish. Aquat. Sci. 50:2404–2420; 1993.

Arts, M.T.; Evans, M.S.; Robarts, R.D. Seasonal patterns of total and energy reserve lipids of dominant zooplanktonic crustaceans from a hyper-eutrophic lake. Oecologia 90:560–571; 1992.

Bell, J.G.; Dick, J.R.; The appearance of rods in the eyes of herring and increased didocosa-hexaenoyl molecular species of phospholipids. J. Mar. Biol. Assn. U.K. 73:679–688; 1993.

Bell, J.G.; Batty, R.S.; Dick, J.R.; Fretwell, K.; Navarro, J.C.; Sargent, J.R. Dietary deficiency of docosahexaenoic acid impairs vision at low light intensities in juvenile herring (*Clupea harengus* L.). Lipids 26:565–573; 1995.

Bell, J.G.; Ghioni, C.; Sargent, J.R. Fatty acid compositions of 10 freshwater invertebrates which are natural food organisms of Atlantic salmon parr (*Salmo salar*): a comparison with commercial diets. Aquaculture 128:301–313; 1994.

Bell, M.V.; Sargent, J.R. Lipid nutrition and fish recruitment. Mar. Ecol. Prog. Ser. 134:315–316; 1996.

Bjorkman, B.; Shapiro, J. Measurement of the lipid-ovary index on stored zooplankton samples. Limnol. Oceanogr. 31:1138–1139; 1986.

Bourdier, G.G.; Amblard, C.A. Lipids in *Acanthodiaptomus denticornis* during starvation and fed on three different algae. J. Plankton Res. 11:1201–1212; 1989.

Bradshaw, S.A.; O'Hara, S.C.M.; Corner, E.D.S.; Eglinton, G. Changes in lipids during simulated herbivorous feeding by the marine crustacean *Neomysis integer.* J. Mar. Biol. Assn. U.K. 70:225–243; 1990.

Capuzzo, J.M.; Lancaster, B.A.; Sasaki, G.C. The effects of petroleum hydrocarbons on lipid metabolism and energetics of larval development and metamorphosis in the American lobster (*Homarus americanus* Milne Edwards). Mar. Environ. Res. 14:201–228; 1984.

Carpenter, S.R.; Kitchell, J.F. Trophic cascade and biomanipulation: interface of research and management—a reply to the comment by DeMelo et al. Limnol. Oceanogr. 37:208–213; 1992.

Cavaletto, J.F.; Vanderploeg, H.A.; Gardner, W.S. Wax esters in two species of freshwater zooplankton. Limnol. Oceanogr. 34:785–789; 1989.

Craig-Schmidt, M.C.; Stieh, K.E.; Lien, E.L. Retinal fatty acids of piglets fed docosahexaenoic and arachidonic acids from microbial sources. Lipids 31:53–59; 1996.

Dahms, H-U. Dormancy in the Copepoda—an overview. Hydrobiologia 306:199–211; 1995.

Dauble, D.D.; Klopfer, D.C.; Carlile, D.W.; Hanf, R.W., Jr. Usefulness of the lipid index for bioaccumulation studies with *Daphnia magna*. In: Bahner, R.C.; Hansen, D.J., eds. Aquatic Toxicology and Hazard Assessment. 8th ed. Philadelphia: American Society for Testing and Materials; 1985: p. 350–358.

Desvilettes, Ch.; Bourdier, G.G.; Breton, J-C. Lipid class and fatty acid composition of planktivorous larval pike *Esox lucius* living in a natural pond. Aquat. Living Resources 7:67–77; 1994.

Désy, J. Fate and effects of triallate in a prairie wetland. M.Sc. thesis, University of Saskatchewan, Saskatoon, Canada; 1996.

Dey, D.B.; Damkaer, D.M.; Heron, G.A. UV-B dose/dose-rate responses of seasonally abundant copepods of Puget Sound. Oecologia 76:321–329; 1988.

Farkas, T. Adaptation of fatty acid compositions to temperature—a study on planktonic crustaceans. Comp. Biochem. Physiol. 64:71–76; 1979.

Farkas, T.; Nemecz, G.Y.; Csengeri, I. Differential response of lipid metabolism and membrane physical state by an actively and passively overwintering planktonic crustacean. Lipids 19:436–442; 1984.

Farkas, T.; Kariko, K.; Csengeri, I. Incorporation of [1-^{14}C] acetate into fatty acids of the crustaceans *Daphnia magna* and *Cyclops strenus* in relation to temperature. Lipids 16:418–422; 1981.

Fraser, A.J. Triacylglycerol content as a condition index for fish, bivalve, and crustacean larvae. Can. J. Fish. Aquat. Sci. 46:1868–1873; 1989.

Gardner, W.S.; Frez, W.A.; Cichocki, E.A.; Parrish, C.C. Micromethod for lipids in aquatic invertebrates. Limnol. Oceanogr. 30:1099–1105; 1985.

Goes, J.I.; Handa, N.; Taguchi, S.; Hama, T. Effect of UV-B radiation on the fatty acid composition of the marine phytoplankton *Tetraselmis* sp: relationship to cellular pigments. Mar. Ecol. Prog. Ser. 114:259–274; 1994.

Goulden, C.E.; Henry, L.L. Lipid energy reserves and their role in Cladocera. In: Meyers, D.G.; Strickler, J.R., eds. AAAS Selected Symposium 85. Boulder, CO: Westview; 1984: p. 167–185.

Goulden, C.E.; Place, A.R. Fatty acid synthesis and accumulation rates in daphnids. J. Exp. Zool. 256:168–178; 1990.

Guckert, J.B.; Nold, S.C.; Boston, H.L.; White, D.C. Periphyton response in an industrial receiving stream: lipid-based physiological stress analysis and pattern recognition of microbial community structure. Can. J. Fish. Aquat. Sci. 49:2579–2587; 1992.

Gurr, M.I.; Harwood, J.L. Lipid Biochemistry—an introduction, 4th ed. London: Chapman & Hall; 1991

Häder, D-P. Effects of enhanced solar UV-B radiation on phytoplankton. Sci. Mar. 60:59–63; 1996.

Hagen, W.; Kattner, G.; Graeve, M. On the lipid biochemistry of polar copepods: compositional differences in the Antarctic calanoids *Euchaeta antarctica* and *Euchirella rostromagna.* Mar. Biol. 123:451–457; 1995.

Hairston, N.G., Jr.; Van Brunt, R.A.; Kearns, C.M.; Engstrom, D.R. Age and survivorship of diapausing eggs in a sediment egg bank. Ecology 76:1706–1711; 1995.

Hall, J.A. Potential adverse effects of long-term consumption of (n-3) fatty acids. Compend. Cont. Educ. Pract. Vet. 18:879–879; 1996.

Hammer, U.T. Saline Lake Ecosystems of the World. Dordrecht, The Netherlands: Dr. W. Junk Publ.; 1986.

Hanson, B.J.; Cummins, K.W.; Cargill, A.S.; Lowry, R.R. Lipid content, fatty acid composition, and the effect of diet of fats of aquatic insects. Comp. Biochem. Physiol. 80B:257–276; 1985.

Hessen, D.O.; DeLange H. J.; Van Donk E. UV-induced changes in phytoplankton cells and its effects on grazers. Freshwat. Biol. 38:513–524; 1997.

Himbeault, K.T. Alternative approaches for monitoring aquatic ecosystems: a study on the effects of municipal sewage effluent on biota in the South Saskatchewan River. M.Sc. thesis, University of Saskatchewan, Saskatoon, Canada; 1995.

Holm, N.P.; Shapiro, J. An examination of lipid reserves and the nutritional status of *Daphnia pulex* fed *Aphanizomenon flos-aquae.* Limnol. Oceanogr. 29:1137–1140; 1984.

Jónasdóttir, S.H.; Kiorboe, T. Copepod recruitment and food composition: do diatoms affect hatching success? Mar. Biol. 125:743–750; 1996.

Kakisawa, H.; Asari, F.; Kusumi, T.; Toma, T.; Sakurai, T.; Oohusa, T.; Hara, Y.; Chihara, M. An allelopathic fatty acid from the brown alga *Cladosiphon okamuranus.* Phytochemistry 27:731–735; 1988.

Laura, D.; DeSocio, G.; Frassanito, R.; Rotilio, D. Effects of atrazine on *Ochrobactrum anthropi* membrane fatty acids. Appl. Environ. Microbiol. 62:2644–2646; 1996.

Layne, K.S.; Goh, Y.K.; Jumpsen, J.A.; Ryan, E.A.; Chow, P.; Clandinin, M.T. Normal subjects consuming physiological levels of 18:3(n-3) and 20:5(n-3) from flaxseed or fish oils have characteristic differences in plasma lipid and lipoprotein fatty acid levels. J. Nutr. 126:2130–2140; 1996.

Locke, A.; Sprules, W.G. Effects of experimental acidification on zooplankton population and community dynamics. Can. J. Fish. Aquat. Sci. 50:1238–1247; 1993.

Madronich, S.; McKenzie, R.L.; Caldwell, M.M.; Björn, L.O. Changes in ultraviolet radiation reaching the earth's surface. Ambio 24:143–152; 1995.

Mooney, H.M.; Cooney, J.J.; Baisden, C.M.; Patching, J.W. Fatty acids of *Dunaliella tertiolecta* and *Skeletonema costatum* grown in the presence of phenyltin compounds. Bot. Mar. 38:423–429; 1995.

Morris, R.J. Absence of wax esters in pelagic Lake Baikal fauna. Lipids 18:149–150; 1983.

Müller-Navarra, D. Evidence that a highly unsaturated fatty acid limits *Daphnia* growth in nature. Arch. Hydrobiol. 132:297–307; 1995a.

Müller-Navarra, D. Biochemical versus mineral limitation in *Daphnia*. Limnol. Oceanogr. 40:1209–1214; 1995b.

Napolitano, G.E.; Hill, W.R.; Guckert, J.B.; Stewart, A.J.; Nold, S.C.; White, D.C. Changes in periphyton fatty acid composition in chlorine-polluted streams. J. North Am. Benthol. Soc. 13:237–249; 1994.

Nishida, I.; Murata, N. Chilling sensitivity in plants and cyanobacteria: the crucial contribution of membrane lipids. Annu. Rev. Plant Physiol. 47:541–568; 1996.

Norsker, N-H.; Stottrup, J.G. The importance of dietary HUFAs for fecundity and HUFA content in the harpacticoid, *Tisbe holothuriae* Humes. Aquaculture 125:155–166; 1994.

Ohman, M.D. Sources of variability in measurements of copepod lipids and gut fluorescence in the California current coastal zone. Mar. Ecol. Prog. Ser. 42:143–153; 1988.

Osterud, B.; Elvevoll, E.; Barstad, H.; Brox, J.; Halvorsen, H.; Lia, K.; Olsen, J.O.; Olsen, R.L.; Sissener, C.; Rekdal, O.; Vognild, E. Effect of marine oils supplementation on coagulation and cellular activation in whole blood. Lipids 30:1111–1118; 1995.

Pond, D.W.; Priddle, J.; Sargent, J.R.; Watkins, J.L. Laboratory studies of assimilation and egestion of algal lipid by Antarctic krill—methods and initial results. J. Exp. Mar. Biol. Ecol. 187:253–268; 1995.

Reis, A.; Gouveia, L.; Veloso, V.; Fernandes, H.L.; Empis, J.A.; Novais, J.M. Eicosapentaenoic acid-rich biomass production by the microalga *Phaeodactylum tricornutum* in a continuous-flow reactor. Bioresource Technol. 55:83–88; 1996.

Renaud, S.; Lorgeril, M.; Delaye, J.; Guidollet, J.; Jacquard, F.; Mamelle, N.; Martin, J-L.; Monjaud, I.; Salen, P.; Touboul, P. Cretan Mediterranean diet for prevention of coronary heart disease. Am. J. Clin. Nutr. 61:1360S-1366S; 1995.

Santer, B.; Lampert, W. Summer diapause in cyclopoid copepods: adaptive response to a food bottleneck. J. Anim. Ecol. 64:600–613; 1995.

Sargent, J.R.; Bell, J.G.; Bell, M.V.; Henderson, R.J.; Tocher, D.R. Requirement criteria for essential fatty acids. J. Appl. Ichthyol. 11:183–198; 1995.

Sawai, Y.; Fujita, Y.; Sakata, K.; Tamashiro, E. Chemical studies of feeding inhibitors for marine herbivores. 3. 20-hydroxy-4, 8, 13, 17-tetramethyl-4, 8, 12, 16-eicosapentanoic acid, a new feeding deterrent against herbivorous gastropods, from the subtropical brown algae *Turbinaria ornata*. Fish Sci. 60:199–201; 1994.

Schindler, D.W.; Curtis, P.J.; Parker, B.R.; Stainton, M.P. Consequences of climate warming and lake acidification for UV-B penetration in North American boreal lakes. Nature 379:705–708; 1996.

Schindler, D.W.; Beaty, K.G.; Fee, E.J.; Cruikshank, D.R.; Debruyn, E.R.; Findlay, D.L.; Linsey, G.A.; Shearer, J.A.; Stainton, M.P.; Turner, M.A. Effects of climatic warming on lakes of the central boreal forest. Science 250:967–970; 1990.

Schwalme, K. Reproductive and overwintering adaptations in northern pike (*Esox lucius* L): balancing essential fatty acid requirements with dietary supply. Physiol. Zool. 67:1507–1522; 1994.

Shaw, B.A.; Anderson, R.J.; Harrison, P.J. Feeding deterrence properties of apo-fucoxanthinoids from marine diatoms. I. Chemical structures of apo-fucoxanthinoids produced by *Phaeodactylum tricornutum*. Mar. Biol. 124:467–472; 1995a.

Shaw, B.A.; Harrison, P.J.; Anderson, R.J. Feeding deterrence properties of apo-fucoxanthinoids from marine diatoms. II. Physiology of production of apo-fucoxanthinoids by the marine diatoms *Phaeodactylum tricornutum* and *Thalassiosira pseudonana,* and their feeding deterrent effects on the copepod *Tigriopus californicus*. Mar. Biol. 124:473–481; 1995b.

Sicko-Goad, L.; Evans, M.S.; Lazinsky, D.; Hall, J.; Simmons, M.S. Effects of chlorinated benzenes on diatom fatty acid composition and quantitative morphology. IV. Pentachlorobenzene and comparison with trichlorobenzene isomers. Arch. Environ. Contam. Toxicol. 18:656–668; 1989.

Siebeck, O.; Böhm, U. Challenges for an appraisal of UV-B effects upon crustaceans under natural radiation conditions with a non-migrating (*Daphnia pulex obtusa*) and a migrating cladoceran (*Daphnia galeata*). Arch. Hydrobiol. Beih. Ergebn. Limnol. 43:197–206; 1994.

Suzuki, M.; Wakana, I.; Denboh, T.; Tatewaki, M. An allelopathic polyunsaturated fatty acid from red algae. Phytochemistry 43:63–65; 1996.

Tatsuzawa, H.; Takizawa, E.; Wada, M.; Yamamoto, Y. Fatty acid and lipid composition of the acidophilic green alga *Chlamydomonas* sp. J. Phycol. 32:598–601; 1996.

Tessier, A.J.; Goulden, C.E. Estimating food limitation in Cladoceran populations. Limnol. Oceanogr. 27:707–717; 1982.

Threlkeld, S.T. Estimating cladoceran birth rates: the importance of egg mortality and the egg age distribution. Limnol. Oceanogr. 24:601–612; 1979.

van Houwelingen, A.C.; Sorensen, J.D.; Hornstra, G.; Simonis, M.M.G.; Boris, J.; Olsen, S.F.; Secher, N.J. Essential fatty acid status in neonates after fish-oil supplementation during late pregnancy. Br. J. Nutr. 74:723–731; 1995.

Wang, K.S.; Chai, T.J. Reduction in omega-3 fatty acids by UV-B irradiation in microalgae. J. Appl. Phycol. 6:415–421; 1994.

Williamson, C.E.; Zagarese, H.E.; Schulze, P.C.; Hargreaves, R.; Seva, J. The impact of short-term exposure to UV-B radiation on zooplankton communities in north temperate lakes. J. Plankton Res. 16:205–218; 1994.

Wyngaard, G.A.; Goulden, C.E.; Nourbakhsh, A. Life history traits of the tropical freshwater copepod *Mesocyclops longisetus* (Crustacea: Copepoda). Hydrobiologia 292/293: 423–427; 1994.

Yamada, N.; Murakami, N.; Morimoto, T.; Sakakibara, J. Auto-growth inhibitory substance from the fresh-water cyanobacterium *Phormidium tenue*. Chem. Pharmacol. Bull. 41:1863–1865; 1993.

Yan, N.D.; Keller, W.; Scully, N.M.; Lean, D.R.S.; Dillon, P.J. Increased UV-B penetration in a lake owing to drought-induced acidification. Nature 381:141–143; 1996.

5

Lipid Dietary Dependencies in Zooplankton

Clyde E. Goulden, Robert E. Moeller, James N. McNair, and Allen R. Place

5.1. Introduction

Zooplankton accumulate large amounts of lipids, as much as 40–70% of the dry mass of their body (Goulden and Henry, 1985; Lee et al., 1972). Some polyunsaturated fatty acids and sterols are essential but required in trace amounts. By contrast, storage and membrane lipids are generally composed of nonessential lipids and make up an important energy reserve.

The known essential fatty acids (EFA) of zooplankton and other metazoans appear to be linoleic, $18:2\omega6$ and linolenic, $18:3\omega3$ acids (Conklin and Provasoli, 1977) and perhaps the polyunsaturated fatty acid, eicosapentaenoic acid (EPA; $20:5\omega3$). However, studies of animals exposed to controlled diets lacking EPA for multiple generations have not been performed. This approach would demonstrate the essential need of EPA for many metazoans. Experimental studies by D'Abramo (1979) suggest that the absence or low concentrations of this polyunsaturated fatty acid may affect clutch sizes and the rate of reproduction in the cladoceran *Moina*. Some sterols, particularly cholesterol, are essential to all crustaceans, although little has been done to study them in zooplankton.

Lipids also play a vital role in the ecology of zooplankton. Storage lipids, such as triacylglycerols (TAG) and wax esters, do represent the primary source of metabolic energy in copepods and Cladocera during periods of low ambient food levels (Goulden and Henry, 1985; Lee et al., 1972). Lipids are also extremely important in egg yolk, again serving as a major part of the metabolic energy of developing embryos and neonates (Goulden et al., 1987).

Although it is known that lipids are critical to survival of zooplankton, many questions remain to explain this observation within the context of the animal's ecology. For example, do the large amounts of lipid accumulated by zooplankton originate from the diet directly as lipid (or fatty acids), or are they biosynthesized by the animals? If accumulated directly from the diet, are there sufficient lipids in natural zooplankton diets to support this large dependency?

The purpose of this research has been to determine whether "food limitations" may stem from deficiencies of particular food components, either as single essential compounds or as groups of compounds such as lipids. An understanding of deficiencies should ultimately enable us to better explain changes in food quality and whether they affect feeding behavior (e.g., omnivory, excess feeding) that could have substantial impacts on ecosystem structure and processes.

The primary food of most zooplankton is algae. It is well established that algal biochemistry is diverse and changeable. Fundamental differences exist among the various taxonomic groups of prokaryotic and eukaryotic algae (e.g., differences in fatty acid compositions among algal phyla) (Wood, 1974; Erwin, 1973; also see chapters that discuss variability of lipids in aquatic systems by Wainman et al., this volume; Napolitano and Cicerone, this volume; Olsen, this volume). In a review of laboratory studies demonstrating poor survival and reproduction of freshwater herbivores on single algal species diets, Provasoli et al. (1959) argued that such differences in biochemistry of apparently edible algae were major determinants of food quality and potentially induced severe dietary deficiencies in herbivores (see Olsen, this volume). Algal biochemistry also changes depending on growth conditions. Limitations by light or by inorganic nutrients in the medium affect biosynthetic pathways. Under growth-limiting conditions, certain algal species accumulate large amounts of energy-rich compounds composed largely of saturated fatty acids or carbohydrate storage products. Biosynthesis of proteins and polyunsaturated fatty acids usually declines, and these changes are evident in the biochemical composition of the algae (Thompson et al., 1996, 1992; Siron et al., 1989; Shifrin and Chisholm, 1981). These biochemical differences in the algae have been demonstrated in laboratory experiments to affect feeding rates and food selectivity (Butler et al., 1989; Cowles et al., 1988; Houde and Roman, 1987), growth, and reproduction of zooplankton (Kiorboe, 1989; Scott, 1980).

We assume that similar effects on growth and reproduction should occur in natural populations of zooplankton in both freshwater and marine habitats. However, tests of food quality effects are very difficult to design and perform in natural habitats. Recently, Müller-Navarra (1995) has found evidence indicating a polyunsaturate fatty acid limitation in natural populations of *Daphnia*. This approach is promising.

Defining food quality is complex because it can be affected by the presence of toxic compounds, chelators, nutrients, and morphological characters that interfere with ingestion or digestion. For the purposes of this study, we refer to a resource as low in quality if the animal is unable to maximize growth and reproduction, per unit of resource carbon mass, by selectively modifying its rate of ingestion of food species or its assimilation of ingested foods. Poor food quality therefore is determined by the presence of deficiencies in the biochemical composition of the food relative to the consumer's requirements. A deficiency should be detectable by supplementing the diet with specific compounds that will increase a consumer's rate of growth and reproduction.

The basis of our interests in lipids as limiting nutrients in daphniids has been stimulated by two studies previously performed by Goulden and Place (1993, 1990). Adult daphniids accumulate lipid preferentially, relative to other biochemical components of the diet (Goulden and Place, 1993). This was determined by measuring total lipids accumulated in adult animals during a single instar. The instar begins after an exoskeleton molt and, in daphniids, is coincident with the release of newly formed eggs into the brood pouch. These eggs develop during the instar and are released just before the molt at the end of the instar. For the determination of the amount of lipids accumulated during an instar, newly formed eggs were removed from part of a cohort of clonal adult females at the beginning of the instar and the adults then analyzed for total lipids and lipid classes. A second set of adult females of the same cohort and age was cultured until the beginning of the next instar and was sacrificed without removal of the new eggs, and their lipid content was analyzed.

The results indicated that in adults at the beginning of an instar (with eggs removed), lipid contents ranged from 10–15% of dry mass of the body. At the beginning of the following instar, adult Cladocera (with fresh eggs) contained up to 35–40% of their dry mass as lipids. Total body mass of the animals increased, as did protein and carbohydrate content, but on a relative basis, the biochemical composition of these adults was most altered by the substantial increases in their lipid composition.

To explain this, we must assume that animals (1) preferentially assimilate lipids, (2) preferentially metabolize assimilated carbohydrate or protein while storing lipids, or (3) biosynthesize lipids from carbohydrate and possibly protein.

The second set of experiments measured fatty acid synthesis rates in daphniids. In a series of tracer experiments using ^{14}C-acetate or ^{3}H-water, Goulden and Place (1990) measured incorporation rates of the isotopes into extractable fatty acids and compared synthesis rates with fatty acid accumulation rates in two daphniid species. They found that given either high food levels or low food levels, the daphniids (*Daphnia pulicaria* and *D. magna*) synthesized <2% of their accumulated fatty acids. So, >98% of the fatty acids was derived directly from the diet. Most of the incorporated isotope appeared in the fatty acids of phospholipids, and the authors suggested that most of the incorporation resulted from chain elongation of dietary acids.

This low rate of synthesis relative to the amount of lipid accumulated indicates that zooplankton are quite dependent on the total lipid and the quality of the lipid (fatty acid composition) in their food.

Because total lipids may represent a small fraction of the biochemical composition of exponential growth-phase algae under laboratory conditions (Cowgill et al., 1984; Lee et al., 1971), lipids or their constituent fatty acids may represent a primary cause of food limitation in daphniids.

In this chapter, we focus primarily on whether lipids may limit reproduction in daphniid Cladocera in two north-temperate freshwater lakes. Although the focus is on lipids, protein was also studied. We do not emphasize the protein results in this chapter, although the data are included and results briefly described.

Our hypotheses of food limitation to be tested in this study are the following:

• Total food hypothesis: Food limitation is due simply to the total quantity of food; specific dietary components always occur in adequate ratios so that none acts as a limiting nutrient.

- Total lipid hypothesis: Food limitation is due to total storage lipid; dietary supplements of triglycerides or non-EFAs will increase reproduction in adult daphniids, but other dietary supplements will not.
- Essential fatty acid hypothesis: Food limitation is due to specific EFAs (linoleic or linolenic); supplements of these will increase reproduction in adult daphniids, but other dietary supplements will not.
- Protein hypothesis: Food limitation is due to protein or essential amino acids; supplements of these will increase reproduction in adult daphniids, but other dietary supplements will not.

We supplemented natural zooplankton diets with microcapsules or particles formulated from purified components and measured subsequent egg production. The daphniids we studied are prominent grazers in many freshwater systems. Adult females allocate most of their growth to reproduction (Taylor and Gabriel, 1985). We performed a series of experiments in two lakes in northeastern Pennsylvania, with emphasis on an experiment conducted in a eutrophic lake (Lake Waynewood) in autumn. In this study, the natural diet was concurrently analyzed in terms of ingestion rates and nutrient-limited status of the common algal food items.

5.2. Methods

Chemicals: The following compounds (with their purity, when stated) were purchased from Sigma Chemical Co. (St. Louis, MO): casein (bovine milk) purified powder; gum arabic; cholesterol (99% +); ergocalciferol; linoleic acid (99% by capillary gas chromatography [GC]); linolenic acid (98% by capillary GC); 1-monooleoyl-rac-glycerol (99%); palmitic acid (99% by capillary GC); *L*-a-phosphatidylcholine (type V-E) from frozen egg yolk (~99%, prepared chromatographically); retinol palmitate, type IV; alpha-tocopherol (from vegetable oil); and triolein (99%). $Na_2^{14}CO_3$ solution was made up from a carrier free crystalline salt (New England Nuclear, Boston, MA) and diluted with ACS grade Na_2HCO_3. NaH_2PO_4 and NH_4NO_3.

5.2.1. Microparticle Preparation

Two types of lipid microcapsules were made by encapsulation in gum arabic (gum arabic primarily consists of a polymer of galactose, rhamnose, arabinose, and glycuronic acid): a TAG microcapsule, and a microcapsule that contained EFA. The EFA microcapsule contained phospholipid, cholesterol, vitamins A, D, and E, monoolein, and three free fatty acids (palmitic and the two EFAs, linoleic and linolenic). Phospholipid and monoolein both served as emulsifiers, and the monoolein would also improve absorption of free fatty acids. The vitamins served as antioxidants. This particle composition was the same as the "FV" particle developed by Conklin and Provasoli (1977), except that an equivalent amount of monoolein was substituted for oleic acid and egg albumin was omitted. In some tests, an equivalent amount of oleic acid was substituted for the two EFAs to make a "non-EFA" microcapsule, or microcapsules were synthesized exclusively from the triglyceride triolein. Lipids were encapsulated by the procedures of Langdon and Waldock (1981). The lipid mixture was emulsified in a 4% solution of gum arabic (40°C) with a high-speed homogenizer. After heating and adjusting pH, the microcapsules were refrigerated to harden the coats, then centrifuged and rinsed twice, and finally autoclaved. Microcapsule concentrations were determined by hemacytometer counts. The capsules averaged 2 μm in diameter. Samples of microcapsules were analyzed and estimated to contain, on average, 1 pg of free fatty acids, by thin-layer chromatography with an Iatroscan flame ionization detector system (Parrish and Ackman, 1983).

Protein was usually added as microparticles of casein. Casein is commonly used in aquaculture because of its wide complement of essential amino acids. To make microparticles, casein suspended in distilled water was initially dissolved by elevating pH to 11.5 with 0.1 Eq · L^{-1} NaOH, precipitated by lowering pH to 4.8 (0.1 Eq · L^{-1} HCl), and then homogenized with a high-speed homogenizer. Particles averaged ~10 μm in diameter but ranged from <1 to >50 μm. Protein concentrations of the microparticle slurry were determined by the modified Lowry assay (Sigma Kit) after sonication. A protein microcapsule was also prepared by following the methods of Langdon (1989) and Langdon and DeBevoise (1990) but with casein substituted for the protein mix suggested by Langdon (1989). However, we were unable to use these because they retained a slight toxic effect on the daphniids.

Prior experiments in our laboratory (C.E.G. and J.N.M.) had demonstrated that daphniids would ingest lipid microcapsules containing either TAG, EFA, or only non-EFA (oleic and palmitic acids). In preliminary experiments, laboratory animals did respond to EFA microcapsules by increased reproduction, but they did not respond to the same microcapsule containing comparable amounts of non-EFAs alone. The lack of response to the non-EFA microcapsule that contained both monoolein and the vitamin additions indicated that the original response to the EFA microcapsule was not a response to the monoolein or the vitamin additions.

5.2.2. Dietary Supplement Experiments

Dietary supplement experiments were performed in the summers of 1988–1989 using *Daphnia catawba* from mesotrophic Lake Lacawac and *Daphnia pulicaria* and *Daphnia laevis* from eutrophic Lake Waynewood. Both lakes are located in Wayne County, in the Pocono region of northeastern Pennsylvania. Spring (April–May) and fall (late September–October) populations of daphniids in these two lakes average 20–30 individuals per liter. Summertime densities seldom exceed 2–5 individuals per liter in Lacawac due to size-selective predation or food limitation, or both (Tessier, 1986).

TABLE 5.1. Detail of supplement experiments.[a]

Date/lake/species	Treatment supplements[b]
May 24, 1988 Lake Lacawac *Daphnia catawba*	TAG μ-capsules ($100 \cdot 10^3 \cdot$ ml^{-1}) *Ankistrodesmus* ($5 \cdot 10^3$ cells $\cdot$ ml^{-1})
June 30,1989 Lake Lacawac *D. catawba*	EFA μ-capsules ($50 \cdot 10^3 \cdot$ ml^{-1}) Non-EFA microcapsules ($50 \cdot 10^3 \cdot$ ml^{-1}) Casein μ-particles (1.25 μg protein $\cdot$ ml^{-1}) *Ankistrodesmus* ($5 \cdot 10^3$ cells $\cdot$ ml^{-1})
July 13, 1989 Lake Lacawac *D. catawba*	EFA μ-capsules ($50 \cdot 10^3 \cdot$ ml^{-1}), Casein μ-particles (1.0 μg protein $\cdot$ ml^{-1}) EFA + Casein *Ankistrodesmus* ($5 \cdot 10^3$ cells $\cdot$ ml^{-1})
July 29, 1989 Lake Waynewood *Daphnia pulicaria*	EFA μ-capsules ($50 \cdot 10^3 \cdot$ ml^{-1}) Casein μ-particles (1.0 μg and 1.5 μg protein $\cdot$ ml^{-1}) *Ankistrodesmus* ($5 \cdot 10^3$ cells $\cdot$ ml^{-1})
October 14, 1989 Lake Waynewood *Daphnia laevis* *D. pulicaria*	EFA μ-capsules ($50 \cdot 10^3 \cdot$ ml^{-1}) Casein μ-particles (1.4 μg protein $\cdot$ ml^{-1}) EFA + Casein *Ankistrodesmus* ($10 \cdot 10^3$ cells $\cdot$ ml^{-1})

[a]EFA, essential fatty acids; TAG, triacylglycerol.
[b]Lipid microcapsules average 1 pg lipid per microcapsule.

Details of daphniid species studied and the supplements used in each experiment are given in Table 5.1. For the field experiments, integrated epilimnetic water samples were collected from central locations in each lake. The samples were filtered through an 80-μm mesh net to remove crustacean zooplankton and were combined into large bottles; this water was thoroughly mixed and divided randomly among treatments. Dietary supplements (microcapsules) containing either TAG or EFA, or additional algae, were mixed in prior to filling replicate biological oxygen demand (BOD) bottles (generally 15 for each treatment; each bottle contained ~300 ml of lake water).

The composition of the dietary supplement differed among experiments (Table 5.1). EFA or the TAG treatments contained $25-100 \cdot 10^3$ microcapsules $\cdot$ ml^{-1} ($0.025-0.1$ μg $\cdot$ ml^{-1} of total TAG or free fatty acids) and were established after preliminary tests with animals to determine that negative effects would not occur during the exposure period (free fatty acids provided in excess can be toxic to organisms). TAG treatments were $50-100 \cdot 10^3$ microcapsules $\cdot$ ml^{-1} ($0.05-0.1$ μg $\cdot$ ml^{-1}). Ambient concentrations of lipids or proteins were not measured because of the difficulty of distinguishing the biochemical composition of the actual diets from detritus and nonfood algae, which compose the bulk of suspended particles in these lakes. Protein microparticles were added at 0.7 or 1.4 μg $\cdot$ ml^{-1} of lake water, but because the larger microparticles could not be ingested, the amount available to the animals was less than this. A treatment of added cultured algae ($5-10 \cdot 10^3$ cells $\cdot$ ml^{-1} of *Ankistrodesmus falcatus* from laboratory cultures; maintained on ASM medium [Carmichael and Gorham, 1974] in a semicontinuous flow culture system and kept in a growth phase) served as a "complete" diet to test for food limitation in the broad sense.

Similar-sized adult animals (1.2–1.4 mm for *Daphnia catawba;* 1.8–2.2 mm for *D. pulicaria;* 1.7–1.9 mm for *D. laevis*) collected from the lake were selected for the experiments to minimize size differences in egg number. Individual animals were transferred to each BOD bottle (one per bottle in the Waynewood experiments; one or three per bottle at Lacawac). The bottles were then placed on a plankton wheel and rotated at 1 rpm, in a room receiving low ambient light during daytime and with temperatures similar to the lake's epilimnion (19°C during the late June and July experiments and 17°C during mid-October). Animals were maintained on experimental diets for 6–7 d (two to three instars) with medium changes (water and supplements) at 1–2-d intervals. Body size, clutch size, and lipid-ovary index (Tessier and Goulden, 1982) were recorded for each animal when the experiment was terminated. Clutch size in *Daphnia* responds within a few days to changes in diet (Frank, 1960). Differences in clutch size were tested by ANOVA, followed by linear contrasts with the control or by two-way ANOVA using SYSTAT (Wilkinson, 1988) or SAS (Anonymous, 1985). Log transformations were usually used to equalize variances. Among treatment differences in lipid-ovary index were determined by using Fisher's Exact test (Sokal and Rohlf, 1981).

5.2.3. Characterization of the Natural Algal Diet

At the time of the October supplementation experiment in Lake Waynewood, the natural algal diet of the *Daphnia* was analyzed in terms of grazing selectivity, biovolume of food ingested, and nutrient limitations of the algae. A composite water sample from 0–6 m depths in Lake Waynewood was screened (153 μm) to remove large zooplankton, distributed among 1-L transparent polyethylene containers (Cubitainers), and treated by adding *Daphnia* (0 or 20 · L^{-1}, mostly *D. laevis*), nitrogen (0 or 30 μmol · L^{-1} NH$_4$NO$_3$), and phosphorus (0 or 6 μmol · L^{-1} NaH$_2$PO$_4$) in all combinations with four replicates. Not all treatment results are included in the Results section. Containers were suspended at 2-m depth in Lake Lacawac for 3 cloudy days, providing light equivalent to that at 1 m in Lake Waynewood. They were shaken twice daily. On day 3, 40-ml subsamples from the containers were incubated with NaH^{14}CO$_3$ (111 Bq · ml^{-1}) in the laboratory under "cool-white" fluorescent illumination (photosynthetically active radiation at 150 μmol · m^{-2} · s^{-1}) for 4 h to measure photosynthetic rate. A freshly collected water sample was incubated along with the day 3 container samples ("lake" treatment). Aliquots were counted by liquid scintillation after being acidified and bubbled to remove unincorporated ^{14}CO$_2$. Parallel subsamples from all containers and the lake treatment were incubated with much lower ^{14}C activity (2.6 Bq · ml^{-1}) for autoradiography of algae. Total inorganic carbon concentration was determined by gas chromatography (Stainton, 1973) and was uniform across all containers (0.40 mM · L^{-1}). Samples were preserved with acid Lugol's solution at 1% (vol/vol) for algal enumeration and autoradiography.

Grazing rates were calculated from the individual clearance rates (Peters, 1984) of 16 food items, representing >95% of the algal biomass present. We used analysis of covariance (ANOCOVA; Snedecor and Cochran, 1967) to calculate clearance rates and significance levels from differences in ln (cell concentration) between treatments with and without *Daphnia*. The ANOCOVA simultaneously provided estimates of nutrient effects (for the N plus P coaddition treatment) and their significance levels. ANOCOVA was an appropriate analytical model because the variances were homogeneous across all treatments and, with the exception of *Chrysochromulina,* slopes were homogeneous (all $P >.1$) for treatments with and without added N and P. For *Chrysochromulina,* the clearance rate was estimated from the NP treatments alone. Average length and biovolume (e.g., cell contents excluding walls, colonial mucilage) were estimated for each food item. Total biovolume is the product of average cell volume and cell concentration. Ingestion rate is the product of total biovolume of a food item (10^3 $\mu m^3 \cdot ml^{-1}$) and clearance rate (ml $\cdot$ *Daphnia*$^{-1} \cdot d^{-1}$).

Parallel samples of the original lake water used for the microparticle supplements were screened (40 μm) to remove the largest, mostly inedible algae, collected on precombusted glass fiber filters (Whatman GF/F), and frozen. These were analyzed for organic carbon and nitrogen (Carlo-Erba Elemental Analyzer). Total phosphorus was determined as phosphate (John, 1970) following combustion and extraction in boiling dilute HCl.

5.2.4. *Algal Counting and Autoradiography*

Preserved algal samples (5 ml) were filtered onto 1-μm Nucleopore polycarbonate membranes. These were immediately inverted onto gelatin-coated slides and quick-frozen (cryogenic aerosol), and the filters were stripped away (Hewes and Holm-Hansen, 1983). After drying ($\sim 35°C$), slides were mounted in 30% glycerol (for cell counting) or dipped in Kodak NTB3 emulsion for ^{14}C track autoradiography (Knoechel and DeNoyelles, 1980; Knoechel and Kalff, 1979, 1976). Following exposure (14 d at 4°C) and processing (7 minutes Kodak D-19, 10 minutes Kodak Fixer, 15 minutes 10% [mass $\cdot$ vol^{-1}] sodium thiosulfate pentahydrate, with intermediate rinses in tap water), slides were dried and then mounted with 30% glycerol. Track data were compiled per cell (*Synura petersenii, Cryptomonas marssonii*) or per unit filament length (*Aphanizomenon flosaquae*) and are rescaled for plotting as a proportion of the mean rate for the freshly collected lake treatment.

5.3. Results

5.3.1. *Supplement Experiments with Natural Populations*

Responses to dietary supplementations were judged from the magnitude and statistical significance of any increase in clutch size, relative to the control.

Animals maintained on the ambient epilimnial phytoplankton without supplementations were considered controls. A clutch size response to the addition of algae (*Ankistrodesmus falcatus*) would imply food limitation, which could be due either to food quantity (total food hypothesis) or to food quality. Response to TAG supplements but not to EFAs or protein would suggest that energy reserves limited reproduction (total lipid hypothesis). Response to the EFA supplements (linoleic and linolenic) but not to TAG or protein would suggest that these fatty acids were limiting in the diet (EFA hypothesis). A response to protein but not to TAG or EFAs suggests that protein may be the limiting factor (protein hypothesis). We also monitored clutch sizes of the ambient populations throughout the experiment, and averages of these data are given in the table legends for each experiment.

The results of the first experiment with algae and TAG supplements suggest *Daphnia catawba* from Lake Lacawac were food limited, as both clutch size and lipid reserves increased with increased food concentration (Table 5.2). TAG microcapsule supplements caused a substantial increase in visible lipid in the body but did not enhance fecundity. The results of this experiment prompted us to develop the EFA microcapsules and protein microparticles used in subsequent experiments.

As in the previous experiment, *Daphnia catawba* collected from Lake Lacawac in late June 1989 responded to algal supplements by increasing clutch size (Table 5.3). Protein alone produced a statistically significant response. In July (Table 5.4), a slight but statistically significant food limitation was apparent, but the protein effect was not statistically significant. The lower fecundities in the control and treatment bottles versus the lake population may have been caused by the use of three animals per vessel, a decision made to counteract increased clutch size when single *Daphnia catawba* were placed in vessels.

Daphnia pulicaria was studied in Lake Waynewood in August 1989, when it was the dominant large zooplankter (three to four individuals per liter) and cyanobacteria were abundant, especially *Anabaena planktonica* and *Aphanizomenon flos-aquae*. *D. pulicaria* responded to protein supplements with a slight but statistically significant increased clutch size (Table 5.5). The algal supplement, repre-

TABLE 5.2. Supplement experiment started on May 24, 1988, with *Daphnia catawba* from Lake Lacawac, PA.[a]

| Treatment | n | Clutch size | | | Lipid index | | |
		X	SD	P[b]	X	SD	P[c]
Control	19	1.2	0.4	—	0.8	0.3	—
+TAG	20	1.5[d]	1.2	NS	2.1	0.5	<.05
+Algae	6	5.1	1.9	<.001	1.67	0.4	NS

[a]+TAG, triacylglycerol microcapsules; +Algae, *A. falcatus;* NS, not significant. Mean body length of lake adults during experiment is 1.5 mm (SD, 0.12).
[b]Linear contrasts with the control following ANOVA.
[c]Fisher's exact test of medians.
[d]Includes degenerate eggs.

TABLE 5.3. Effect of dietary supplements on *Daphnia catawba* clutch size (log transformed), Lake Lacawac, June 30, 1989.[a]

Treatment		Clutch size	
	n	*X*	SD
Control	15	0.88	0.46
EFA	15	0.73	0.40
Non-EFA	15	0.99	0.76
+Protein	15	2.06	1.18
+Algae	14	2.19	0.80
Source of variance	df	SS	*P*
Treatment	4	4.570	<.001
Error	69	6.748	
EFA versus control	1	0.045	NS
Non-EFA versus control	1	0.007	NS
Protein versus Control	1	1.444	<.001
Algae versus control	1	2.058	<.001

[a]+Algae, *A. falcatus;* EFA, essential fatty acids; NS, not significant. One-way ANOVA followed by linear contrasts with the control. Mean length of lake animals during experiment is 1.3 mm (SD, 0.06 mm); mean clutch size is 2.58 eggs per female (SD, 1.57 eggs per female).

TABLE 5.4. Results of dietary supplement experiments in Lake Lacawac (July 13, 1989) with *Daphnia catawba.*[a]

Species		Clutch size	
	n	*X*	SD
Control	15	1.10	0.60
+EFA	15	1.20	0.92
+Protein	14	1.18	0.95
+EFA +Protein	15	0.69	0.48
+Algae	15	1.93	0.86
Source of variance	df	SS	*P*
EFA	1	0.075	NS
Protein	1	0.045	NS
EFA * Protein	1	0.023	NS
Error	55	40.704	
+Algae[b]	1	0.812	.015
Error	28	3.402	

[a]+Algae, *A. falcatus;* EFA, essential fatty acids; NS, not significant. Data analyzed by two-way ANOVA with main effects. Lake adults at time of experiment had a mean length of 1.45 mm (SD, 0.1 mm) and a mean clutch size of 1.1 eggs per female (SD, 0.28 eggs per female).
[b]One-way ANOVA comparing control and +Algae treatment.

TABLE 5.5. Results of dietary supplement experiments with *Daphnia pulicaria* from Lake Waynewood (August 1989).[a]

Treatment	n	Clutch size	
		X	SD
Control	14	4.7	2.0
+Lo-EFA	12	4.7	2.6
+Hi-EFA	14	4.3	2.6
+Lo-Protein	14	4.4	2.6
+Hi-Protein	14	6.8	2.7
+Hi-EFA +Lo-Protein	15	5.9	2.4
+Algae	15	5.6	3.2
Source of variance	df	SS	P
EFA	2	0.595	NS
Protein	2	44.846	.04
EFA * Protein	1	3.124	NS
+Algae	1	5.143	NS
Error	87	563.917	

[a]+Algae, *A. falcatus;* EFA, essential fatty acids; NS, not significant. Clutch size analyzed by two-way ANOVA (SAS). Interaction with all lipid or protein treatments. The lake adult population at the time of the experiment had a mean length of 1.8 mm and a mean clutch size of 2.06 (SD, 0.68 eggs).

senting a smaller protein addition (Lowry assays), was intermediate between the control and the protein treatments and not statistically distinguishable from either.

5.3.2. Lake Waynewood Experiment: October 1989

5.3.2.1. Algal Diet

At the time of the October microcapsule experiment in Lake Waynewood, the algae in that eutrophic lake were dominated by two colonial cyanobacteria, *Gomphosphaeria nageliana* and *Aphanizomenon flos-aquae,* which were not perceptibly grazed (Table 5.6). Several other colonial forms, including *Synura petersenii,* were grazed at low rates. Grazing rates were highest on small flagellates ($\sim$20–50 ml $\cdot$ daphniid^{-1} $\cdot$ d^{-1}), including *Cryptomonas marssonii* (18 ml $\cdot$ daphniid^{-1} $\cdot$ d^{-1}) (Fig. 5.1). *Cryptomonas* was relatively abundant and contributed the largest part (40%) to the total volume of ingested algae.

Nutrient enrichment assays suggested that the phytoplankton as a whole was not limited by N or P, or the combination of N with P. After 3 d, algal chlorophyll *a* and 4-h ^{14}C uptake were uniform across all nutrient treatments lacking *Daphnia* ($P >$.2 from one-way ANOVA) as well as across treatments with *Daphnia* (separate one-way ANOVA, $P >$.2). Grazing slightly reduced both chlorophyll *a* and ^{14}C uptake. These results for the whole algal community, which was dominated by inedible types, are not presented in detail. Cell counts and autoradiography

TABLE 5.6. Algal diet of *Daphnia* from Lake Waynewood, October 1989.[a]

Food item[b]	Biovolume (10^3 $\mu m^3 \cdot ml^{-1}$)	Ingestion[c] (10^4 $\mu m^3 \cdot day^{-1}$)
Gomphosphaeria nageliana	790	(0)
Aphanizomenon flos-aquae	340	(0)
*Cryptomonas marssonia***	200	360
*Synura petersenii**	110	66
Chrysochromulina parva	73	131
Eudorina elegans	52	31
Rhodomonas lacustris	36	120
Fragilaria crotonensis	34	31
Phytoflagellates <4 μm	29	61
Mallomonas akrokomos	18	25
Cosmarium sp.	14	(0)
Sphaerocystis schroeteri	13	14
Anabaena planktonica	6	6
Pseudopedinella erkensis	6	23
Trachelomonas volvocina	4	(0)
Kathablepharis sp.	4	19

[a]Algal abundance is presented as biovolume, the cell-content volume of a species per milliliter lake water. Ingestion represents consumption of algae by an individual daphniid per day (the product of biovolume and clearance rate).
[b]Significant growth response to N+P: ** P <.001, * P <.05.
[c]Ingestion is taken as zero for items with clearance rates not significantly greater than zero.

of three selected species provided details about the edible algal subcomponent (Fig. 5.1). Cell counts revealed a small but statistically significant stimulation by the N+P coaddition in *Cryptomonas marssonii* (P <.001) and in *Synura petersenii* (P <.05) (Table 5.6). Counts of the single-element additions (composite sample, data not shown) suggested that P alone was primarily responsible for the effect. Autoradiography of these same species from day 3 of the enrichment did not reveal statistically significant nutrient effects on short-term photosynthetic rates, however, suggesting that the nitrogen and/or phosphorus limitations detected by cell counts must not have been severe (Fig. 5.1). Chemical analysis of <40 μm seston gave C/N/P molar ratios of 113:14.5:1, close to the Redfield ratio (106:16:1), also not indicative of severe nutrient deficiency (Vollenweider, 1985).

5.3.2.2. Response by *Daphnia*

The pattern of response seen in the previous experiments—positive to algae and protein but not to EFA—was repeated in the October 1989 investigation of Lake Waynewood daphniids. Both *Daphnia laevis* and *D. pulicaria* responded to the protein addition as well as to the extra algae (Table 5.7). The inclusion of *D. pulicaria* at low and uneven replication was fortuitous; we mistakenly assumed that only one species was present when setting up the treatments. Neither daphniid responded to the lipid (EFA) microcapsules alone. EFA added with protein did

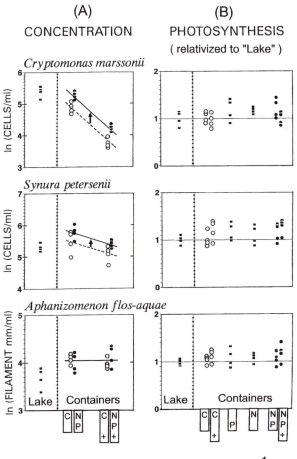

TREATMENTS   ("+" : 20 *Daphnia*·L⁻¹)

FIGURE 5.1. Response of algal concentrations (A) and relative photosynthetic rates (B) to grazing and inorganic nutrient additions (Lake Waynewood, October 1989). Algal counts from the control (C) and nitrogen plus phosphorus treatments (NP), each with ("+") and without *Daphnia,* demonstrate significant stimulation of growth in *Cryptomonas* ($P < .001$ from ANOCOVA) and *Synura* ($P < .05$), which were both significantly grazed ($P < .001$ and $< .01$, respectively). The cyanobacterium *Aphanizomenon* was neither grazed nor nutrient limited. The homogeneity of photosynthetic rates across all nutrient and grazer treatments suggests an absence of strong nutrient limitations. In each graph, container results for the third day of incubation are compared with cell counts and photosynthetic rates of algae in water freshly collected on day 3 ("lake" treatment to left of vertical divider).

TABLE 5.7. Reproductive response of *Daphnia laevis* and *D. pulicaria* to dietary supplements (Lake Waynewood, October 1989).[a]

| | Daphnia laevis | | | Daphnia pulicaria | | |
| | Clutch size | | | Clutch size | | |
Treatment	n	X	SD	n	X	SD
Control	8	3.2	1.2	5	6.0	0.7
+Protein	8	6.0	1.8	5	8.4	3.2
+EFA	11	3.0	1.0	3	4.3	0.6
+Protein + EFA	9	5.7	1.9	5	11.8	3.0
+Algae[b]	12	8.1	2.8	2	11.5	0.7
Source of variance	SS		P	SS		P
EFA	0.75		NS	0.003		NS
Protein	64.9		<.001	1.683		<.001
Interaction	0.02		NS	0.529		.022
Error	71.50			1.126		

[a]+Algae, *A. falcatus;* EFA, essential fatty acids; NS, not significant. Lake animals at time of experiment had 2.3 eggs per egg-bearing female (SD, 0.94 eggs per egg-bearing female) at the beginning of the experiment (species not separated). Initial size: *D. laevis* length, 1.7 mm (SD, 0.06 mm); *D. pulicaria* length, 2.2 mm (SD, 0.1 mm).
[b]One-way ANOVA comparing control and +algae treatment gave P <.001 for both species. *D. pulicaria* data were log-transformed before statistical analyses.

increase clutch size in *D. pulicaria* above that produced by the protein alone, to a level statistically indistinguishable from that produced by the algal addition.

5.4. Discussion

The natural *Daphnia* populations we studied were generally food-limited, as evidenced by increases in clutch size of animals receiving supplements of *Ankistrodesmus falcatus* to the lake phytoplankton. In four of five cases, clutch size also increased when the diet was supplemented with protein microparticles. Lipid, as TAG or fatty acid supplements containing two EFAs, was never limiting to clutch size (rejecting the EFA hypothesis), although in one case, protein supplementation apparently shifted the diet to a secondary EFA limitation (Table 5.7; see the *Daphnia pulicaria* treatment interaction $P = .022$). When clutch size was artificially increased with algal or protein supplementation, we continued to find clutches in all stages of development. This implies that we were seeing increased production of progeny, not formation of nonviable eggs, which cease development at an early egg development stage.

Food limitation does appear to require a more complex explanation than just a shortage of total food quantity (total food hypothesis). Our experimental results cause us to reject total storage lipid as a factor limiting growth and reproduction (total lipid hypothesis). However, our experiments did not examine impacts on survival. Survival would definitely be affected by the amount of TAG (Goulden and Henry, 1985), and daphniids appear to depend on the diet for TAG (Goulden and Place, 1990). In the 1988 experiment, the lipid-ovary index of *D. catawba* increased from 0.3 to 2 as triacylglycerol was assimilated (Table 5.2). By contrast, the fatty acid microcapsules without TAG produced no discernible change in the lipid status of animals in these experiments (data not shown). Although the fatty acid microcapsules were introduced at a lower concentration than the triacylglycerol microcapsules ($25 \cdot 10^3 \cdot ml^{-1}$ versus $100 \cdot 10^3 \cdot ml^{-1}$), this experiment and a later laboratory experiment (Goulden, unpublished data) suggest that visible lipid accumulation depends in large part on the composition of the lipid classes in the diet, particularly the presence and amount of triacylglycerol.

The analysis of the algal diet during the October experiment from Lake Waynewood, when the *Daphnia* responded strongly to protein supplementation, established two important facts regarding the conditions during that experiment: first, that the concentration of readily ingested algae was not very high, despite the high algal biomass in the lake; and second, that these edible algae were not strongly limited by nitrogen or phosphorus. Circulation of the water column to 6–8 m at that time meant that nutrients that had accumulated in the anoxic hypolimnion during the summer were being advected into the epilimnion. At the same time, light availability was decreasing (only 1% of surface irradiance penetrated to 3 m, the middle of the mixed layer).

The edible fraction of algae amounted to $0.36 \cdot 10^6 \ \mu m^3 \cdot ml^{-1}$ of lake water, equivalent to about 80 ng C $\cdot ml^{-1}$ (Reynolds, 1984). This was well below 400 ng C $\cdot ml^{-1}$, the level that saturates assimilation rate (Lampert, 1977) and maximum clutch size (Goulden et al., 1982) of similar-sized daphniids.

EFA deficiencies were not identified in this study, but we suspect that such deficiencies, especially of the polyunsaturated EFAs, can occur wherever bacteria and single-celled coccoid cyanobacteria form a major part of the diet of zooplankton. Recent studies by Müller-Navarra (1995) do suggest a more widespread occurrence of limitation by EPA. Bacteria and coccoid cyanobacteria are deficient in polyunsaturated fatty acids (Murata and Nishida, 1987; Wood, 1974; Erwin, 1973).

Lipids in lake seston of <50-μm diameter can be 30–50% of dry weight, depending on the time of the year (Kreeger et al., 1997). This may suggest that lake seston (assuming that the seston is primarily algae) reflects nutrient limitation of the phytoplankton because nutrient limitation often stimulates lipid accumulation. However, these measurements need to be repeated in other lakes and during different years to characterize what is happening in the phytoplankton and seston of lakes. If these values do typify the composition of the edible fraction of the seston consumed and digested by zooplankton, it would indicate that adequate lipid is present for zooplankton reproduction and survival. Further, this suggests that a dependency on dietary lipids is not a risky strategy for zooplankton.

Zooplankton populations are often found to be food-limited during the summer months (Tessier, 1986; Lampert, 1985; Threlkeld, 1979). Under low food conditions, lipid reserves are crucial to the survival of *Daphnia* adults (Goulden and Henry, 1985; Lampert and Bohrer, 1984) as well as embryos and neonates

(Goulden et al., 1987). This role of storage lipids in population regulation must be distinguished from growth and reproductive limitations set by suboptimal concentrations of protein, essential amino acids, or fatty acids in the diet. Lipids and protein conceivably play distinctly different roles in the regulation of *Daphnia* populations. The experimental results reported here suggest that dietary lipids are adequate to support good zooplankton growth and reproduction under the conditions we studied. When food quality is poor, in the absence of toxins, dietary proteins may be found to be limiting in ambient diets. However, we cannot exclude limitation by other lipids not tested in this study (Müller-Navarra, 1995).

Acknowledgments. The research reported here was supported by NSF grants BSR-86-15328 to C. E. Goulden and R. E. Moeller, BSR-82-14820 to Goulden, and DCB-87-09340 to A. Place. Funding from the Environmental Associates of the Academy of Natural Sciences and a Ruth Patrick Fellowship to Moeller stimulated the collaborative project. Moeller acknowledges support from the Andrew Mellon Foundation through the Pocono Comparative Lakes Project at Lehigh University. We are grateful to the Lacawac Sanctuary and its former director, Sally Jones, for access to field and laboratory facilities there and to Alan Everett for his help in the field and laboratory throughout the project. We express our sincere appreciation to Dr. Ruth Patrick and the late Dr. Luigi Provasoli for their encouragement of this research.

References

Anonymous. SAS user's guide: statistics. Version 5. SAS Institute, Cary, North Carolina, USA; 1985.

Butler, N.M.; Suttle, C.A.; Neill, W.E. Discrimination by freshwater zooplankton between single algal cells differing in nutritional status. Oecologia 78:368–372; 1989.

Carmichael, W.W.; Gorham, P. An improved method for obtaining axenic clones of planktonic blue-green algae. J. Phycol. 10:238–240; 1974.

Conklin, D.E.; Provasoli, L. Nutritional requirements of the water flea, *Moina macrocopa.* Biol. Bull. 152:337–350; 1977.

Cowgill, U.M.; Williams, D.M.; Esquivel, J.B. Effect of maternal nutrition on fat content and longevity of neonates of *Daphnia magna.* J. Crust. Biol. 4:173–190; 1984.

Cowles, T.J.; Olson, R.J.; Chisholm, S.W. Food selection by copepods: discrimination on the basis of food quality. Mar. Biol. 100:41–49; 1988.

D'Abramo, L.R. Dietary fatty acid and temperature effects on the productivity of the cladoceran, *Moina macrocopa.* Biol. Bull. 157:234–248; 1979.

Erwin, J.A. Comparative biochemistry of fatty acids in eukaryotic microorganisms. In: Erwin, J.A., ed. Lipids and Biomembranes of Eukaryotic Microorganisms. New York: Academic Press; 1973:p. 41–143.

Frank, P.W. Prediction of population growth form in *Daphnia pulex* cultures. Am. Nat. 94:357–372; 1960.

Goulden, C.E.; Henry, L. Lipid energy reserves and their role in Cladocera. In: Meyers, D.G.; Strickler, J.R., eds. Trophic Interactions within Aquatic Ecosystems. Boulder, CO: Selected Symposium Volume, American Association for the Advancement of Science, Westview Press; 1985:p. 167–185.

Goulden, C.E.; Place, A. Lipid accumulation and allocation in daphniid Cladocera. Bull. Mar. Res. 53:106–114; 1993.

Goulden, C.E.; Place, A.R. Fatty acid synthesis and accumulation rates in daphniids. J. Exp. Zool. 256:168–178; 1990.

Goulden, C.E.; Henry, L.; Berrigan, D. Egg size, post-embryonic yolk, and survival ability. Oecologia 72:28–31; 1987.

Goulden, C.E.; Henry, L.; Tessier, A.; Durand, M. Body size, energy reserves, and competitive ability in three species of Cladocera. Ecology 63:1780–1789; 1982.

Hewes, C.D.; Holm-Hansen, O. A method for recovering nanoplankton from filters for identification with the microscope: the filter-transfer-freeze (FTF) technique. Limnol. Oceanogr. 28:389–394; 1983.

Houde, S.E.L.; Roman, M.R. Effects of food quality on the functional ingestion response of the copepod *Acartia tonsa*. Mar. Ecol. Prog. Ser. 40:69–77; 1987.

John, M.K. Colorimetric determination of phosphorus in soil and plant material with ascorbic acid. Soil Sci. 109:214–220; 1970.

Kiorboe, T. Phytoplankton growth rate and nitrogen content: implications for feeding and fecundity in a herbivorous copepod. Mar. Ecol. Prog. Ser. 55:229–234; 1989.

Knoechel, R.; DeNoyelles, F., Jr. Analysis of the response of hypolimnetic phytoplankton in continuous culture to increased light or phosphorus using track autoradiography. Can. J. Fish. Aquat. Sci. 37:434–441; 1980.

Knoechel, R.; Kalff, J. The advantages and disadvantages of grain density and track autoradiography. Limnol. Oceanogr. 24:1170–1171; 1979.

Knoechel, R.; Kalff, J. Track autoradiography: a method for the determination of phytoplankton species productivity. Limnol. Oceanogr. 21:590–596; 1976.

Kreeger, D.A.; Goulden, C.E.; Kilham, S.S.; Lynn, S.G.; Datta, S.; Interlandi, S.J. Seasonal changes in the biochemistry of lake seston. J. Freshwat. Biol. 38:535–554; 1997.

Lampert, W., ed. Food limitation in zooplankton. Arch. Hydrobiol. Suppl. Ergebn. Limnol. 21:497; 1985.

Lampert, W. Studies on the carbon balance of *Daphnia pulex* De Geer as related to environmental conditions. II. The dependence of carbon assimilation on animal size, temperature, food concentration and diet species. Arch. Hydrobiol. Suppl. 48:310–335; 1977.

Lampert, W.; Bohrer, R. Effect of food availability on the respiratory quotient of *Daphnia magna*. Comp. Biochem. Physiol. 78A:221–223; 1984.

Langdon, C.J. Preparation and evaluation of protein microcapsules for a marine suspension-feeder, the Pacific oyster *Crassostrea gigas*. Mar. Biol. 102:217–224; 1989.

Langdon, C.J.; DeBevoise, A.E. Effect of microcapsule type on delivery of dietary protein to a marine suspension-feeder, the oyster *Crassostrea gigas*. Mar. Biol. 105:437–443; 1990.

Langdon, C.J.; Waldock, M.J. The effect of algal and artificial diets on the growth and fatty acid composition of *Crassostrea gigas* spat. J. Mar. Biol. Assn. U.K. 61:431–447; 1981.

Lee, R.F.; Nevenzel, J.C.; Paffenhofer, G.A. The presence of wax esters in marine planktonic copepods. Naturwissenschaften 59:406–411; 1972.

Lee, R.F.; Nevenzel, J.C.; Paffenhofer, G.A. Importance of wax esters and other lipids in the marine food chain: phytoplankton and copepods. Mar. Biol. 9:99–108; 1971.

Müller-Navarra, D. Evidence that a highly unsaturated fatty acid limits growth in nature. Arch. Hydrobiol. 132:297–307; 1995.

Murata, N.; Nishida, I. Lipids of blue-green algae (Cyanobacteria). In: Stumpf, P. K., Conn,

E.E., eds. The Biochemistry of Plants. vol. 9. New York: Academic Press; 1987:p. 315–347.

Parrish, C.C.; Ackman, R.G. Chromarod separations for the analysis of marine lipid classes by Iatroscan thin-layer chromatography-flame ionization detection. J. Chromatogr. 262:103–112; 1983.

Peters, R.H. Methods for the study of feeding, grazing and assimilation by zooplankton. In: Downing, J.A.; Rigler, F.H., eds. A Manual on Methods for the Assessment of Secondary Productivity in Fresh Waters. Oxford: Blackwell Scientific; 1984:p. 336–412.

Provasoli, L.; Shiriashi, K.; Lance, J.R. Nutritional idiosyncrasies of *Artemia* and *Tigriopus* in monoxenic culture. Ann. N.Y. Acad. Sci. 77:250–261; 1959.

Reynolds, C.S. The Ecology of Freshwater Phytoplankton. Cambridge, U.K., Cambridge University Press; 1984.

Shifrin, N.S.; Chisholm, S.W. Phytoplankton lipids: Interspecific differences and effects on nitrate, silicate and light-dark cycles. J. Phycol. 17:374–384; 1981.

Scott, J.M. Effect of growth rate of the food alga on the growth/ingestion efficiency of a marine herbivore. J. Mar. Biol. Assn. U.K. 60:681–702; 1980.

Siron, R.; Giusti, G.; Berland, B. Changes in the fatty acid composition of *Phaeodactylum tricornutum* and *Dunaliella teriolecta* during growth and under phosphorus deficiency. Mar. Ecol. Prog. Ser. 55:95–100; 1989.

Snedecor, G.W.; Cochran, W.G. Statistical Methods, 6th ed. Ames, IA: Iowa State University Press; 1967.

Sokal, R.R.; Rohlf, F.J. Biometry. New York: W. H. Freeman. 1981.

Stainton, M.P. A syringe gas-stripping procedure for gas chromatographic determination of dissolved inorganic and organic carbon in fresh water and carbonate in sediments. J. Fish. Res. B. Can. 30:1441–1445; 1973.

Taylor, B.; Gabriel, W. Reproductive strategies of two similar *Daphnia* species. Verh. Int. Verein. Limnol. 22:3047–3050; 1985.

Tessier, A.J. Comparative population regulation of two planktonic Cladocera (*Holopedium gibberum* and *Daphnia catawba*). Ecology 67:285–302; 1986.

Tessier, A.J.; Goulden, C.E. Estimating food limitation in cladoceran populations. Limnol. Oceanogr. 27:707–717; 1982.

Thompson, P.A.; Guo, M-X.; Harrison, P.J. The influence of irradiance on the biochemical composition of three phytoplankton species and their nutritional value to larval Pacific oysters (*Crassostrea gigas*). Aquaculture 143:379–391; 1993.

Thompson, P.A.; Guo, M-X.; Harrison, P.J. The influence of temperature. I. On the biochemical composition of eight species of marine phytoplankton. J. Phycol. 28:481–488; 1992.

Threlkeld, S.T. The midsummer dynamics of two *Daphnia* species in Wintergreen Lake, Michigan. Ecology 60:165–179; 1979.

Vollenweider, R.A. Elemental and biochemical composition of plankton biomass; some comments and explorations. Arch. Hydrobiol. 105:11–29; 1985.

Wilkinson, L. SYSTAT: The System for Statistics. Evanston, IL: SYSTAT, Inc.; 1988.

Wood, B.J.B. Fatty acids and saponifiable lipids. In: Stewart, W.D.P., ed. Algal Physiology and Biochemistry. Botanical Monographs. vol. 10. Berkeley, CA: University of California Press; 1974:p. 236–265.

6

Seasonal Dynamics of Lipids in Freshwater Benthic Invertebrates

Joann F. Cavaletto and Wayne S. Gardner

6.1. Introduction

Although there is a wealth of information on lipids in marine organisms, especially zooplankton (Kattner and Hagen, 1995; Tande and Henderson, 1988; Clarke et al., 1987; Lee, 1975; Lee and Hirota, 1973), relatively little information is available on freshwater benthic organisms. The benthos of large temperate lakes often experience a seasonal variation in food supply that is similar to the high to midlatitudes of the ocean (Lee, 1975). For instance, in temperate large lakes the diatom bloom that eventually settles to the profundal zone of the lake may be the main source of high-quality food for the benthos during the year (Fitzgerald and Gardner, 1993; Johnson and Wiederholm, 1992; Gardner et al., 1990; Fahnenstiel and Scavia, 1987).

The seasonal variation in lipid content of benthic organisms can provide clues to the ecology of the organisms. The high energy content of lipids (39.35 J · mg^{-1}) over proteins (23.63 J · mg^{-1}) and carbohydrates (17.18 J · mg^{-1}) make it the most efficient energy-storing compound for most freshwater benthic organisms. To survive periods when high-quality food is not reaching the benthic community, cold stenothermic benthic invertebrates often store energy as lipids (i.e., triacylglycerols [TAG]) when food is abundant (Johnson and Wiederholm, 1992; Gardner et al., 1990; Gauvin et al., 1989; Gardner et al., 1985a). The seasonal lipid dynamics of an organism may also depend on their feeding strategy. For example, some benthic animals feed at the sediment surface whereas others feed below the surface. Consequently, animals may feed on particles at various phases of decomposition and nutritional quality depending on their feeding mode. In addition, the trophic state of the water body that the benthic invertebrates inhabit may affect their lipid dynamics. Besides availability of food resources, morphology, life history, habitat, and the reproductive state of a benthic animal may contribute to its seasonal lipid profile.

In this chapter, we explore the seasonal changes of lipids in different benthic invertebrates with emphasis on large temperate lakes. Seasonal lipid dynamics in benthos from lotic environments is limited, so only minimal comparative information is provided on stream invertebrates. Because the location of benthic invertebrates in the lake is in part responsible for their seasonal lipid content, this chapter is organized according to habitats. Furthermore, research gaps in lipid dynamics of freshwater benthic invertebrates are explored.

6.2. Results and Discussion

6.2.1. Slope and Profundal Zones

6.2.1.1. Crustacea: Amphipoda

The benthic amphipods *Diporeia* spp. are the most abundant benthic invertebrates in the slope and profundal regions of the upper Laurentian Great Lakes (Nalepa, 1987); they also live in many other glacial lakes from temperate to subarctic zones (Moore, 1979; Dadswell, 1974). In North America, *Diporeia* spp. were formerly known as one species, *Pontoporeia hoyi;* however, recent changes in the taxonomy of these amphipods have identified at least two and perhaps as many as eight species (Bousfield, 1989). Until the taxonomy of these benthic amphipods is resolved, we will refer to the species formerly known as *Pontoporeia hoyi* as *Diporeia.*

Diporeia inhabits the upper 2 cm of sediment and thoroughly mix this upper sediment layer during burrowing and feeding (Robbins et al., 1979). In addition, *Diporeia* prefers feeding in fine silty sediments over coarse ones (Mozley and Howmiller, 1977). The slope region (30–90 m) of the Great Lakes, with its cold temperatures (4–8°C) and an adequate food supply, supports the highest *Diporeia* densities throughout the year (Evans et al., 1990; Nalepa, 1987; Johannsson et al., 1985; Winnell and White, 1984). *Diporeia* is by far the most dominant organism in profundal zones (>90 m); however, overall densities are lower than those found in the slope region, likely because of diminished food supplies (Nalepa, 1991; Evans et al., 1990).

Diporeia lipid levels vary seasonally, mainly because of changes in TAG, the lipid reserve (Cavaletto et al., 1996; Gardner et al., 1985b). A seasonal increase in *Diporeia* lipid reserves occurs sometime following a spring bloom of large diatoms that settle to the bottom before the lake stratifies. However, various lag times occur between the diatom bloom event and accumulation of lipid reserves in the animals that appears to depend on the lake and depth (Fig. 6.1). Evidence of this diatom sedimentation has been shown in the slope region of Lake Michigan by examining alkane composition and biogenic silica fluxes in material collected in a sediment trap located just below the thermocline at a 45-m-deep site in Lake Michigan (Gardner et al., 1989). The ratio of aquatic plankton markers (short alkane length; C_{17}) to terrestrial markers (long alkane length; C_{29}) indicated a large input of plankton/autochthonous material in the spring (late April–May), whereas the ratio during other times of the year indicated a terrestrial/allochthonous input. In addition, high biogenic silica flux in early spring also supported the occurrence of this event (Johengen et al., 1994; Gardner et al.,

FIGURE 6.1. Seasonal variation of total lipid levels in *Diporeia* spp. from 45-m and 100-m sites in Lake Michigan and 35-m and 125-m sites in Lake Ontario. Bars represent standard errors. (Data from Cavaletto et al., 1996.)

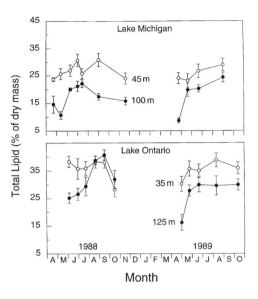

1989). It appears that *Diporeia* depends on diatom sedimentation for a high-quality food source. *Diporeia* from the slope has a higher incident of full guts in the spring, when diatom sedimentation to the benthos occurs, than it does during other times of the year (Quigley, 1988).

In the profundal zone of the lake, the percentage of *Diporeia* with full guts is more constant through the seasons, possibly indicating that food resources are relatively consistent throughout the year (Evans et al., 1990). Likely, the relatively long time required for diatoms to settle to the deeper profundal zone results in more decomposition and zooplankton grazing of the diatoms than occurs in the shallower slope region. Thus, the quality of food sources reaching the benthic community may be inversely related to depth. These food differences are evidenced by lower *Diporeia* lipid levels in individuals from the profundal zone compared with those from the slope region (Fig. 6.1) (Cavaletto et al., 1996). However, lipid levels in *Diporeia* are not always lower in the profundal region, and this difference can vary by season. For instance, in Lake Ontario during 1988, *Diporeia* lipid levels were higher in the slope region in the spring, but there was no difference in lipid levels in the slope and profundal regions in late summer (Fig. 6.1) (Cavaletto et al., 1996).

Diporeia from Lake Ontario accumulated lipid levels up to 12% higher than *Diporeia* from Lake Michigan (Cavaletto et al., 1996). This result reflects ecological differences in the two lakes or possibly genetic differences between the two *Diporeia* populations (Bousfield, 1989). One ecological difference between the lakes is that Lake Ontario is more productive than Lake Michigan. This may result in more algal and detrital material provided to the benthos in Lake Ontario (Johengen et al., 1994). In addition, Lake Ontario is dominated by small zooplankton that may not be able to consume large diatoms that are usually dominant in spring blooms, thus allowing a greater proportion to settle to the bottom of the lake and become available to the benthic community (Mazumder et al., 1992; Johannsson et al., 1991). In Lake Michigan, hypolimnetic calanoid copepods are dominant in the spring and readily feed on the diatom bloom (Vanderploeg et al., 1992).

Lipid levels in *Diporeia* are similar to lipid levels found in two species of related amphipods from a Baltic archipelago. *Monoporeia affinis* had a maximum lipid level in late summer of 27% of dry mass as compared with the less seasonally variable lipid levels of *Pontoporeia femorata* that were 20–23% of dry mass (Hill et al., 1992). Another study of Baltic Sea *M. affinis* revealed higher seasonal lipids that had a maximum of 42% of dry mass (Lehtonen, 1996). The maximum lipid level recorded for *Diporeia* from Lake Michigan was 46% of dry mass in 1984 (Gardner et al., 1985a). However, annual variation occurs as demonstrated by a decline in *Diporeia* maximum lipid levels to 31% of dry mass in 1988 and 1989 (Fig. 6.1). Generally, *Diporeia* is an important pelagic benthic coupling link and fish food source (McDonald et al., 1990; Wells, 1980). Due to its high lipid content, *Diporeia* contributed about 71% of the mean energy content of all macroinvertebrates averaged over southern Lake Michigan as compared with 66% of the biomass (Gardner et al., 1985a).

In addition to providing metabolic fuel when food is scarce, TAGs are important to the reproductive process of *Diporeia*. Adult female *Diporeia* have the highest TAG levels, males have low levels of TAG, and juvenile TAG levels are intermediate (Fig. 6.2) (Quigley et al., 1989). Males are highly active and swim above the sediment surface in search of mates. This high level of activity in males probably rapidly depletes lipid reserves that were accumulated in the juvenile stage. The final molt of the adult male may also be demanding energetically (Hill, 1992). By contrast, the high lipid levels of the females are likely conserved and

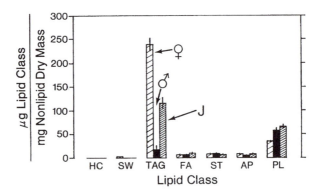

FIGURE 6.2. Mean micrograms ($\pm$ SE) of lipid class per milligram nonlipid dry mass among female, male, and juvenile *Diporeia*. HC, hydrocarbon; SW, sterol ester/wax ester; TAG, triacylglycerol; FA, free fatty acid; ST, sterol; AP, acetone mobile polar lipid; PL, phospholipid. (From Quigley et al., 1989.)

then transferred to their brooded eggs or young to provide them with energy during their first few days of life. In addition, females with high lipid reserves may be able to produce larger broods (Cavaletto et al., 1996).

Lipid levels may also be important to the duration of the life cycle in *Diporeia.* The length of the *Diporeia* life cycle increases with increasing depth. For example, in Lake Michigan, *Diporeia* has a 1-year life cycle in the shelf region (10–20 m), in the slope (30–90 m) it has a 2-year life cycle, and in the profundal zone (>90 m) the life cycle is 2.5–3 years (Winnell and White, 1984). Food availability and increased temperatures during the summer in the shelf region are probably responsible for these varying life cycles; however, a certain lipid level may need to be achieved before completion of the life cycle can occur in the different regions (Hill et al., 1992; Quigley et al., 1989).

6.2.1.2. Mysidacea

Mysis relicta, commonly known as the opossum shrimp, usually inhabit deep, cold, glacial lakes, although they can adapt to a wide range of environmental conditions including temperatures up to 21°C, low oxygen, and eutrophication (Beeton and Gannon, 1991). *Mysis* usually inhabit depths >30 m, but in the Great Lakes they may also be found in shallow regions when the lake is isothermal or during upwelling events (Mozley and Howmiller, 1977). *Mysis* are considered to be epibenthic because they spend the day in the sediments and migrate up into the water column at night (Beeton, 1960). However, in deep regions of a lake (100 m) mysids can occur 7 m above the sediments during the day (Beeton, 1960). *Mysis* are omnivorous, feeding on detrital particles in sediments, algae, zooplankton, and small *Diporeia* (Parker, 1980; Bowers and Grossnickle, 1978; Lasenby and Langford, 1972). *Mysis* can take 1–4 years to reach sexual maturity, depending on food availability (Adare and Lasenby, 1994; Morgan, 1980; Lasenby and Langford, 1972) and temperature (Berrill and Lasenby, 1983). After a female *Mysis* deposits her eggs in the marsupium and breeds with a male, she continues to grow as the embryos and larvae develop. Following release of the young, the female will usually die; however, in some instances *Mysis* have been documented to produce a second brood (Morgan and Beeton, 1978).

Mysis from four Canadian lakes accumulated lipids as they matured. Life cycle length depended on growth rate along with lipid accumulation rate and it varied in the *Mysis* populations from the different lakes. For example, in the first 6 months of life, *Mysis* with a 2-year life cycle grew at a rate of 1.0 mm $\cdot$ month^{-1} and accumulated 43.5 μg of lipid, whereas *Mysis* with a 1-year life cycle grew 1.8 mm $\cdot$ month^{-1} and accumulated 125 μg of lipid (Adare and Lasenby, 1994). On maturity, *Mysis* attained their peak lipid levels in summer (Fig. 6.3). Maximum mysid lipid levels ranged from 30 to 40% of dry mass in Canadian temperate lakes (Adare and Lasenby, 1994). These levels are comparable with lipid levels of 33% of dry mass for Lake Michigan *Mysis* (Fig. 6.4) (Gardner et al., 1985a). TAG is the most abundant lipid class (75.6% of total lipid) in *Mysis* from Lake Michigan during late spring (Table 6.1).

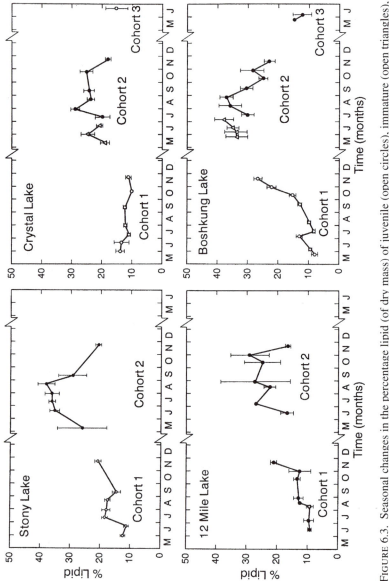
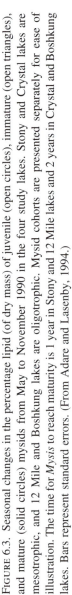

FIGURE 6.3. Seasonal changes in the percentage lipid (of dry mass) of juvenile (open circles), immature (open triangles), and mature (solid circles) mysids from May to November 1990 in the four study lakes. Stony and Crystal lakes are mesotrophic, and 12 Mile and Boshkung lakes are oligotrophic. Mysid cohorts are presented separately for ease of illustration. The time for *Mysis* to reach maturity is 1 year in Stony and 12 Mile lakes and 2 years in Crystal and Boshkung lakes. Bars represent standard errors. (From Adare and Lasenby, 1994.)

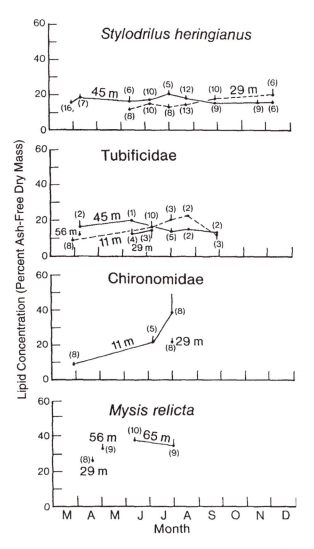

FIGURE 6.4. Total lipid content, expressed as a percentage of ash-free dry mass (ash-free dry mass is approximately 90% of dry mass for oligochaetes and chironomids, and 85% of dry mass for *Mysis relicta;* Johnson and Brinkhurst, 1971) in major species of Lake Michigan macroinvertebrates sampled in southeastern Lake Michigan in 1984. Bars represent standard errors. Numbers in parentheses represent the number of replicate animals analyzed. (From Gardner et al., 1985a.)

The life cycle of mysids in the Canadian lakes study was 1 or 2 years in oligotrophic and mesotrophic lakes, suggesting that lipid accumulation rate was independent of lake trophic status (Fig. 6.3) (Adare and Lasenby, 1994). However, this situation seems unusual because in other *Mysis* studies, life cycle length is considered to be inversely related to lake trophic state (Morgan, 1980), and

TABLE 6.1. Range of total lipids (% of dry mass) found in the literature, and the lipid class profiles of macrobenthic invertebrates.[a]

	Diporeia	Tubificidae	Chironomidae	Mysis	Dreissena
Total lipid range	9–46	8–20	8–35	10–41	6–18
Lake	Michigan	Michigan	Michigan	Michigan	St. Clair
Collection date for lipid classes	7/12/88	7/22/87	7/22/87	6/10/87	7/24/90
HC	nd	1.4 (0.6)	nd	0.6 (0.3)	nd
SE	nd	nd	nd	1.2 (0.1)	nd
ME	nd	nd	nd	1.0 (0.6)	2.4 (0.8)
TAG	75.6 (4.8)	9.7 (1.6)	29.1 (4.9)	72.1 (2.0)	17.2 (4.7)
FFA	0.9 (0.6)	3.3 (1.0)	1.4 (1.0)	0.9 (0.6)	3.0 (1.9)
AL	nd	1.6 (0.6)	1.6 (1.1)	1.8 (1.1)	nd
ST	9.7 (4.5)	10.3 (1.1)	7.8 (2.2)	3.4 (0.6	8.2 (1.8)
DG	nd	0.6 (0.3)	0.9 (0.6)	nd	nd
AMPL	4.8 (1.1)	13.9 (1.8)	17.9 (2.6)	4.2 (0.8)	6.3 (1.2)
PL	9.0 (2.4)	59.2 (4.4)	41.3 (5.7)	14.8 (2.2)	62.9 (6.8)

[a]Lipid class data are from the following sources; *Diporeia* spp. (Cavaletto et al., 1996); Tubificidae, Chironomidae, and *Mysis relicta* (Cavaletto and Gardner, unpublished data); *Dreissena polymorpha* (Nalepa et al., 1993). Lipid classes were determined with thin-layer chromatography with flame ionization detection (Parrish, 1987). Lipid class abbreviations are HC, hydrocarbon; SE, sterol ester; ME, methyl ester; TAG, triacylglycerol; FFA, free fatty acid; AL, aliphatic alcohol; ST, sterol (alicyclic alcohol); DAG, diacylglycerol; AMPL, acetone mobile polar lipid (this group may include chlorophyll, glycolipid, and monoacylglycerol); and PL, phospholipid. nd, Not detected. Standard errors are in parentheses.

correspondingly, life cycle length is dependent on the rate of lipid accumulation (Adare and Lasenby, 1994). It appears that other biological and physical water-quality parameters are important to the life cycle length of *Mysis* in some of these Canadian lakes. These parameters could include food that is inedible or nutritionally lacking to mysids during a certain period of the year and the average temperature of the hypolimnion (Lasenby, 1991).

Differences in life cycle length also occur in *Mysis* from the Great Lakes. For example, *Mysis* from Lake Michigan reach maturity 8 months faster than *Mysis* from Lake Ontario because the growth rate of *Mysis* from Lake Ontario declines during the winter months, whereas the growth rate of *Mysis* from Lake Michigan remains the same all year long (Johannsson et al., 1994). Furthermore, *Mysis* from Lake Michigan used 11–31% less calories to reach maturity than *Mysis* from Lake Ontario. This result differs from data in a study that compared *Mysis* from an Arctic and a temperate lake. Although the *Mysis* from the Arctic lake had a 2-year life cycle and the *Mysis* from the temperate lake had a 1-year life cycle, the calories used to reach maturity were the same for both populations (Lasenby and Langford, 1972). The lipid level/caloric utilization and life cycle length differences in *Mysis* must be adaptations to differences in food webs between lakes or perhaps a limitation of essential substances in the diet at certain times of the year.

6.2.2. Shelf and Nearshore Zones

6.2.2.1. Crustacea: Amphipoda

Unlike *Diporeia,* the majority of freshwater amphipods inhabit nearshore areas of lakes, ponds, and streams, and they usually live in and around vegetation from the surface to 1 m deep (Pennak, 1978). Seasonal variation in lipid content was determined in two common species of amphipods, *Hyalella azteca* and *Gammarus lacustris,* in conjunction with a contaminant study (Arts et al., 1995). Amphipods were analyzed from three Canadian prairie lakes that are in the vicinity of agricultural herbicide applications.

In *H. azteca,* lipid levels ranged from 2.5 to 16.3% of dry mass and varied seasonally in two of three lakes. In both of these lakes, lipid levels were highest in spring and declined through the summer, with lipid levels increasing again in autumn in *H. azteca* from one of the lakes. In addition, females usually had higher lipid levels than males in spring and early summer during reproduction (Arts et al., 1995). Similarly, total lipid levels in *H. azteca* from a Michigan pond were 17.4% ± 3.7 (SE) of dry mass in autumn (Cavaletto and Gardner, unpublished data). In *G. lacustris,* lipid levels ranged from 2.4 to 15.7% of dry mass. Like *H. azteca,* lipid levels in *G. lacustris* were highest, especially in females, during the reproductive period in spring and early summer (Arts et al., 1995). TAG was the most abundant lipid class in both *H. azteca* and *G. lacustris,* ranging from 22.6 to 67.3% of total lipid. In addition, the contaminant studied (i.e., triallate) was positively correlated with TAG in both amphipods (Arts et al., 1995).

6.2.2.2. Annelida: Oligochaeta

Oligochaete worms are the most abundant benthic macroinvertebrates in many shallow bays, harbors, and river mouths, and they occur from nearshore areas to the profundal zone (Schloesser et al., 1995; Brinkhurst, 1970, 1967). Many species of oligochaetes are used as water-quality indicators. For example, high densities of *Tubifex tubifex* or *Limnodrilus hoffmeisteri* are usually found in polluted environments. The presence of *Stylodrilus heringianus* is an indication of oligotrophic conditions, and they are more commonly found in the profundal zone of lakes (Nalepa, 1991; Mozley and Howmiller, 1977). Benthic oligochaetes usually feed with their anterior ends 4–6 cm below the sediment–water interface and defecate at the sediment surface (Appleby and Brinkhurst, 1970). Oligochaetes are continuous feeders and derive nutrition from the detritus and bacteria present in the sediment (Wavre and Brinkhurst, 1971; Brinkhurst and Chua, 1969). Lipids have been studied in oligochaetes from the family Tubificidae and *S. herigianus* (Lumbriculidae) (Gardner et al., 1985a). These oligochaetes had no apparent seasonal lipid pattern; however, total lipids varied between 8–20% of dry mass (Fig. 6.4).

Phospholipids (PL) are the most abundant lipid class in tubificids. Oligochaetes have large surface areas of unprotected epidermis (no carapace or cuticle) that require a high portion of structural membrane PL. Sterol (ST; most commonly cholesterol) is also an important membrane component, and it is fairly abundant in the worms. Combined PL and ST represent 70% of tubificid total lipid (Table 6.1). The acetone mobile polar lipids (AMPL) are polar lipid compounds that are mobile in 100% acetone; this lipid group includes chlorophyll. AMPL may represent a portion of the worm's gut contents. Lipid reserves (TAG) are not a large component of tubificid lipids (Table 6.1). It is possible that because oligochaetes feed below the sediment surface, they have a more constant but probably a low-quality food resource (i.e., detritus and bacteria) compared with organisms that feed closer to the sediment surface. This relatively constant food supply may allow oligochaetes to maintain its growth, respiration, and reproduction with only a minimum of lipid reserves (Gardner et al., 1985a).

6.2.2.3. Insecta

The aquatic midge larvae, members of the family Chironomidae, are ubiquitous in lentic and lotic environments. They are most common in shallow (<8 m) environments where food is relatively abundant. Chironomid larvae feed on surficial sediment, and many species build feeding tubes where they draw in food particles by self-produced currents. Chironomids are selective feeders, and it has been demonstrated that food items in their gut are at a higher concentration than food items in the surrounding sediments (Johnson et al., 1989). Although they feed on detrital particles and bacteria, algae (especially diatoms) can be an important part of their diet (Pinder, 1992; Johnson et al., 1989). In fact, chironomid larvae that were fed detrital particles alone did not survive or grow as well as larvae that were fed both detritus and diatoms (Pinder, 1992). Chironomids inhabiting the near-shore areas of Lake Michigan likely rely on diatoms as an essential part of their diet. Following the spring diatom bloom, chironomid larvae from Lake Michigan show an increase in total lipid levels by midsummer (Fig. 6.4). Storage lipids (TAG) occur at intermediate levels in chironomids as compared with other Great Lakes benthic invertebrates (Table 6.1). These quantities of lipid classes are comparable with levels found in chironomids from a Scottish river where TAG was 54% and PL was 31% of total lipids (Bell et al., 1994).

Chironomids are holometabolous insects, and their metamorphosis from larvae to adult is a high energy activity. A study of biochemical components (protein, carbohydrates, and lipids) of chironomid larvae revealed that the instar prior to prepupal reorganization had a more rapid increase in carbohydrates and lipids than the earlier instars. When prepupal tissue organization did begin, carbohydrates and lipids were the components that were used most rapidly (Beattie, 1978).

Accumulation of lipid reserves was also necessary for proper metamorphosis and reproduction to occur in the caddis fly *Clisoronia magnifica*. Caddis fly larvae that were fed a high-quality diet (high carbohydrate) accumulated more lipid, spent less time in the larval form, took longer to pupate, and lived longer as adults than individuals on a low-quality diet (Cargill et al., 1985). Like the chironomid larvae, the majority (i.e., 80%) of lipid reserves that accumulated in the caddis fly larvae occurred during the last instar.

Another similar situation occurs in the aquatic soldier fly larvae, *Hedriodiscus truquii,* which was studied for ecological energetics modeling in thermal springs (Stockner, 1971). The algivorous larvae pass through four instars from egg to adult. Total lipid levels of *H. truquii* may double or quadruple from the second to fourth instar (pupae) and finally reach a maximum lipid level of approximately 20% of dry mass (Stockner, 1971). Energy expenditure during metamorphosis is revealed in the adult's lower lipid levels of 11–12% of dry mass. In addition, the short-lived adults (i.e., a few days) probably do not feed, therefore relying on the lipid acquired as larvae to complete their reproductive cycle.

Mayfly nymphs, members of the order Ephemeroptera, live in diverse habitats from fast-running streams to slow-moving rivers and lakes. Most mayflies are herbivores and detritivores, but a few may be true carnivores (Edmunds, 1978). In addition, they are very sensitive to poor water-quality conditions. Thus, many species were eliminated from their natural habitats due to cultural eutrophication.

Lipid content has been analyzed for studies of mayfly toxicokinetics (Landrum and Poore, 1988) and biochemical composition (Meyer, 1990). Lipid content of the mayfly nymphs varies seasonally. The stream mayfly *Epeorus sylvicola* had its highest values in lipid level, dry mass, and body length in June, just prior to mature nymph emergence to the subimago or preadult stage. The maximum mean lipid level was 12% of dry mass, and the range throughout the year for *E. sylvicola* was from 2 to 20% of dry mass (Meyer, 1990). *Hexagenia limbata,* a burrowing mayfly nymph, had a similar lipid range from 2.4 to 17.7% of dry mass. Like *E. sylvicola,* the mean maximum lipid level in *H. limbata* occurred in June (Landrum and Poore, 1988). In several species of mayflies from a Scottish stream, TAG was determined to be the most abundant lipid class (i.e., 50–60% of total lipid), whereas PL ranked second (i.e., 30% of total lipid) during August (Bell et al., 1994).

Sexual differences in lipid level were found in large nymphs of *E. sylvicola.* Male lipid levels were approximately 15% of dry mass, whereas female lipid levels were approximately 10% of dry mass in individuals >15 mg dry mass (Meyer, 1990). The lower lipid level in females may be due to production of eggs when stored lipid is used for the synthesis of yolk proteins (Meyer, 1990). In compliance with this, protein levels were higher in the >15-mg dry mass female than in male nymphs of the same size.

6.2.2.4. Mollusca: Bivalvia

The mollusk *Diplodon patagonicus* lives in Lake Nahuel Huapi located in the Patagonian Andes Mountains. Total lipid levels varied seasonally in *D. patagonicus.* Lipid values of 4.1% of dry mass (percentage dry mass was converted from wet mass data by assuming that dry mass is 15% of wet mass) occurred in winter, whereas lipid values of 9.7% of dry mass occurred in the spring (Pollero et al., 1981). This lipid increase was likely due to an increase in *D. patagonicus* feeding in the spring but not on a diatom bloom. The spring-melt runoff is low in nutrients in Lake Nahuel Huapi and does not promote a spring diatom bloom. In addition, the lack of seasonal variation in *D. patagonicus* fatty acid (FA) composition indicated no occurrence of a seasonal algal event. However, FA analysis did reveal an unsaturated FA ratio typical of terrestrial plants. Therefore, it appears that land-based detrital material is an important component of the diet of *D. patagonicus* (Pollero and Brenner, 1981; Pollero et al., 1981b). In addition, reproduction appeared to influence lipid composition of a related species, *Diplodon delodontus* (Pollero et al., 1983). For example, TAG levels of *D. delodontus* increased in both sexes prior to gametogenesis, although TAG levels were higher in females than in males. Furthermore, specific unsaturated FAs also increased with the maturing of the oocytes. Like other benthic invertebrates, lipid production in the spring is important for *Diplodon* spp. to successfully reproduce.

Zebra mussels (*Dreissena polymorpha*) invaded the lakes and rivers of central Europe at the beginning of the past century. In the late 1980s, *Dreissena* invaded the Great Lakes ecosystem (Hebert et al., 1989) and continue to spread into water bodies across the south and to the west in North America. *Dreissena* are nuisance organisms due to their mode of attachment and rapid colonization of hard surfaces. We are only beginning to understand the cascading effects that result from the invasion of this exotic species. For example, dreissenids alter native ecosystems due to their tremendous ability to filter phytoplankton from the water (Fanslow et al., 1995) and shift the main use of primary production from the pelagic zone to the benthos (Lowe and Pillsbury, 1995). Many studies have been done to try to better understand and possibly help control the spread of *Dreissena* (Nalepa and Fahnenstiel, 1995; Nalepa and Schloesser, 1993).

Dreissena are short lived for mollusks (3–4 years), but they still produce many offspring. Unlike most freshwater bivalves that have a parasitic larval stage, *Dreissena* has a free-swimming veliger larva like marine bivalves (Sprung, 1989). *Dreissena* mature rather fast and begin to produce gametes when they are approximately 8 mm in length; they may increase the number of gametes produced as they grow (Sprung, 1991).

Biochemical composition of *Dreissena* soft tissues has been examined in several studies in relation to growth, reproduction, and starvation under natural and experimental conditions (Sprung, 1995; Nalepa et al., 1993; Sprung and Borcherding, 1991; Walz, 1979). *Dreissena* are unusual bivalves whose major energy reserve appears to be lipid rather than glycogen (Sprung, 1995; Sprung and Borcherding, 1991). Lipid content of *Dreissena* varies seasonally (Fig. 6.5). *Dreissena* total lipid was usually highest in the spring and declined throughout the summer and fall, although this may vary with food availability (Sprung, 1995; Nalepa et al., 1993). Lipid is concentrated in the digestive gland of *Dreissena,* and on proper conditions for reproduction (i.e., mainly a temperature of 12°C or greater), some lipid is transferred from the digestive gland to the gonad (Sprung, 1995). Like most freshwater organisms, *Dreissena* storage lipid is TAG. Their TAG concentration peaks in the spring and then declines throughout the remaining seasons (Fig. 6.6) (Nalepa et al., 1993). Structural membrane component lipid (i.e., PL) remains relatively stable from spring until fall and is most often the main lipid class in zebra mussels (Fig. 6.6; Table 6.1).

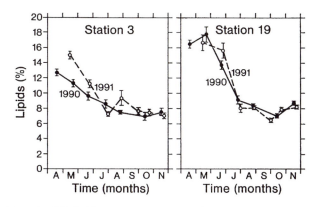

FIGURE 6.5. Mean ($\pm$ SE) lipid content (percentage of dry mass) of *Dreissena polymorpha* from the two sampling sites in southern Lake St. Clair on sampling dates in 1990 and 1991. Station 3 had high mussel densities, and station 19 had low mussel densities. (From Nalepa et al., 1993.)

The seasonal increase in *Dreissena* TAG levels appears to be related to food availability and reproductive state (Garton and Haag, 1993; Walz, 1979). In Lake St. Clair, part of the connecting waterway between Lakes Huron and Erie, *Dreissena*'s lipid composition was studied at two sites (stations 3 and 19) with different zebra mussel densities. At both sites, lipid levels peaked in April or May and then declined. However, spring maximum total lipid and TAG levels were highest at station 19 (Figs. 6.5 and 6.6) (Nalepa et al., 1993). This observation seems to provide some evidence that lipid levels in *Dreissena* are related to food supply because spring chlorophyll concentration was higher at station 19 (i.e., 5.1 $\mu g \cdot L^{-1}$) than at station 3 (i.e., 2.1 $\mu g \cdot L^{-1}$). However, other factors are probably involved. For instance, *Dreissena* densities were higher at station 3 than at station 19, and this could limit the food per mussel at station 3. In addition, the water temperature increased earlier in the spring at station 3 than at station 19, a factor that would increase mussel respiration and possibly induce reproduction sooner at station 3 than at station 19. All these factors exerted stress on the *Dreissena* population at station 3 that may have resulted in lower maximum lipid levels.

Another indicator of *Dreissena* condition is the comparison of respiration with ammonium excretion rates or O/N ratio (Quigley et al., 1993). Although the usual biochemical composition of an organism must be considered, some generalizations can be made regarding O/N ratios in starving animals. O/N ratios of ≥ 60 may indicate relatively high use of lipids, a ratio between 50–60 may indicate equal use of lipid and protein, ratios <50 and >16 may indicate increasingly more catabolism of protein over lipid, whereas low O/N ratios between 3–16 indicate use of protein (Mayzaud and Conover, 1988). Seasonal O/N ratios of Lake St. Clair *Dreissena* revealed a ratio of 48 in April, signifying relatively equal use of

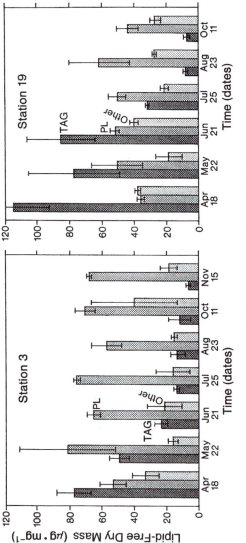

FIGURE 6.6. Seasonal variation (mean ± SE) in major lipid classes in *Dreissena polymorpha* from the two sampling sites in Lake St. Clair in 1990. Station 3 had high mussel densities, and station 19 had low mussel densities. Values are given as μg lipid · mg lipid-free dry mass^{-1}. TAG, triacylglycerol; PL, phospholipid; Other, acetone mobile polar lipid, free fatty acid, sterol. (From Nalepa et al., 1993.)

lipids and protein. An O/N ratio of 50 is considered the lower limit for physiological health in the blue mussel, *Mytilus edulis* (Bayne et al., 1985). In August, *Dreissena* had an O/N ratio of 16 that signified high use of protein, probably from the catabolism of soft body tissues because *Dreissena* mass and lipid declined from April to August (Nalepa et al., 1993). However, a low ratio could also signify the use of a high-protein food source. During the fall months, the zebra mussel's O/N ratio increased to 30, but this number is still below the physiologically healthy ratio (Quigley et al., 1993). Thus, despite *Dreissena's* rapid increase in densities since their successful invasion of North America, they may not be in a healthy state in all locales, likely due to food limitation during certain times of the year.

6.3. Research Needs

FA characterization in marine organisms has enhanced our understanding of food web dynamics in many marine ecosystems, but this approach has not been used as extensively in freshwater environments. FA biomarkers could likely, in freshwater ecosystems, trace food web dynamics as has been done for marine ecosystems (Kattner et al., 1996; St. John and Lund, 1996; Kattner and Hagen, 1995; Parrish et al., 1995; Kattner, 1989; Napolitano and Ackman, 1989). For example, in Jeddore Harbour, Nova Scotia, the appearance of odd-chain FAs in smelt were eventually traced to *Pontoporeia femorata* (a marine relative of *Diporeia* spp.). *Pontoporeia femorata* was a significant part of the smelt diet in the spring when actively reproductive males swam in the water column (Paradis and Ackman, 1976). The odd-chain FAs may have originated from microbes or allochthonous sources in the harbor (Paradis and Ackman, 1977).

FA characterization of freshwater organisms is very limited, especially for benthic invertebrates. However, some FA analysis has been done for *Diplodon* spp. (Pollero et al., 1983, 1981; Pollero and Brenner, 1981), aquatic insects (Bell et al., 1994; Hanson et al., 1985, 1983; Lee et al., 1975), and nectobenthic amphipods from Lake Baikal (Dembitsky et al., 1994; Morris, 1984). In the Great Lakes, identification of FA biomarkers in *Diporeia* or oligochaetes could possibly be used to trace seasonal changes in their diet, and these biomarkers might eventually be traced to their consumers such as mysids and fish.

Several FAs are essential to organisms at various trophic levels. Invertebrates and fish must obtain essential FAs from their diet for proper development and reproduction. Some FAs are important precursors to hormones. For example, the FA compositions of river invertebrates that are the natural food to salmon were examined and compared with the FA composition of commercial diets fed to farmed salmon (Bell et al., 1994). Although essential FAs were present in the commercial diet, important FA ratios were not optimal for salmon undergoing smoltification. A diet with FA ratios closer to that found in the salmon's natural diet of freshwater invertebrates prevented certain pathologies from developing in the fish.

More extensive FA analysis on benthic macroinvertebrates is necessary to develop a complete understanding of seasonal lipid and food web dynamics in freshwater environments and how FAs may alter development and reproduction. As in the case of *Mysis,* it is unclear whether their maturation time is more dependent on food quantity or quality. An examination of FA profiles of mysids and their potential food items in lakes with different mysid maturation times could be beneficial in understanding why life cycle length varies in *Mysis.* For example, improved growth of juvenile scallops was correlated to the occurrence of the essential FA 22:6ω3 in their diet, rather than merely a diet abundant in lipids (Parrish et al. 1995). This essential FA is common to the Cryptophyceae. Seasonal changes in algal composition and the essential FAs that they may provide could contribute to the maturation time of *Mysis* either by direct consumption or indirectly through zooplankton prey.

Long-term monitoring of total lipid content in certain benthic invertebrates could provide additional insight into ecosystem changes. *Diporeia* and other amphipods may be ideal organisms for this approach. In Lake Michigan, seasonal lipid content of *Diporeia* has been analyzed over several years at the same site. In 1984, the yearly mean lipid levels in *Diporeia* were 33.7% $\pm$ 3.3 (SE) of dry mass, in 1988 the levels declined to 28.2% $\pm$ 1.0 (SE) of dry mass, and in 1989 levels were 25.7% $\pm$ 1.2 (SE) of dry mass (Cavaletto et al., 1996; Gardner et al., 1985a). Maximum lipid levels in *Diporeia* reached 46% of dry mass in 1984, and in 1988 maximum levels were only 30% of dry mass. The decline in total lipid could reveal an increase in *Diporeia* population size causing intraspecific competition for food or reflect a decline in the food supply to the benthos. *Diporeia* densities at this site have remained relatively stable through the 1980s (T. Nalepa, personal communication), indicating that food availability to *Diporeia* may have changed from the mid- to late 1980s. However, additional data are necessary to reveal true trends in the lipid levels of *Diporeia.*

Lipid research on freshwater benthic macroinvertebrates has been focused primarily on organisms from temperate to Arctic regions. Little or no information is available for benthic invertebrates in subtropical and tropical lakes. Following results from tropical zooplankton (Lee and Hirota, 1973), lipid energy reserves may not be as important to subtropical/tropical as for temperate and Arctic benthic organisms because the former may have a more continuous food supply. However, it is possible that lipid variation may still occur for reasons other than climatic temperature variation (e.g., high- and low-precipitation periods). It would be interesting to document and compare lipid dynamics in invertebrates from temperate/Arctic regions to those from subtropical/tropical regions. In addition, FA profiles may be interesting to compare in benthic organisms from these two contrasting environments. For instance, to improve membrane fluidity, the degree of unsaturation of membrane FAs appears to be inversely related to temperature (Sargent, 1976). For this reason, FA unsaturation may be less in membrane lipids of benthic species from subtropical/tropical than in those from temperate/Arctic environments.

6.4. Conclusions

The benthic or epibenthic macroinvertebrate species that thrive in cold environments (e.g., slope and profundal zones in the Great Lakes) tend to have higher maximum lipid levels than the benthic invertebrate species that reside in habitats with seasonal changes in water temperatures (e.g., shelf and nearshore zones of lakes) (Fig. 6.7). The high lipid levels may not be from living in cold temperatures per se but from the seasonal variability in food inputs that usually occur in constantly cold environments. For example, *Diporeia* and *Mysis relicta* had the highest maximum lipid levels (>40% of dry mass) of the benthic invertebrates described, and they both reach their highest densities at >30 m or below the thermocline. Food quantity and quality vary seasonally in large temperate lakes, and these seasonal differences are amplified in the hypolimnion. Accumulating lipid reserves is a way for *Diporeia* and *Mysis* to store energy for a period when food is not available and also to support reproduction. Because both *Diporeia* and *Mysis* brood their eggs for many months and release live young, they may have higher reproductive success by transferring more lipid to their young. In addition, both *Diporeia* and *Mysis* have the ability to vary their life cycle length; this appears to be an adaptation to living where food is variable.

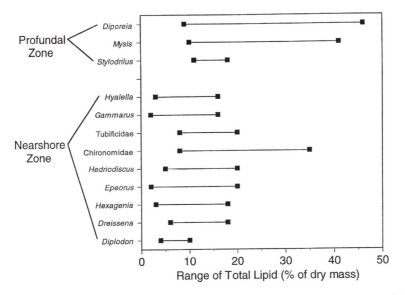

FIGURE 6.7. Summary of freshwater macroinvertebrates ranges of total lipid (% of dry mass). The macroinvertebrates are from the profundal zone (i.e., below the thermocline of deep lakes) and nearshore zones that include the shelf of large lakes and the littoral zone of lakes, ponds, rivers, and streams.

Benthic or epibenthic macroinvertebrate species that generally reside in near-shore zones do not usually accumulate as much reserve lipid or have as great of variation in lipid levels as the macrobenthic invertebrate species that reside in the profundal zone (Fig. 6.7). Usually, nearshore areas are more productive and may provide a more constant food supply to the benthic community. In addition, a reduction in metabolic rate during the "cold" season may conserve energy during this time of the year. Lipid levels often do increase seasonally but do not normally reach the maximum lipid levels found in *Diporeia* and *Mysis* (Fig. 6.7). The increase in lipid levels in chironomids and zebra mussels appears to be related to spring phytoplankton input, and the lipids are then allocated for reproduction. Lipid levels also increased in the amphipods *Hyalella azteca* and *Gammarus lacustris* and in aquatic insect larvae prior to reproduction. The Tubificidae and *Stylodrilus heringianus* differ due to their lack of seasonal variation in lipid reserves. The oligochaete mode of feeding apparently provides them with a rela-tively constant supply of subsurface bacteria and detritus throughout the year in either nearshore or profundal zones.

Acknowledgments. We thank K. Adare, M. Quigley, and T. Nalepa for use of their figures; T. Nalepa and P. Lavrentyev for their helpful comments; C. Darnell for editorial assistance; and J. R. Sargent for reviewing this manuscript. This chapter is GLERL contribution 1027.

References

Adare, K.I.; Lasenby, D.C. Seasonal changes in the total lipid content of the opossum shrimp, *Mysis relicta* (Malacostraca: Mysidacea). Can. J. Fish. Aquat. Sci. 51:1935–1941; 1994.

Appleby, A.G.; Brinkhurst, R.O. Defecation rate of three tubificid oligochaetes found in the sediment of Toronto Harbour, Ontario. J. Fish. Res. Bd. Can. 27:1971–1982; 1970.

Arts, M.T.; Ferguson, M.E.; Glozier, N.E.; Robarts, R.D.; Donald, D.B. Spatial and tem-poral variability in lipid dynamics of common amphipods: assessing the potential for uptake of lipophilic contaminants. Ecotoxicology 4:91–113; 1995.

Bayne, B.L.; Brown, D.A.; Burns, K.; Dixon, D.R.; Ivanovici, A.; Livingstone, D.R.; Lowe, D.M.; Moore, M.N.; Stebbing, A.R.D.; Widdows, J. The Effects of Stress and Pollution on Marine Animals. New York: Praeger; 1985.

Beattie, D.M. Life-cycle and changes in carbohydrates, proteins and lipids of *Pentapedilum uncinatum* Goet. (Diptera; Chironomidae). Freshwat. Biol. 8:109–113; 1978.

Beeton, A.M. The vertical migration of *Mysis relicta* in Lakes Huron and Michigan. J. Fish. Res. Bd. Can. 17:517–539; 1960.

Beeton, A.M.; Gannon, J.E. Effect of environment on reproduction and growth of *Mysis relicta*. Am. Fish. Soc. Symp. 9:144–148; 1991.

Bell, J.G.; Ghioni, C.; Sargent, J.R. Fatty acid compositions of 10 freshwater invertebrates which are natural food organisms of Atlantic salmon parr (*Salmo salar*): a comparison with commercial diets. Aquaculture 128:301–313; 1994.

Berrill, M.; Lasenby, D.C. Life cycle of the freshwater mysid shrimp *Mysis relicta* reared at two temperatures. Trans. Am. Fish. Soc. 112:551–553; 1983.

Bousfield, E.L. Revised morphological and relationships within the amphipod genera *Pontoporeia* and *Gammaracanathus* and the "glacial relict" significance of their postglacial distributions. Can. J. Fish. Aquat. Sci. 46:1714–1725; 1989.

Bowers, J.A.; Grossnickle, N.E. The herbivorous habits of *Mysis relicta* in Lake Michigan. Limnol. Oceanogr. 23:767–776; 1978.

Brinkhurst, R.O. Distribution and abundance of tubificid (Oligochaeta) species in Toronto Harbour. J. Fish. Res. Bd. Can. 27:1961–1969; 1970.

Brinkhurst, R.O. The distribution of aquatic oligochaetes in Saginaw Bay, Lake Huron. Limnol. Oceanogr. 12:137–143; 1967.

Brinkhurst, R.O.; Chua, K.E. Preliminary investigation of the exploitation of some potential nutritional resources by three sympatric tubificid oligochaetes. J. Fish. Res. Bd. Can. 26:2659–2668; 1969.

Cargill, A.S.; Cummins, K.W.; Hanson, B.J.; Lowry, R.P. The role of lipids, fungi, and temperature in the nutrition of a shredder caddisfly, *Clistoronia magnifica*. Freshwat. Invertebr. Biol. 4:64–78; 1985.

Cavaletto, J.F.; Nalepa, T.F.; Dermott, R.; Gardner, W.S.; Quigley, M.A.; Lang, G.L. Seasonal variation of lipid composition, weight and length in juvenile *Diporeia* spp. (Amphipoda) from Lakes Michigan and Ontario. Can. J. Fish. Aquat. Sci. 53:2044–2051; 1996.

Clarke, A.; Holmes, L.J.; Hopkins, C.E.C. Lipid in the arctic food chain: *Calanus, Bolinopsis, Beroe*. Sarsia 72:41–48; 1987.

Dadswell, M.J. Distribution, ecology, and postglacial dispersal of certain crustaceans and fishes in eastern North America. Publ. Zool. No. 11, National Museum of Canada, Ottawa, 110 pp.; 1974.

Dembitsky, V.M.; Kashin, A.G.; Rezanka, T. Comparative study of the endemic freshwater fauna of Lake Baikal-V. Phospholipid and fatty acid composition of the deep-water amphipod crustacean *Acanthogammaraus* (Brachyuropus) *grewingkii*. Comp. Biochem. Physiol. 108B:443–448; 1994.

Edmunds, G.F. Ephemeroptera. In: Merritt, R.W.; Cummins, K.W., eds. An Introduction to the Aquatic Insects of North America. Dubuque: Kendall/Hunt Publishing Co.; 1978:p. 57–80.

Evans, M.S.; Quigley, M.A.; Wojcik, J.A. Comparative ecology of *Pontoporeia hoyi* populations in southern Lake Michigan: the profundal region versus the slope and shelf. J. Great Lakes Res. 16:27–40; 1990.

Fahnenstiel, G.L.; Scavia, D. Dynamics of Lake Michigan phytoplankton: recent changes in surface and deep communities. Can. J. Fish. Aquat. Sci. 44:509–514; 1987.

Fanslow, D.L.; Nalepa, T.F.; Lang, G.A. Filtration rates of the zebra mussel (*Dreissena polymorpha*) on natural seston from Saginaw Bay, Lake Huron. J. Great Lakes Res. 21:489–500; 1995.

Fitzgerald, S.A.; Gardner, W.S. An algal carbon budget for pelagic-benthic coupling in Lake Michigan. Limnol. Oceanogr. 38:547–560; 1993.

Gardner, W.S.; Quigley, M.A.; Fahnenstiel, G.L.; Scavia, D.; Frez, W.A. *Pontoporeia hoyi*, an apparent direct link between spring diatoms and fish in Lake Michigan. In: Tilzer, M.; Serruya, C., eds. Large Lakes: Ecological Structure and Function. Heidelberg: Springer Verlag; 1990:p. 632–644.

Gardner, W.S.; Eadie, B.J.; Chandler, J.F.; Parrish, C.C.; Malczyk, J.M. Mass flux and "nutritional composition" of settling epilimnetic particles in Lake Michigan. Can. J. Fish. Aquat. Sci. 46:1118–1124; 1989.

Gardner, W.S.; Nalepa, T.F.; Frez, W.A.; Cichocki, E.A.; Landrum, P.F. Seasonal patterns in lipid content of Lake Michigan macroinvertebrates. Can. J. Fish. Aquat. Sci. 42:1827–1832; 1985a.

Gardner, W.S.; Frez, W.A.; Cichocki, E.A.; Parrish, C.C. Micromethod for lipids in aquatic invertebrates. Limnol. Oceanogr. 30:1100–1105; 1985b.

Garton, D.W.; Haag, W.R. Seasonal reproductive cycles and settlement patterns of *Dreissena polymorpha* in western Lake Erie. In: Nalepa, T.F.; Schloesser, D.W., eds. Zebra Mussels: Biology, Impacts, and Control. Boca Raton, FL: Lewis Publishers; 1993:p. 111–128.

Gauvin, J.M.; Gardner, W.S.; Quigley, M.A. Effects of food removal on nutrient release rates and lipid content of Lake Michigan *Pontoporeia hoyi*. Can. J. Fish. Aquat. Sci. 46:1125–1130; 1989.

Hanson, B.J.; Cummins, K.W.; Cargill, A.S.; Lowry, R.R. Lipid content, fatty acid composition, and the effect of diet on fats of aquatic insects. Comp. Biochem. Physiol. 80:257–276; 1985.

Hanson, B.J.; Cummins, K.W.; Cargill, A.S.; Lowry, R.R. Dietary effects on lipid and fatty acid composition of *Clistoronia magnifica* (Trichoptera: Limnephilidae). Freshwat. Invertebr. Biol. 2:2–15; 1983.

Hebert, P.D.N.; Muncaster, B.W.; Mackie, G.L. Ecological and genetic studies on *Dreissena polymorpha* (Pallas): a new mollusc in the Great Lakes. Can. J. Fish. Aquat. Sci. 46:1587–1591; 1989.

Hill, C. Interactions between year classes in the benthic amphipod *Monoporeia affinis:* effects on juvenile survival and growth. Oecologia 91:157–162; 1992.

Hill, C.; Quigley, M.A.; Cavaletto, J.F.; Gordon, W. Seasonal changes and composition in the benthic amphipods *Monoporeia affinis* and *Pontoporeia femorata*. Limnol. Oceanogr. 37:1280–1289; 1992.

Johannsson, O.E.; Rudstam, L.G.; Lasenby, D.C. *Mysis relicta:* assessment of metalimnetic feeding and implications for competition with fish in Lakes Michigan and Ontario. Can. J. Fish. Aquat. Sci. 51:2591–2602; 1994.

Johannsson, O.E.; Mills, E.L; O'Gorman, R. Changes in the nearshore and offshore zooplankton communities in Lake Ontario: 1981–88. Can. J. Fish. Aquat. Sci. 48:1546–1557; 1991.

Johannsson, O.E.; Dermott, R.M.; Feldcamp, R.; Moore, J.E. Lake Ontario long term biological monitoring program: report for 1981 and 1982 database. Can. Data Rep. Fish. Aquat. Sci. 552, 103 pp.; 1985.

Johengen, T.; Johannsson, O.; Pernie, G.; Millard, E.S. Temporal and seasonal trends in nutrient dynamics and biomass measures in Lakes Michigan and Ontario in response to phosphorus control. Can. J. Fish. Aquat. Sci. 51:2570–2578; 1994.

Johnson, M.G.; Brinkhurst, R.O. Production of benthic macroinvertebrates of Bay of Quinte and Lake Ontario. J. Fish. Res. Bd. Can. 28:1699–1714; 1971.

Johnson, R.K.; Wiederholm, T. Pelagic-benthic coupling—The importance of diatom interannual variability for population oscillations of *Monoporeia affinis*. Limnol. Oceanogr. 37:1596–1607; 1992.

Johnson, R.K.; Bostrom, B.; van de Bund, W. Interactions between *Chironomus plumosus* and the microbial community in surficial sediments of a shallow, eutrophic lake. Limnol. Oceanogr. 34:992–1003; 1989.

Kattner, G. Lipid composition of *Calanus finmarchicus* from the North Sea and the arctic. A comparative study. Comp. Biochem. Physiol. 94B:185–188; 1989.

Kattner, G.; Hagen, W. Polar herbivorous copepods—different pathways in lipid biosynthesis. ICES J. Mar. Sci. 52:329–335; 1995.

Kattner, G.; Hagen, W.; Falk-Petersen, S.; Sargent, J.R.; Henderson, R.J. Antarctic krill *Thysanoessa macrura* fills a major gap in marine lipogenic pathways. Mar. Ecol. Prog. Ser. 134:295–298; 1996.

Landrum, P.F.; Poore, R. Toxicokenetics of selected xenobiotics in *Hexagenia limbata.* J. Great Lakes Res. 14:427–437; 1988.

Lasenby, D.C. Comments on the roles of native and introduced *Mysis relicta* in aquatic ecosystems. Am. Fish. Soc. Symp. 9:17–22; 1991.

Lasenby, D.C.; Langford, R.R. Growth, life history, and respiration of *Mysis relicta* in an arctic and temperate lake. J. Fish. Res. Bd. Can. 29:1701–1708; 1972.

Lee, R.F. Lipids of arctic zooplankton. Comp. Biochem. Physiol. 51B:263–266; 1975.

Lee, R.F.; Hirota, J. Wax esters in tropical zooplankton and nekton and the geographical distribution of wax esters in marine copepods. Limnol. Oceanogr. 18:227–239; 1973.

Lee, R.F.; Polhemus, J.T.; Cheng, L. Lipids of the water strider *Gerris remigis* Say (Heteroptera: Gerridae). Seasonal and developmental variations. Comp. Biochem. Physiol. 51B:451–456; 1975.

Lehtonen, K.K. Ecophysiology of the benthic amphipod *Monoporeia affinis* in a northern Baltic open-sea area: seasonal variations in body composition, with bioenergetic considerations. Mar. Ecol. Prog. Ser. 143:87–98; 1996.

Lowe, R.L.; Pillsbury, R.W. Shifts in benthic algal community structure and function following the appearance of zebra mussels (*Dreissena polymorpha*) in Saginaw Bay, Lake Huron. J. Great Lakes Res. 21:558–566; 1995.

McDonald, M.E.; Crowder, L.B.; Brandt, S.B. Changes in *Mysis* and *Pontoporeia* populations in southeastern Lake Michigan: a response to shifts in the fish community. Limnol. Oceanogr. 35:220–227; 1990.

Mayzaud, P.; Conover, R.J. O:N atomic ratio as a tool to describe zooplankton metabolism. Mar. Ecol. Prog. Ser. 45:289–302; 1988.

Mazumder, A.; Lean, D.R.S.; Taylor, W.D. Dominance of small filter feeding zooplankton in the Lake Ontario food-web. J. Great Lakes Res. 18:456–466; 1992.

Meyer, E. Levels of major body compounds in nymphs of the stream mayfly *Epeorus sylvicola* (PICT.) (Ephemeroptera: Heptageniidae). Arch. Hydrobiol. 117:497–510; 1990.

Moore, J.W. Ecology of a subarctic population of *Pontoporeia affinis* Lindstrom (Amphipoda). Crustaceana 36:264–276; 1979.

Morgan, M.D. Life history characteristics of two introduced populations of *Mysis relicta*. Ecology. 61:551–561; 1980.

Morgan, M.D.; Beeton, A.M. Life history and abundance of *Mysis relicta* in Lake Michigan. J. Fish. Res. Bd. Can. 35:1165–1170; 1978.

Morris, R.J. The endemic faunae of Lake Baikal: their general biochemistry and detailed lipid composition. Proc. R. Soc. Lond. B222:51–78; 1984.

Mozley, S.C.; Howmiller, R.P. Environmental status of the Lake Michigan region: zoobenthos of Lake Michigan. Argonne National Lab. Rep. ANL/ES-40. Vol. 6. Argonne, IL: U.S. Energy Research and Development Administration; 1977.

Nalepa, T.F. Status and trends of the Lake Ontario macrobenthos. Can. J. Fish. Aquat. Sci. 48:1558–1567; 1991.

Nalepa, T.F. Longterm changes in the macrobenthos of southern Lake Michigan. Can. J. Fish. Aquat. Sci. 44:515–524; 1987.

Nalepa, T.F.; Fahnenstiel, G.L. *Dreissena polymorpha* in Saginaw Bay, Lake Huron ecosystem: overview and perspective. J. Great Lakes Res. 21:411–416; 1995.

Nalepa, T.F.; Schloesser, D.W., eds. Zebra Mussels: Biology, Impacts, and Control. Boca Raton, FL: Lewis Publishers; 1993.

Nalepa, T.F.; Cavaletto, J.F.; Ford, M.; Gordon, W.M.; Wimmer, M. Seasonal and annual variation in weight and biochemical content of the zebra mussel, *Dreissena polymorpha,* in Lake St. Clair. J. Great Lakes Res. 19:541–552; 1993.

Napolitano, G.E.; Ackman, R.G. Lipids and hydrocarbons in *Corophium volutator* from Minas Basin, Nova Scotia. Mar. Biol. 100:333–338; 1989.

Paradis, M.; Ackman, R.G. Influence of ice cover and man on the odd-chain hydrocarbons and fatty acids in the waters of Jeddore Harbour, Nova Scotia. J. Fish. Res. Bd. Can. 34:2156–2163; 1977.

Paradis, M.; Ackman, R.G. Localization of a marine source of odd chain-length fatty acids. I. The amphipod *Pontoporeia femorata* (Kroyer). Lipids 11:863–870; 1976.

Parker, J.I. Predation by *Mysis relicta* on *Pontoporeia hoyi:* a food chain link of potential importance in the Great Lakes. J. Great Lakes Res. 6:164–166; 1980.

Parrish, C.C. Separation of aquatic lipid classes by chromarod thin-layer chromatography with measurement by Iatroscan flame ionization detection. Can. J. Fish. Aquat. Sci. 44:350–356; 1987.

Parrish, C.C.; McKenzie, C.H.; MacDonald, B.A.; Hatfield, E.A. Seasonal studies of seston lipids in relation to microplankton species composition and scallop growth in South Broad Cove, Newfoundland. Mar. Ecol. Prog. Ser. 129:151–164; 1995.

Pennak, R.W. Freshwater Invertebrates of the United States. New York: John Wiley & Sons; 1978:p. 451–463.

Pinder, L.C.V. Biology of epiphytic Chironomidae (Diptera: Nematocera) in chalk streams. Hydrobiologia 248:39–51; 1992.

Pollero, R.J.; Brenner, R.R. Effect of the environment and fasting on the lipid and fatty acid composition of *Diplodom patagonicus.* Lipids 16:685–690; 1981.

Pollero, R.J.; Irazu, C.E.; Brenner, R.R. Effect of sexual stages on lipids and fatty acids of *Diplodon delodontus.* Comp. Biochem. Physiol. 76B:927–931; 1983.

Pollero, R.J.; Brenner, R.R.; Gros, E.G. Seasonal changes in lipid and fatty acid composition of the freshwater mollusk, *Diplodom patagonicus.* Lipids 16:109–113; 1981.

Quigley, M.A. Gut fullness of deposit-feeding amphipod, *Pontoporeia hoyi,* in southeastern Lake Michigan. J. Great Lakes Res. 14:178–187; 1988.

Quigley, M.A.; Gardner, W. S.; Gordon, W.M. Metabolism of the zebra mussel (*Dreissena polymorpha*) in Lake St. Clair of the Great Lakes. In: Nalepa, T.F.; Schloesser, D.W., eds. Zebra Mussels: Biology, Impacts, and Control. Boca Raton, FL: Lewis Publishers; 1993:p. 295–306.

Quigley, M.A.; Cavaletto, J.F.; Gardner, W.S. Lipid composition related to size and maturity of the amphipod *Pontoporeia hoyi.* J. Great Lakes Res. 15:601–610; 1989.

Robbins, J.A.; McCall, P.L.; Fisher, J.B.; Kresowski, J.R. Effects of deposit feeding on migration of ^{137}Cs in lake sediments. Earth Planet. Sci. Lett. 42:277–287; 1979.

Sargent, J.R. The structure, metabolism and function of lipids in marine organisms. In: Malins, D.C.; Sargent, J.R., eds. Biochemical and biophysical perspectives in marine biology, volume 3. London: Academic Press; 1976:p. 151–212.

Schloesser, D.W.; Reynoldson, T.B.; Manny, B.A. Oligochaete fauna of western Lake Erie 1961 and 1982: signs of sediment quality recovery. J. Great Lakes Res. 21:294–306; 1995.

Sprung, M. Physiological energetics of the zebra mussel *Dreissena polymorpha* in lakes I. Growth and reproductive effort. Hydrobiologia 304:117–132; 1995.

Sprung, M. Costs of reproduction: a study on metabolic requirements of the gonads and fecundity of the bivalve *Dreissena polymorpha*. Malacologia 33:63–70; 1991.

Sprung, M. Field and laboratory observations of *Dreissena polymorpha* larva: abundance, growth, mortality and food demands. Arch. Hydrobiol. 115:537–561; 1989.

Sprung, M.; Borcherding, J. Physiological and morphometric changes in *Dreissena polymorpha* (Mollusca; Bivalvia) during a starvation period. Malacologia 33:179–191; 1991.

St. John, M.A.; Lund, T. Lipid biomarkers: linking the utilization of frontal plankton biomass to enhanced condition of juvenile North Sea cod. Mar. Ecol. Prog. Ser. 131:75–85; 1996.

Stockner, J.G. Ecological energetics and natural history of *Hedriodiscus truquii* (Diptera) in two thermal spring communities. J. Fish. Res. Bd. Can. 28:73–94; 1971.

Tande, K.S.; Henderson, R.J. Lipid composition of copepodite stages and adult females of *Calanus glacialis* in arctic waters of the Barents Sea. Polar Biol. 8:333–339; 1988.

Vanderploeg, H.A.; Bolsenga, S.J.; Fahnenstiel, G.L.; Leibig, J.R.; Gardner, W.S. Plankton ecology in an ice-covered bay of Lake Michigan: utilization of a winter phytoplankton bloom by reproducing copepods. Hydrobiologia 243/244:175–183; 1992.

Walz, N. The energy balance of the freshwater mussel *Dreissena polymorpha* PALLAS in laboratory experiments and in Lake Constance. Arch. Hydrobiol. Suppl. 55.3/4:235–254; 1979.

Wavre, M.; Brinkhurst, R.O. Interactions between some tubificid Oligochaetes and bacteria found in the sediments of Toronto Harbour, Ontario. J. Fish. Res. Bd. Can. 28:335–341; 1971.

Wells, L. Food of alewives, yellow perch, spot-tail shiners, trout-perch, slimy and fourhorn sculpins in southeastern Lake Michigan. U.S. Fish Wildl. Serv. Tech. Paper 98. Washington, D.C.: U.S. Fish and Wildlife Service; 1980.

Winnell, M.H.; White, D.S. Ecology of shallow and deep water populations of *Pontoporeia hoyi* (Smith) (Amphipoda) in Lake Michigan. Freshwat. Invertebr. Biol. 5:118–138; 1984.

7
Ecological Role of Lipids in the Health and Success of Fish Populations

S. Marshall Adams

7.1. Introduction

The health or fitness of fish populations in aquatic ecosystems can be assessed by a variety of criteria ranging from individual biochemical and physiological measures to population level attributes such as growth, reproductive success, or year-class strength. Lipid storage and dynamics within the organism, however, are a particular important attribute of fish health and future population success. Lipids are critical not only to the survival and fitness of individuals but also to the success of reproduction and recruitment of future year classes within the population. This chapter discusses four areas in which lipids play an important role in fish health including (1) overwinter starvation and survival, (2) energy allocation strategies, (3) reproductive performance and early life history, and (4) environmental stress.

The primary purpose of this chapter is to evaluate and discuss the importance of the energetic lipids (triacylglycerols [TAG]) from an ecological perspective, with less emphasis on the more specific biochemical aspects of lipid dynamics in fish such as the role of fatty acids and lipid classes in fish health. These latter aspects of fish lipid dynamics are usually more related to feeding and nutritional aspects, which are presented in a variety of publications that address these specific areas including Halver (1989), Henderson and Tocher (1987), Cowey et al. (1985), and Love (1980). Furthermore, the purpose of this chapter is not to provide an exhaustive coverage of the literature relative to the ecological role of lipids in fish health but to give representative examples that illustrate the many and diverse ways that lipids are important in regulating and maintaining population stability and overall fitness of fish. Even though the objective of this book is to address the role of lipids in freshwater organisms, much of what we understand about the importance of lipids in fish comes from marine studies. Therefore, in areas where information is generally lacking for freshwater fish, discussions are supplemented with examples from marine systems that are relevant to freshwater environments. Reproductive development and early life history, lipid cycling, and energy allocation are the primary areas where studies of marine fish appear to be more complete compared with freshwater species.

7.2. Results and Discussion

7.2.1. Overwinter Starvation and Survival

The energetic demands of fish are met primarily by lipid oxidation (Cowey and Sargent, 1979). Depending on the species, TAG, or the energy-based lipids, are stored in various tissues including muscle, liver, subdermal tissue, and the mesenteries. During starvation, TAG is mobilized preferentially to the structural (mostly phospholipids) lipids, but these membrane lipids can also be mobilized if starvation is prolonged (Henderson and Tocher, 1987; Love, 1980). At temperate and north- temperate latitudes, many fish species display seasonal cycles in energy storage (Dygert, 1990; Sheridan et al., 1983; Adams et al., 1982; Medford and Mackay, 1978; Foltz and Nordon, 1977; Newsome and Leduc, 1975; MacKinnon, 1972). During the winter nonfeeding periods, fish use these lipid stores for basic maintenance and other metabolic needs. Beyond these maintenance needs, however, substantial quantities of both energetic lipids (i.e., TAG) and phospholipids for cell membranes must be supplied to the gonads, especially the ovaries, during reproductive development. Because energy for both metabolic and reproductive needs are drawn from stored lipids (Fig. 7.1) and depots are not usually re-

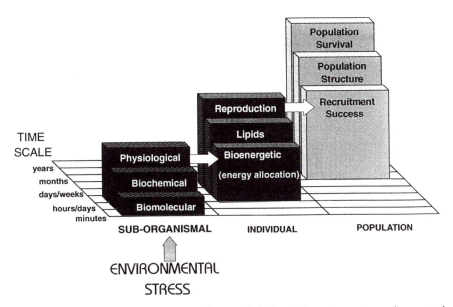

FIGURE 7.1. Hierarchical response scheme of biological systems to environmental stressors illustrating the relationship between type of response and time scale of response at different levels of biological organization. The allocation of available energy into lipids and reproduction is influenced at the suborganismal level by biochemical and physiological processes with effects being ultimately manifested at the population level through recruitment success and population dynamics. (Modified from Munkittrick and McCarty, 1995.)

plenished over the nonfeeding periods, lipid levels are typically reduced to very low levels by the end of the winter. Exhaustion of lipid reserves can have a variety of negative effects on fish condition and health and compromise individual survival and ultimately population success.

Mortality due to exhaustion of stored energy has been reported for a number of fish species (Henderson et al., 1988; Adams et al., 1985, 1982; Isley, 1981; Oliver et al., 1979). Fish must maintain a critical or minimal level of fat to survive (Hoar, 1983). Minimal fat levels for survival have been reported for various species as 1.4% wet wt. in herring (*Clupea harengus*) (Wilkins, 1967), 2.2% dry wt. in yellow perch (*Perca flavescens*) (Newsome and Leduc, 1975), 3.2% dry wt. in sand smelt (*Atherina boyeri*) (Henderson et al., 1988), and 1% wet wt. in gizzard shad (*Dorosoma cepedianum*) (Adams et al., 1985). Several studies also indicate that overwinter starvation mortality occurs when a fixed proportion (approximately 50%; i.e., the 50% rule) of the initial body weight is lost (Kleiber, 1961). For example, weight loss percentages in the 50% range for mortality resulting from starvation have been reported for several species (57%, Johnson and Evans, 1996; 55%, Shuter et al., 1980; 56%, Savitz, 1971). The assumption for the 50% rule is that a starved fish depletes the usable portion of its body energy (mainly TAGs), leaving only phospholipids, protein, and small amounts of carbohydrates.

The association between critical lipid (TAG) levels and survival may be related to the period when all the neutral lipids have been exhausted and fish begin to use phospholipids as an energy source. Even though phospholipids can be metabolized to provide some minimal energy during starvation (Love, 1980; Wilkens, 1967), the biochemical and physiological consequences for doing so appear to be severe for the organism. These structural lipids play a key role in the activity of various enzyme complexes that oxidize intermediates in the tricarboxylic acid cycle.

Structural lipids are also essential in cellular metabolism as they relate to osmotic and electrolyte homeostasis in the cell. Alterations in membrane permeability and ion transport mechanisms can occur under starvation, reducing the ability of fish to osmoregulate, particularly at low temperatures (Stanley and Colby, 1971; Morris and Bull, 1968). Nordlie and Leffler (1975) have suggested that reduced survival of fish during the winter may be due to this impaired osmoregulatory ability.

Over the winter nonfeeding period, most temperate zone fish undergo physiological changes that result in declining body condition and depletion of energy reserves (Cunjak, 1988). Many authors have shown the close relationship between the lipid content of fish and their ability to survive the winter. To survive through the winter, fish either accumulate a critical amount of fat or feed periodically to maintain their reserves. Fish that can switch their allocation of available energy from somatic growth to fat reserves before winter have a higher probability of overwinter and early spring survival (Thompson et al., 1991). In populations of temperate zone fish near the northern limit of their range, overwinter mortality is often high (Shuter and Post, 1990; Oliver et al., 1979). Freshwater fish at higher latitudes are particularly more vulnerable to overwintering mortality not only because of shorter periods for feeding and accumulation of energy stores, but also because of extended nonfeeding periods when any remaining lipids are used. The ability of fish to store energy and survive the winter generally decreases, therefore, from south to north as the growing season becomes shorter and the starvation period is extended (Miranda and Hubbard, 1994). The likelihood that young fish will die from exhaustion of energy stores is reduced at low latitudes where the growing season is longer and winters are shorter and milder. Shuter and Post (1990) surmised that overwinter starvation mortality of young fish may be the critical factor limiting the northern distribution of both yellow perch and smallmouth bass (*Micropterus dolomieui*) throughout large parts of their ranges. The northern distribution limits for other species may similarly be influenced by winter starvation dynamics.

Overwinter mortality appears to be a size-dependent phenomenon, with the smaller fish in a cohort experiencing higher levels of mortality than larger individuals (Hutchings, 1994; Adams et al., 1982; Shuter et al., 1980; Toneys and Coble, 1979). Size-dependent overwinter mortality is based, in part, on metabolic allometry (Miranda and Hubbard, 1994; Thompson et al., 1991; Post and Evans, 1989). Smaller fish, with lower energy stores and relatively higher metabolic rates, generally display higher mortality rates during winter when energy stores are often depleted and fish cannot meet basic metabolic demands (Post and Evans, 1989). Small individuals use proportionally more of their fat stores to maintain basal metabolic rate during periods of low food than do larger individuals (Schmidt-Nielsen, 1984; Shuter et al., 1980). A positive relationship between body size and lipid content has been found for yellow perch (Toneys and Coble, 1979), smallmouth bass (Shuter et al., 1980; Oliver et al., 1979), sand smelt (Henderson et al., 1988), and largemouth bass, *Micropterus salmoides* (Adams et al., 1982; Isley, 1981). Also, fish appear to have a compensatory reaction that raises the metabolic rate when temperatures approach critically low levels (Shul'man, 1974), with the rate of fat metabolism increasing at very low temperatures (Prosser, 1973). Thus, metabolism plays a significant role in lipid dynamics of the fish during the winter and has a major influence on survival rate.

Impairment of osmoregulatory ability can occur not only as a result of phospholipid utilization during periods of severe starvation but also as a result of biochemical changes in membrane structure at low temperatures. Lipids have a large influence on membrane permeability with fatty acid chain length, percentage unsaturated fatty acids, and number of double bonds per chain regulating fluidity, ion fluxes, and enzyme kinetics (Friedman et al., 1986). As temperatures decrease, the amount of unsaturated fatty acids in the plasma membrane increases along with concomitant changes in osmotic sensitivity, fluidity, and permeability. At low temperatures, replacement of native phosphatidylcholine by more saturated species considerably modifies membrane properties, resulting in a progressive increase in osmotic fragility and permeability (Leray et al., 1986). Changes in membrane permeability and impaired functioning of ion transport mechanisms at lower temperatures generally reduce the ability of fish to osmoregulate, which either results in death or increases the organism's vulnerability to additional stressors (Smit et al., 1981; Toneys and Coble, 1980; Davis and Simco, 1976). Because electrolyte regulation is an energy-consuming process (Nordlie and Leffler, 1975; Morris and Bull, 1968), the proportion of energy used for such regulation may be higher at lower than at higher temperatures and may therefore result in an additional drain on already depleted energy stores during the winter. For example, Johnson and Evans (1996) found that the overwinter mortality of white perch (*Morone americana*) at 4°C occurred primarily because of starvation but at 2.5°C resulted from both ionic osmoregulatory imbalance as well as starvation.

Starvation and depletion of lipids to critically low levels can result in increased susceptibility of fish to a variety of environmental stressors (Roff, 1982; Shul'man, 1974). Both Walkley and Meakins (1970) and Ghittino (1989) reported that poor nutrition and low energy reserves from starvation can result in increased susceptibility to disease and parasites. Increased susceptibility to disease is probably related to deficiencies in the immune system induced under starvation conditions (Gurr, 1983). Reduction of energy stores due to spawning can also result in mortality from parasitic infection (Roff, 1982).

Starvation due to exhaustion of energy reserves, particularly in smaller individuals, weakens fish and also renders them more susceptible to predation (Minton and McLean, 1982; Roff, 1982; Herting and Witt, 1967). Increased growth in young piscivores as a result of forging on high-energy prey reduces not only predation risk but increases their overwinter survival (Buijse and Houthuijzen, 1992; Maceina and Isley, 1986; Shelton et al., 1979; Chevalier, 1973; Kramer and Smith, 1962). Shortage of prey results in reduced growth, which in turn increases the vulnerability of smaller largemouth bass to predation and leads to increased overwinter mortality (Wicker and Johnson, 1987). Mild winters may also result in a prolonged period of predation risk because temperatures may be high enough to increase metabolic demands for limited fat reserves but just low enough to depress feeding, resulting in a negative energy balance. Even if temperatures were mild enough during the winter such that predators could occasionally feed, the foraging costs per unit of energy intake could be significantly higher when food availability is relatively low. This situation would led to an additional energy expenditure and would further deplete fat reserves (MacKinnon, 1972).

Species or individuals that spawn early in the year increase their probability of overwinter survival and escaping predation by storing proportionally more fat reserves prior to the onset of winter. Later-spawning fish may accumulate insufficient fat reserves and starve to death over a normal winter. For example, sand smelt display density-dependent control on population size by limiting the number of fish that can spawn at the optimum time (Henderson et al., 1988). Restricted availability of spawning ground results in overall population regulation because some individuals are forced to spawn later in the spring, leaving them insufficient time to build up fat reserves to survive the winter. This density-dependent control acts to limit the number of individuals that can breed simultaneously at the most desirable time and leads to a size-selective winter mortality, ultimately resulting in a balance between spawning site availability and population size. For largemouth bass, Adams and DeAngelis (1987) demonstrated that late-spawned individuals were too small to ingest their primary prey of young-of-the-year shad. These fish, therefore, failed to reach the critical size needed to accumulate sufficient lipid stores by the end of their first growing season, resulting in a decreased probability of overwinter survival. Increased risk of predation may arise because of the weakened condition of the fish due to insufficient energy (body lipids). Increased predation risk could also occur due to an increase in foraging activity after spawning when the lipid reserves are depleted (Roff, 1982). Small fish, by diverting energy from maintenance into reproduction, which has a higher energy cost, greatly increase their vulnerability to predation.

For several species living in the temperate and north-temperate zones, their first winter is considered to be the final survival bottleneck before year-class strength is set (Thompson et al., 1991; Henderson et al., 1988). Overwinter mortality can cause substantial variability in year-class strength of some species that is independent of adult stock sizes (Post and Evans, 1989). Size-dependent predation and overwinter mortality, which are influenced by lipid stores, have been implicated as possible regulators of year-class strength in yellow perch. An abundant young-of-the-year perch population can suppress their own food resources resulting in utilization of suboptimum prey sizes. This situation in turn causes density-dependent starvation and mortality to increase (Post and Evans, 1989). Winter starvation mortality, therefore, is maximal at northern latitudes, and at more southerly latitudes with longer growing seasons and shorter winters, overwinter mortality is lower and less variable with population sizes being more stable.

Summary and Future Initiatives. Availability of lipid reserves can have major consequences on the health of individual fish and influence survival and population success. Fish, particularly the smaller individuals in a population, that do not enter the winter nonfeeding period with sufficient fat reserves not only decrease their probability of overwinter survival but usually have reduced reproductive performance the following spring. Impaired osmoregulatory ability, possibly due to utilization of structural membrane lipids and concomitant changes in membrane permeability, may be a major factor for increased mortality of smaller fish over the winter. Starvation and depletion of lipids to critically low levels can result in increased susceptibility of fish to disease, predation, and other environmental stressors. The dynamics of overwinter survival as dictated by lipid stores can be a major regulator of year-class strength in many fish populations, particularly those from north-temperate areas. Species from higher latitudes are more vulnerable to starvation and overwinter mortality because of shorter time periods for feeding and energy accumulation and the longer periods over which dependence on stored lipids is required. Density-dependent controls can also operate in fish populations to regulate the number of individuals competing for limited food resources, thus optimizing the amount of energy available to individuals for lipid storage and for increasing their probability of overwinter survival.

The seasonal availability of lipid stores is critical to the health and future success of many fish species. It is therefore important to have a better understanding of how food availability and lipid reserves ultimately dictate important life processes such as overwinter survival and reproductive performance. For example, for the major economic and ecological species of concern, and particularly those from temperate and north-temperate areas that experience seasonal pulses in energy and environmental parameters, a better understanding is needed concerning the relationships between minimal levels of fat reserves and the probability of overwinter survival and future reproductive success. In addition, the relationships between fish food availability during the growing season, levels of lipid stores in the late fall, and future year-class strength should be more clearly defined. Within the context of food availability, knowledge of the temporal patterns in total lipids and fatty acid composition of the major prey species would also be useful in helping to predict overwinter survival and future reproductive success. Even though they are more difficult to quantify, it is also important to identify density-dependent processes that are instrumental in optimizing lipid levels in fish during the growing season for overwinter survival and reproductive success. For example, the relative importance of changes in competition for food through changes in intraspecific and interspecific competition, the role of density-dependent breeding and spawning behavior (which could extend or reduce the feeding and growing period), and competition for preferred habitats for feeding, spawning, and shelter are all important for maximizing the amount of available energy ultimately allocated to lipid storage. Understanding these relationships between environmental factors, lipid reserves, survival, and reproductive success is key to effective management of fishery resources and for maximizing the future success of fishery populations.

7.2.2. Energy Allocation Strategies

Many fish species display annual cycles in growth and lipid storage that are tightly coupled to levels of ecosystem production, particularly lipid production in the phytoplankton (see Arts, this volume). An organism without an energy storage capacity is constantly dependent on energy flow from its food source for survival. Any interruption in this flow will result in decreased condition and health of the organism and possibly death. When energy is stored, the organism attains increased homeostatic capacity or stability, which allows it to survive periods of interruption in its food supply (MacKinnon, 1972). Energy storage as a resource allocation strategy represents a way of buying a degree of independence from the environment (Reznick and Braun, 1987). By storing energy, individuals can establish a "savings account" to draw on when environmental resources are scarce. Energy reserves, therefore, provide a means of ameliorating temporal mismatches between food supply and energy demand (King and Murphy, 1985; Shul'man, 1974). When food is constantly available throughout the year, there is little energy conservation advantage in adapting a cyclic feeding or storage strategy. In systems with large seasonal variations in food availability, however, cyclic feeding and storage strategies have a definite energy conservation advantage.

The deposition and use of lipid reserves are important in the overall allocation of energy in fish and thus have an ultimate function in the demography and life history of a species (Larson, 1991). There are both costs and benefits to the organism, however, in regard to not only how energy allocated to reserves is applied to other immediate uses such as growth and reproduction, but how these reserves are ultimately used (Slobodkin, 1962). Costs are balanced against the benefits of reserve utilization, and the trade-offs are evaluated in the context of resource allocation to current and future reproduction (Fisher, 1958). For example, in most poikilothermic species, fecundity increases with body size (Munro, 1990; Bell, 1980), and hence, optimizing the age at first reproduction means trading off future increased egg production associated with growth against the advantages of early maturity (Roff, 1981, 1980).

Timing of breeding may be dependent on the quantity and quality of available energy. In years of low food availability, retaining lipids in mesenteries and other tissues increases the prospect of maternal survival and greater production of progeny in future, more optimal (food production) years, perhaps at the expense of the present year class (MacFarlane et al., 1993). Some fish may defer reproductive maturity, rather than maturing during the present year, to store the necessary reserves for overwinter survival, particularly in years of low food supply. One advantage of shifting reproductive effort from the end of the growing season in the fall to the following spring is that of optimizing both fecundity and offspring fitness. Lipid storage allows the production of a large clutch early in the season compared with smaller clutches that may have resulted from a later spawning (Reznick and Braun, 1987). Young fish that are born earlier in the season have a longer time interval for growth during favorable periods when prey abundance is the highest. Early breeding, therefore, enhances the probability of individual survival and thus increases the recruitment of individuals into the next year class. Overwinter lipid reserves, therefore, permit a useful temporal shift in reproductive effort, allowing fish to deter energy allocation to a time that may be more propitious for reproduction (Reznick and Braun, 1987).

Resource allocation strategies in fish involve trade-offs between available energy partitioned among growth, reproduction, maintenance, and storage (Fig. 7.1). The optimization of energy allocation strategies is a relatively long-term phenomenon operating over many generations of a species. The optimal level of reproductive effort is a trade-off between current reproduction and survival coupled with future reproduction (Stearns, 1992). Surplus energy can be used for reproductive tissue, which represents an investment in current reproductive success, or it can be used to build up somatic tissue, which is an investment in future reproduction at the expense of current reproductive success.

There appears to be two conflicting or alternative strategies of energy resource allocation in fish. Either a fish may maintain a constant body weight over time and adjust gamete quantity and quality accordingly, or it may produce a constant number of eggs and sacrifice energy reserves and somatic tissue to meet reproductive requirements. Very few species appear to adopt either of these extreme reproductive strategies, rather they typically employ some type of intermediate strategy. The range of energy allocation strategies used by fish populations is discussed below and represents strategies near the end of these extremes and some that demonstrate intermediate patterns. In all these examples, food resources are assumed to be limited, which forces fish to use different energy allocation strategies depending on the short- and long-term availability of these food resources.

The three-spined stickleback, *Gasterosteus aculeatus,* is an example of a species that sacrifices body weight and condition to maintain egg production in the face of food restriction. Relatively high levels of hydration of the carcass and liver and the comparatively small size of the liver in spawning females, even at high rations, provide evidence that egg production has a priority over somatic growth (Wootton, 1977). Both the observed increases of water in body tissues and the smaller liver size are indicative of lipid utilization and mobilization to support gonad development. Several other aspects of the stickleback life history suggest that investment in reproduction is more profitable than growth. For example, small size at maturity, short life expectancy, multiple spawnings, and poor postspawning survival indicate emphasis on short-term reproductive success. Decreased postspawning survival, in particular, indicates mobilization of lipids from body tissues, which weakens fish and renders them vulnerable to mortality from numerous sources (Shul'man, 1974).

Another extreme in energy allocation strategy is represented by the winter flounder, *Pseudopleuronectes americanus,* which typically sacrifices gonad development or egg production to maintain body weight (Tyler and Dunn, 1976). This strategy channels energy that is not used for maintenance or lipid storage into increasing body size so that when a good year of food resources occurs, fish will be able to develop larger ovaries. Evidence for use of this strategy in this species includes positive correlations between calories consumed and body condition, liver weight, percentage of lipid in the liver, percentage of fish with yolk-bearing ovaries, and ovary weight. Under reduced rations, fewer fish initiate yolk development, and for those fish that do produce yolk, fewer eggs actually receive yolk. Therefore, the fact that fish maintained body condition including levels of fat in the liver and that reproductive development was hindered during reduced feeding indicate that this species placed a higher priority on maintaining body size than on producing reproductive products.

The bluegill sunfish, *Lepomis macrochirus,* employs an energy allocation strategy that is somewhat intermediate between that of the stickleback and the winter flounder. For juvenile bluegill the strategy for maximizing survival, at least in the northern part of its range, is to divert as much energy as possible into somatic growth while maintaining some lipid reserves for overwinter survival (Booth and Keast, 1986). Rapid growth of juveniles is important because female fecundity and male success in competition for breeding space are weight-dependent (Gross, 1982). Maximizing growth in length is also critical for enhancing abilities to capture prey (Werner, 1979) and maintaining growth refuge from predators (Keast and Eadie, 1984). The energy allocation dilemma for bluegill, therefore, is that enough energy has to be channeled into growth to maximize survival during the growing season for minimizing the risk of predation while conserving enough energy as lipids to maximize overwinter survival.

Many fish species balance somatic growth with reproductive output. In the dwarf surfperch, *Micrometrus minimus,* small females postpone reproduction as an adaptive size-specific tactic to maximize fecundity (Schultz et al., 1991). Late breeding places less demand on stored energy, and postponement enables individuals with low initial lipid reserves to breed without starving. By delaying reproduction, females breed at a larger size and are more fecund (Bell, 1980). By delaying reproduction, greater numbers of offspring are produced; however, offspring survival may be decreased because the young would not have sufficient time and food resources to grow and store adequate lipid reserves for overwinter survival. The yellowtail rockfish, *Sebastes flavidus,* also uses an intermediate resource allocation strategy. In lean years this species retains lipids in mesenteries until the time of breeding or even until later in the year. This strategy increases the prospects of maternal survival and greater production of progeny in future, more optimum (higher food production) years, perhaps at the expense of the present year class (MacFarlane et al., 1993). In walleye, *Stizostedion vitreum,* females elaborate gonadal tissue only if there is sufficient fat to do so. Energy for gonadal development may cease, however, if visceral fat supplies drop below a critical level (Henderson et al., 1996).

Among most fish species there is a strong selection pressure to make the appropriate gonadal responses at the correct times of the year in predictable or more stable environments (Munro, 1990). Such pressure occurs because gonadal growth and maturation are associated with considerable energy investment at the expense of somatic growth, and this high energy drain can ultimately result in increased mortality and a reduction in future fecundity (Myers, 1984; Lamon and Ward, 1983). Thus, even after maturation, natural selection should favor those genotypes that respond to the appropriate environmental factors in a manner that permits the most efficient sequential distribution of resources between somatic and gonadal compartments. Ideally, somatic growth should be continued for as long as possible to maximize the potential fecundity for that season and before diverting energy to the gonads (Munro, 1990). If an individual diverts, therefore, energy from somatic growth to gonadal investment too early, then it must either breed early and jeopardize survival of that brood or wait until optimal breeding conditions. Conversely, fish that postpone gonad growth mature later in the spawning season and must either undergo gonad regression and recycle gonadal material back to the body (with associated energy losses) or attempt a later and probably unsuccessful spawning. Individuals that attempt to increase potential fecundity by postponing gonad development too long in a particular season will also be selected against.

Summary and Future Initiatives. Reproduction and growth in fishes do not occur without some compromises. Fish employ a variety of strategies for allocating available energy such that the trade-offs between these basic functions tend toward optimization. At the core of these allocation strategies are lipid reserves that function as a connector or a "common currency" between environmental resource availability and various uses by the organism such as reproduction. If reproductive function is maximized, some other component such as growth is usually compromised. The energy allocation strategies used by each species are the product of its life history adaptation and natural selection, which allows it to compete successfully in its environment.

Allocation of available energy for growth, metabolism, reproduction, and lipid storage usually follows species-specific patterns or strategies. If the environment is significantly altered through natural or anthropogenic changes, then until a new steady state is reached, allocation patterns may be disrupted and the continued success of a species in that particular environment may be compromised. Alterations in energy allocation patterns from the expected or normal state could be a useful indicator of aquatic system dysfunction. Studies are needed that focus on how changes in allocation patterns can ultimately affect ecologically significant functions such as reproduction, survival, and growth dynamics. For example, in aquatic systems that become increasingly affected by a variety of stressors, increasing amounts of assimilated energy would be diverted to metabolic processes for maintaining or repairing damaged biological systems. In this situation, less energy would be available for growth, reproduction, or lipid storage. Under these circumstances, it would be worthwhile to know the relationship between environmental perturbations, changes in allocation patterns, and the effect on lipid storage, survival, and reproductive performance.

7.2.3. Reproductive Development and Early Life History

Population failure in many fish species is highly variable and may occur, in part, because of reduced reproductive fitness of adult populations during years of low food supply and inadequate fat reserves. During the winter fasting period, fish gonads are developed by using lipid reserves (Love, 1980) which are the main source of energy for the synthesis of generative tissue (Shul'man, 1974) (Fig. 7.1). Poor reproductive success have been attributed to both the reduced condition of the gonad during maturation and decreased larval success following hatching. When maternal diets are deficient, insufficient transfer of lipids to developing oocyte may reduce fecundity and the viability of the progeny (Heming and Buddington, 1988; Watanabe, 1985). Lipids that are transferred to the gonads are incorporated as nutritive material in the yolk of the oocyte and serve as the principal endogenous food source for the developing embryo. The level of fat reserves in the body and the rate of their utilization in gonad maturation, therefore, is a major determinant of reproductive success.

Inadequate fat reserves have been implicated in the reduced reproductive success of several fish species. Because of a lack of summer food, mature yellow perch were unable to build sufficient fat reserves to meet the demands of oogenesis and winter maintenance requirements (Newsome and Leduc, 1975). In walleye, reproductive success and variations in year-class strength have been partially explained by the energy condition of the female and the proportion of the stock able to spawn successfully each year (Henderson et al., 1996). Thus, recruitment success in this species is dependent on the energy acquired before spawning, including that energy obtained from reabsorption of previous gonadal tissue (Henderson and Nepszy, 1994). Tyler and Dunn (1974) also reported that in winter flounder the reduced fecundity and rate of oocyte recruitment was due to low food supply and low fat reserves prior to the maturation period. Reduced fecundity in this species was also linked to low energy availability through a mechanism of increased oocyte resorption. In addition to oocyte resorption, follicular atresia has also been observed in fish under conditions of reduced food supply (Hester, 1964; Scott, 1962). Both Bagenal (1966) and DeVlaming (1971) found that low energy reserves reduced fecundity by effecting a decrease in the rate of oocyte recruitment. Maturation of male Atlantic salmon (*Salmo salar*) parr was suppressed when mesenteric fat failed to reach a critical level in May (Rowe et al., 1991). Shul'man (1974) introduced the concept that fish must attain a minimum fat content before maturation can be initiated. In this regard, Atlantic salmon returning from sea required a minimum fat content of 12% in the spring if they were likely to spawn the following autumn. Likewise, the onset of maturation in the arctic char, *Salvelinus alpinus,* appears to be triggered by a critical level of lipids during maturation (Rowe et al., 1991).

The source of fat reserves for gonad growth and maturation depends on the fish species and life history. In general, the primary energy reserve controlling maturation appears to be carcass or muscle fat in pelagic species such as clupeids; liver fat in benthic species such as flatfish and the Gadidae (Cowey and Sargent, 1972); and mesenteric fat in salmonids (Rowe et al., 1991) and many of the Centrachids (Adams et al., 1982), Percids (Henderson et al., 1996), and Cyprinids. Exceptions to these generalizations may exist such as for plaice (a benthic species), in which Dawson and Grimm (1980) found that the carcass was the main source of lipid reserves. Some members of the family Scorpaenidae (scorpionfish) such as the yellowtail rockfish, however, use both liver and mesenteric fat in ovarian development.

The high energy demands that reproductive development places on stored lipids is illustrated by a unique case of gizzard shad mortality in a southeast U.S. reservoir during the mid-1980s. An unusually large die-off of mature female gizzard occurred in the late spring and was attributed to severe starvation (Adams et al., 1985). These shad, which normally reproduce in the spring, failed to spawn in the spring of the previous year because of a series of unusual environmental conditions, primarily temperature related, during that period. Consequently, those individuals that failed to spawn in the spring spawned the next fall and, in the process, depleted their remaining lipid reserves. Delayed spawning in this situation eventually proved to be the main cause of mortality because the fall is typically the period when energy reserves are accumulated and stored for over-winter survival and gonad maturation. Therefore, the late spawning not only depleted available fat reserves, but also did not allow sufficient time for replenishment of lipid stores before the winter fasting period. The late-spawning individuals thus entered the winter period with depleted fat levels that were insufficient for basic maintenance needs throughout the winter.

Lipids influence reproductive success not only through gonad development and maturation but through viability of the progeny. Larval survival and success has been linked to egg quality as expressed by stored energy in the egg (Kamler, 1992;

Docker et al., 1986). When external food is scarce, larger larvae from larger eggs survive longer than those hatched from smaller eggs (Ware, 1975). Higher survival of these individuals resulting from larger eggs has been noted, particularly for several salmonid species (Gall, 1974). A positive relationship between viability of offspring and total lipid concentration in the eggs has been demonstrated for a number of species (Brown and Taylor, 1992; Kamler, 1992; Thorpe et al., 1984). Eggs characterized by higher lipid content tend to produce larger larvae, which provides the larvae with several survival advantages (Kamler, 1992; Thorpe et al., 1984). Higher lipid reserves in the maternal parent and the eggs improve the ability of larvae to delay initial feeding for longer periods than larvae with less energy resources (Brown and Taylor, 1992), with larger larvae having more energy reserves for seeking and capturing prey.

Body size, which is based on endogenous energy available from the egg, is an important variable also influencing the physiology, ecology, and behavior of larval fish (Miller et al., 1988). Mechanisms that regulate larval success and growth operate in a size-dependent fashion. For example, body size influences the vulnerability of larval fish to predation through differential encounter rates and predator escape abilities. Also, the ability of larger larvae to successfully feed is enhanced because they have longer reaction distances to prey and increased swimming abilities. Swimming performance also influences the ability of larval fish to maintain their position in optimal areas for growth and survival (Brown and Taylor, 1992). Before feeding occurred, Knutsen and Tilseth (1985) demonstrated that mouth gape size was correlated to egg size. Therefore, the importance of larval size to feeding is that a broader spectra of prey sizes are available to the larger larvae compared with smaller individuals.

Relatively small differences in larval size at the time of hatch may translate into substantial differences at the population level in regard to long-term survival, subsequent recruitment, and year-class strength (Adams and DeAngelis, 1987). Fish that incur very small size advantages early in their life history proportionally increase their probability of survival because of slight advantages in foraging and swimming abilities, escape from predators, and maximization of the period between depletion of endogenous energy stores and feeding. This later point becomes particularly important if the production of food resources is mismatched in time with the initiation of larval feeding.

The trade-offs between number and size of offspring is an important factor determining reproductive potential (Pepin and Myers, 1991). Properties of larvae as related to egg size have counterbalancing advantages, however, depending on the reproductive strategy employed (Kamler, 1992). In species with r strategies (i.e., those that produce many small eggs) from the total biomass of eggs in the ovary, a larger number of smaller individuals develop that use food resources more efficiently and exhibit a high potential for rapid growth. In species with a k strategy (i.e., those that produce a few large eggs), fewer but larger larvae are produced, and their larger size increases their chances of survival because the risk of predation is reduced (Miller et al., 1988) and their prey capture and foraging abilities are increased (Blaxter and Hempel, 1963). According to Brown and Taylor (1992), female spawning fish face two major trade-offs when allocating available energy for reproductive products. Available energy must be allocated not only between egg size and number but among carbohydrate, protein, and lipid components within the individual eggs. Life history theory suggests that natural selection favors females allocating reproductive energy in a manner maximizing fitness return per unit of ovarian resource invested (Fleming and Gross, 1990; Sargent et al., 1987). Total egg number, therefore, should vary in response to selection pressures on egg size, egg composition, and total investment in egg production (Fleming and Gross, 1990).

Summary and Future Initiatives. Lipids as energy reserves in fish are instrumental in dictating reproductive success and ultimately year-class strength and the

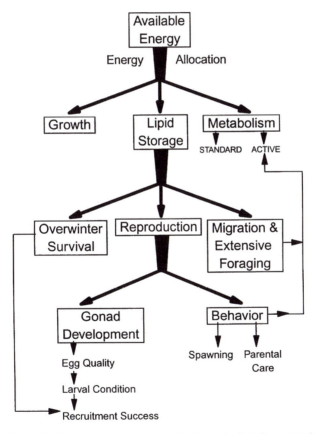

FIGURE 7.2. General allocation strategy of available (assimilated) energy into the main functional processes of growth, metabolism, lipid storage, and reproduction for a typical temperate or north-temperate fish species.

structure of fish populations (Fig. 7.2). In many species, lipids that are built up and stored in various compartments of the body are used over the winter and early spring for basic maintenance needs and for gonadal development. Numerous studies have shown that fish that fail to build up adequate lipid reserves before winter generally experience some level of reproductive failure the following spring. Because the quality of eggs is related to lipid quantity, the success of the hatched larvae is a size-dependent phenomenon also related to maternal lipid availability. Larger larvae hatched from larger eggs generally experience higher survival and success because of enhanced feeding ability, decreased vulnerability to predation, and a longer period of subsistence on endogenous energy before exogenous resources become available. Relatively small differences in larval size at the time of hatch may ultimately translate into substantial differences at the population level in regards to long-term survival and success of the population.

Successful reproductive performance is intricately linked to the availability of lipid reserves for gonadal development. Several studies have documented that the health and level of lipid depots in the adult female in the fall prior to the overwinter nonfeeding period is a key indicator of reproductive success the following spring. Related to this, studies are needed to better establish, for important economic and ecological species, the relationship between minimum or critical lipid levels in the maternal parent and the ultimate effect of these levels on gonad development, egg quality, and larval hatching success and survival. In addition, the environmental factors that can affect the health and condition of the adult for the period preceding the cessation of feeding and initiation of gonadal development should also be studied. For example, much is known about striped bass and how environmental conditions in certain aquatic systems can place various stresses on the adults, causing loss of body condition including reduced storage of lipids. By understanding the dynamics of physicochemical factors, such as temperature and dissolved oxygen, and processes within the food chain that can have negative impacts on adults, fishery stocks can be better managed to achieve higher reproductive and population success.

7.2.4. Lipids and Environmental Stress

Exposure to environmental stressors is an inescapable aspect of an aquatic organism's life. Fluctuating thermal and hydrological regimes, pH, oxygen, siltation, and contaminants impose constraints on the ability of an organism to cope with its environment (Dehn and Schirf, 1986). Most aquatic species are adapted to their environment, but this adaptation does not imply an absence of energetic expenditure to maintain homeostasis in the face of sublethal environmental stressors (Wedemeyer et al., 1984).

Environmental stressors first initiate the primary stress response system by affecting biomolecular, biochemical, and physiological processes (Fig. 7.2). These primary stress responses may be followed by secondary stress responses in which alterations in energy allocation and metabolism occur (Leatherland and Sonstegard, 1984; Mazeaud and Mazeaud, 1981), including changes in lipids, protein, or glycogen dynamics within the body. If the stress is of sufficient magnitude or duration, then tertiary changes may also be manifested as alterations in growth, reproduction, or even survival of individuals and populations (Adams, 1990; Wedemeyer et al., 1984) (Fig. 7.2).

Environmental stressors such as contaminants and unfavorable temperature regimes are known to alter both the quantity and quality of lipids in fish. Studies of lipids as an energy source in fish have revealed their importance during periods of stress (Jezierska et al., 1982). Stress responses are energy-draining processes that use triglycerides as the main energy source, although phospholipids may also be used under severe conditions (Benton et al., 1994). Energy supplied to the organism by breakdown of lipids during such increased metabolic demand comes primarily from oxidation of fatty acids. The free fatty acids yield additional energy for mitigating or ameliorating the stress condition caused by the impact of contaminants or other environmental stressors (Tulasi et al., 1992). Thus, stressors can deplete the lipid reserves over time, resulting in changes to important physiological processes in the organism.

7.2.4.1. Contaminant Effects

Both increases and decreases in total lipids and triglycerides have been observed when fish are exposed to contaminants such as pesticides, heavy metals, and PCBs. A number of studies, mostly involving laboratory exposures, have reported on the effects of pesticides on lipid levels in the liver and muscle tissue of various fish species. Benton et al. (1994) observed that the sailfin molly, *Poecilia latipinna,* exposed to DDT for 21 d demonstrated decreases in total body lipids and triglycerides that were attributed to increased metabolism (catabolism) following DDT exposure. Conversely, in juvenile coho and chinook salmon fed DDT in their diets, carcass lipid content increased (Buhler et al., 1969). A decrease in liver lipids was observed in the freshwater fish, *Channa punctata,* exposed to increasing concentrations of the organochlorine insecticide endosulfan for 96 h (Murty and Devi, 1982). Exposure of the freshwater catfish, *Clarias batrachus,* to sublethal concentrations of the pesticide malathion for 4 weeks during the vitellogenic phase of its reproductive cycle resulted in elevated liver lipids (LaL and Singh, 1987). In this case, elevated liver lipids may have been caused by the pesticide stimulating lipogenic activity in the liver and increasing fatty acid synthesis. Lipid increases may have also occurred because oxidation and utilization of fatty acids were inhibited. In female catfish, however, malathion inhibited mobilization of hepatic phospholipids and free fatty acids to the gonads. Because the teleostean ovary imports lipids and phospholipids from the liver during the vitellogenic phase under the influence of sex steroids and gonadotrophin, the inability of the exposed fish to mobilize these lipids to the gonads may have been caused by the pesticide reducing circulating levels of sex steroids and gonadotrophins. Freeman and Idler (1975) also reported that organochlorine and organophosphorus compounds suppressed gonadal growth by this mechanism, and Rao and Rao (1984) found that organophosphorus insecticides inhibited the synthesis and metabolism of steroid compounds. When tilapia were exposed to the pesticide methyl parathion for only 48 h, the total lipids and phospholipids decreased in the muscle and liver, but the free fatty acid levels increased (Rao and Rao, 1984). The decreased lipid levels indicated glyconeogenesis in these tissues, whereas the increased free fatty acids along with the decreased tissue lipids suggested increased lipolysis. Increase in fatty acids at the expense of utilization of body lipids indicated that these compounds were used for extra energy to mitigate the stress effects of pesticide exposure. Other investigators have also observed fatty livers following exposure to organochlorine compounds such as PCB (Klaunig et al., 1979; Lipsky et al., 1978). Lipid accumulation in this organ appears to result from an inability to convert lipids in hepatocytes to a form suitable for use by the organism (Runnells et al., 1965).

Lipid stores can serve a protective function from the toxicant effects of contaminants. The herbicide atrazine caused liver lipid degeneration in grey mullet after a 21-d exposure (Biagianti-Risbourg and Bastide, 1995). Following exposure, the size of lipid droplets increased in the liver. According to Fabacher and Chambers (1971), lipid droplets can sequester fat-soluble contaminants such as pesticides and herbicides, providing protection for the organism from the toxic effects of these chemicals by a dilution effect of the contaminant in the body tissue. Geyer et al. (1994) also reported that lipids in aquatic organisms serve as a protective reservoir against the toxic effects of lipophilic, relatively persistent organic chemicals because they bioconcentrate mainly in the body lipids. Therefore, in organisms with high lipid content, only a relatively small fraction of hydrophobic chemicals may actually reach target organs. The disadvantage of this protective effect, however, is that lipids can bioconcentrate high levels of these contaminants and may ultimately cause toxicological problems once lipids are mobilized for the high energy demands of overwinter survival, gonadal development, spawning activities, or migration. In their DDT exposure experiments, Buhler et al. (1969) speculated that increases in lipid levels under certain types of contaminant exposure may represent a compensatory protective response by the fish to provide additional pesticide storage capacity. Following exposure of mosquitofish in drainage ditches containing insecticides, Fabacher and Chambers (1971) also concluded that high lipid concentration served as a protective reservoir to the organism against lipophilic contaminants. A disadvantage, however, of allocating contaminants to lipid stores is that these contaminants are usually biomagnified through the food chain, increasing the potential toxic effects on predators and even to human consumers.

Heavy metal exposure also has been found to have variable effects on lipid dynamics in the organism. Tulasi et al. (1992) exposed the freshwater fish *Anabas testudinus* to sublethal concentrations of lead nitrate for 30 d and found that total lipids and phospholipids in the ovary and liver decreased while free fatty acids increased. Lead apparently affects lipid metabolism of fish and may impair fecundity and reduce reproductive success. The effects of mercuric chloride on freshwater catfish exposed to 0.2 mg $\cdot$ L^{-1} for 10, 20, and 30 d were reported by Bano and Hasan (1989). Total lipid and phospholipid increased in both liver and muscle tissue, whereas peroxidation of endogenous lipids in these tissues also increased.

Katti and Sathyanesan (1984) also found that liver lipids increased when the catfish *Clarias batrachus* was exposed to cadmium chloride for up to 135 d. In contrast to these two studies, Ram and Sathyanesan (1984) reported that freshwater fish subjected to 0.01 ppm mercuric chloride displayed decreases in liver and ovarian lipids. Decreases in ovarian lipids and absence of yolk formation in the oocyte could have been due to toxic effects of $HgCl_2$ on gonadotrophic secretion that impaired liver vitellogenin synthesis. Freshwater fish subjected to 1–4 ppm copper sulfate for 60–84 d demonstrated a gradual decrease in liver and muscle lipids that was attributed to either oxidation or hydrolysis of these lipids (Pandey and Pandey, 1994; Ansari, 1983). However, the lipid content of Atlantic salmon exposed to four concentrations of aluminum for 60 d was not changed (McKee et al., 1989).

7.2.4.2. Thermal Effects

Varying thermal regimes can also alter the lipid dynamics and the health of fish by (1) increasing metabolic activity, thereby increasing utilization of lipid stores; (2) restricting fish from preferred feeding or breeding areas; (3) altering foraging behavior; and (4) increasing susceptibility of fish to disease or parasites, which could result in even greater metabolic costs.

The overall condition, lipid levels, and reproductive performance of striped bass, *Morone saxatilus,* has been compromised in certain aquatic systems in which preferred habitat for this species has been constricted by a combination of high temperatures and low dissolved oxygen. Coutant (1987) hypothesized that poor body condition of adult striped bass caused by habitat limitations early in ovarian development ultimately reduced the reproductive potential of this species. For some eutrophic systems that stratify during the summer, such as Cherokee Reservoir in east Tennessee and Chesapeake Bay, warm upper-layer water combined with low dissolved oxygen in deeper cooler water forces or squeezes fish into a narrow band of the water column. Temperatures in these constricted zones are typically higher than preferred for the species, resulting in increased metabolic demands, higher competition for limited food resources, and increased susceptibility to disease and other environmental stressors (Coutant and Benson, 1990). This type of habitat restriction often results in poor performance as evidenced by a reduction in the percentage of successfully spawning females and in embryo survival. Even though the number of eggs was not affected, the survivorship of eggs and larvae was reduced after spawning. Larval survival, as discussed in the previous section, may have been impaired as a result of the lower concentrations of yolk and lipids derived from the malnourished parent. The survivorship and vigor of larvae and reproductive success of the population can, therefore, be linked to the environmental experience of the parent when the eggs were developing.

Although summer habitat constriction was not as dramatic as that observed for striped bass, northern pike, *Esox lucius,* also experienced habitat restriction and a loss of condition over the summer in a stratified Ohio reservoir (Headrick and Carline, 1993). This species was also squeezed into a narrow zone of unfavorable temperatures because of warm epilimnetic water and anoxic bottom water. In addition, lipid dynamics and recruitment success of striped bass were compromised in a thermally enriched reservoir in Virginia. In comparison with the condition of the fish in the reference reservoir, ovary weights, condition factor, and concentrations of vitello-triglyceride of the impacted fish were significantly reduced in the thermal reservoir (Grimes, 1993). Results of this study suggest that environmental stress, such as altered thermal regimes, affects the condition of ovarian tissue possibly through the catabolism of vitelline substrates. Reproductive success and recruitment were affected by thermal conditions, which qualitatively altered profiles of vitello-compounds important in egg hatching and fry survival.

Direct effects of increased temperature have also been reported for a variety of species living in aquatic systems heated by power station effluents and other industrial facilities discharging heated water to the environment. Largemouth bass from a thermally enriched power plant cooling reservoir had lower body condition (condition factor) and lower fat reserves than fish from unheated areas (Gibbons et al., 1978). From this same system, Eure and Esch (1974) found that the number of parasites per host were significantly higher in fish from heated water sites than in fish from cooler areas. Increased loading of parasites and other disease organisms in fish living in conditions above their range of optimal temperatures occurs not only because such organisms are more prolific under warmer conditions but also because the immune system of fish can be more easily compromised when lipid levels are low (Gurr, 1983; Shul'man, 1974). Chronic malnutrition and reduced lipid reserves, as reflected by decreased condition, were also noted by Graham (1974) for four sunfish species residing in a power plant cooling impoundment. Total body lipids and serum triglycerides were drastically reduced in sunfish residing downstream of an industrial complex discharging thermal effluents and a mixture of contaminants (Adams et al., 1992). Lower energy reserves were attributed to increased metabolic demands due to both contaminant exposure and thermal additions. These fish also had increased parasite loads and infection areas (macrophage aggregates) in the liver and spleen, which were explained by the weakened condition of the fish as a result of low energy reserves and a compromised immune system (Rice et al., 1996).

Temperature regimes can also affect the condition, energy dynamics, and survival of fish during their migration run. Glebe and Leggett (1981) studied depletion of energy reserves during the freshwater migration of American shad, *Alosa sapidissima,* and found that the magnitude of postspawning mortality was linked to the migratory energy demands. Shad migrating late in the run at significantly higher temperatures expanded energy at a greater rate and thereby experienced more extensive overall lipid depletion. Previous studies had found that 56% of the annual variability in postspawning survival of shad was due to differences in adverse river water temperature during the migration. Because temperature itself was not directly lethal, the greater energy use at higher temperatures was assumed to be an important survival regulator for this species.

7.2.4.3. Other Stressors

A wealth of literature is available on the effects of mixed contaminant effluents on fish health, particularly fish populations exposed to pulp and paper mill discharges. Few of these studies, however, have reported on lipid dynamics in fish following exposure to contaminants in these effluents. Notably, Hodson et al. (1992) found that white sucker, *Catostomus commersoni*, residing downstream of pulp mill discharges had increased lipid levels as reflected by total carcass lipid. Similarly, Sandstrom et al. (1988) found elevated lipids in perch, *Perca fluviatilis*, from Baltic coastal areas but observed reduced gonad growth in the exposed fish. Both these studies attributed altered energy metabolism in these fish to disturbances or changes in the way that energy was allocated within the organism. In contrast to these findings, however, McMaster et al. (1991) observed reduced lipid stores, smaller gonads, and decreased energetic commitment to growth in white sucker exposed to paper mill effluents. Adams et al. (1996) also found significantly reduced lipid stores compared with the reference as reflected by the visceral-somatic index in redbreast sunfish, *Lepomis auritus*, below a pulp mill discharge. In this case, however, growth was elevated in these fish as was the condition factor. The somewhat contrasting results among these four studies relative to energy allocation to growth, lipid storage, and reproduction probably reflect site-specific differences in energy allocation strategies that may be related not only to the nature and magnitude of contaminant exposure but also to the influence of other environmental and ecological factors such as food and habitat availability (Adams et al., 1996).

The sublethal effects of textile mill effluents on lipid composition in different tissues of fish are similar to those reported for paper mill discharges. A decline in lipids occurred in both muscle and liver, which was attributed to utilization of this energy source for the purpose of mitigating stress caused by the textile mill effluents (Rajan, 1990). Tannic acid, which is a major constituent of tannery effluents, caused a decrease in the total lipid content of liver and muscle when fish were exposed to 59 ppm for only 96 h (Somanath, 1991). As other investigators working with a variety of contaminant stressors have concluded, the decrease of lipids in the liver and muscle could have been caused by fatty acid oxidation to meet the additional energy requirements under stress conditions.

The indirect effects of contaminants on fish health and lipid dynamics can be illustrated by the study of Munkittrick and Dixon (1988), in which lakes containing elevated levels of heavy metals impaired lipid dynamics and reproduction in a population of white sucker. Fish from the contaminated lakes had decreased muscle lipids, serum lipid concentrations, and visceral lipid reserves. Reproduction was impaired as evidenced by decreased egg size and fecundity, suggesting that less energy was committed to gonad development in fish from the contaminated system than in fish at nonaffected sites. Most of the alterations in lipid dynamics and reproduction in these fish were attributed to nutritional deficiencies as a result of the chronic effects of elevated metals in the sediment on the food base of the sucker. Thus, in this case, contaminants had an indirect effect on lipid dynamics in fish by suppressing food availability, resulting in nutritional and fat storage deficiencies.

Changes in salinity regimes have also been implicated in causing lipid changes in fish. Roche et al. (1983) exposed sea dace, *Dicentrarchus labrax,* to salinities ranging from 4 to 40 ppt and noted changes in total lipids and triglycerides in liver and muscle. As salinity increased from 4 to 40 ppt, total lipids increased in both liver and muscle. Dramatic increases were observed as salinity was raised from 4 to 18 ppt, but no increases were noted after 18 ppt. Salinity changes caused alterations in the phospholipids, particularly phosphatidylcholine and phosphatidylethanolamine. Thus, a hypo-osmotic medium seems to cause decreased biosynthesis of some lipid compounds in this species.

The ability of fish to tolerate increased acidification stress in aquatic systems affected by acid rain is largely dependent on the starvation and lipid status of the organism. Cunningham and Shuter (1986) found that starvation and low fat levels led to progressive weakening of the osmoregulatory system, thereby reducing the tolerance of young smallmouth bass to reduced pH. Not only did starvation and low pH have an interactive effect on total body lipid composition, but chronic exposure to low pH tended to significantly increase metabolic costs that further reduced lipid levels. Thus, the toxicity of reduced pH on juvenile smallmouth bass appears to be related to the level of emaciation as expressed by lipid stores (Kwain et al., 1984). The occurrence of starvation at the end of winter is a critical factor in the life cycle and survival of smallmouth bass in acidified systems.

Summary and Future Initiatives. Environmental stressors can alter both the quantity and quality of lipids in fish. Exposure to contaminants such as pesticides, heavy metals, and PCBs can increase or decrease total lipids and triglycerides, depending on the species of fish and the concentrations and duration of contaminant exposure. The decrease in lipid levels of fish experiencing many types of stress suggests that free fatty acids are mobilized to yield extra energy for the purpose of mitigating or ameliorating the effects of that stress. For lipophilic contaminants, lipids can serve as a protective reservoir helping to prevent toxicants from reaching and affecting target organs. Varying thermal regimes can affect lipid dynamics in fish by increasing metabolism, by altering foraging and breeding behavior, or by increasing susceptibility to disease and parasites. Energy allocation patterns in fish have also been shown to be influenced by not only contaminants such as paper mill effluents but also by ecological factors such as food and habitat availability. Therefore, when evaluating the effects of environmental stressors such as contaminants on lipid dynamics, the influence of other environmental factors should also be accounted for. Although some stressors exert direct effects on lipid dynamics in fish, other stressors such as contaminants can have indirect effects on lipid dynamics by suppressing food availability for predatory species.

Stressors of all kinds, including natural and anthropogenic sources, have been shown to affect lipid dynamics in various fish species. Much of what we understand, however, about lipid response to specific stressors such as contaminants is derived from controlled laboratory or microcosm studies in which a single stressor is usually involved. In nature, however, fish may experience simultaneously a variety of stressors such as various types of contaminants, fluctuating temperature or hydrologic regimes, or variations in food or habitat availability. All these factors, individually or in concert, can influence the health and lipid dynamics of fish. Studies are needed, therefore, that focus on the incremental role or contribution of influential ecological factors in altering lipid dynamics in fish. For example, even if fish are exposed to high levels of contaminants in a field situation, factors such as food availability and competition can both modulate or exacerbate the influence of the stressor of concern. Studies are needed that address the influence of multiple stressors on lipid dynamics of fish and how available energy is partitioned within the organism under different types of stressors such as organic contaminants, heavy metals, and temperature. Within this context, it is important to distinguish between the direct effects of environmental stress on fish that can be realized through metabolic pathways and the indirect effects of stressors that can be mediated through the food chain affecting the quality and quantity of energy available to consumer species. In addition, research effort should be directed toward determining causal relationships between stress responses in fish at the lower levels of biological organization (i.e., biochemical/physiological) and lipid storage and dynamics (i.e., Fig. 7.2). Although more efforts have focused on this aspect of establishing causal relationships between stress and fish health, additional studies are necessary to refine our understanding of the relationship between seasonal levels of lipid reserves and ecological significant responses such as reproductive performance and population success. Just as knowledge of critical lipid levels is needed to help define the success of overwinter survival and reproductive performance, a better understanding of the influence of stressors on critical lipid levels in relation to increased susceptibility to disease, predation, and reduced foraging ability is also needed. Environmental stressors that alter the quality (i.e., phospholipids and cell membrane integrity) or quantity of lipids weaken fish, render them more susceptible to disease and predation, and also decrease behavioral and reproductive performance. In summary, understanding the lipid dynamics of fish as influenced by different types and levels of environmental stressors should help provide effective environmental regulation and management guidelines for maintaining healthy and successful fish populations in various types of aquatic ecosystems.

Acknowledgment. This chapter is publication 4671, Environmental Sciences Division, ORNL.

References

Adams, S.M. Status and use of bioindicators for evaluating effects of chronic stress on fish. Am. Fish. Soc. Sym. 8:1–8; 1990.

Adams, S.M.; DeAngelis, D.L. Indirect effects of early bass-shad interactions on predator population structure and food web dynamics. In: Kerfoot, W.C.; Sih, A., eds. Predation in Aquatic Communities. Hanover, CT: Univ. Press of New England; 1987:p. 103–117.

Adams, S.M.; Ham, K.D.; Greeley, M.S.; LeHew, R.F.; Hinton, D.E.; Saylor, C.F. Downstream gradients in bioindicator responses: point source contaminant effects on fish health. Can. J. Fish. Aquat. Sci. 53:2177–2187; 1996.

Adams, S.M.; Crumby, W.D.; Greeley, M.S.; Ryon, M.G.; Schilling, E.M. Relationships between physiological and fish population responses in a contaminated stream. Environ. Toxicol. Chem. 11:1549–1557; 1992.

Adams, S.M.; Breck, J.E.; McLean, R.B. Cumulative stress-induced mortality of gizzard shad in a southeastern U.S. reservoir. Environ. Biol. Fish. 13:103–112; 1985.

Adams, S.M.; McLean, R.B.; Huffman, M.M. Structuring of a predator population through temperature-mediated effects on prey availability. Can. J. Fish. Aquat. Sci. 39:1175–1184; 1982.

Ansari, I.A. Effects of copper sulphate on the lipid content of *Channa punctatus* (Block). J. Adv. Zool. 4:109–111; 1983.

Bagenal, T.B. The ecological and geographic aspects of the fecundity of the plaice. J. Mar. Biol. Assn. U.K. 46:161–185; 1966.

Bano, Y.; Hasan, M. Mercury induced time-dependent alterations in lipid profiles and lipid peroxidation in different body organs of catfish *Heteropneustes fossilis*. J. Environ. Sci. Health B24:145–166; 1989.

Bell, G. The costs of reproduction and their consequences. Am. Nat. 116:45–76; 1980.

Benton, M.J.; Nimrod, A.C.; Benson, W.H. Evaluation of growth and energy storage as biological markers of DDT exposure in sailfin mollies. Ecotoxicol. Environ. Saf. 29:1–12; 1994.

Biagianti-Risbourg, S.; Bastide, J. Hepatic perturbations induced by a herbicide (atrazine) in juvenile grey mullet *Liza ramada* (Mugilidae, Teleostei): an untrastructural study. Aquat. Toxicol. 31:217–229; 1995.

Blaxter, J.H.S.; Hempel, G. The influence of egg size on herring larvae (*Clupea harengus* L.). J. Cons. Int. Explor. Mer. 28:211–240; 1963.

Booth, D.J.; Keast, J.A. Growth energy partitioning by juvenile bluegill sunfish, *Lepomis macrochirus* Rafinesque. J. Fish. Biol. 28:37–45; 1986.

Brown, R.W.; Taylor, W.W. Effects of egg composition and prey density on the larval growth and survival of lake whitefish (*Coregonus clupeaformis* Mitchill). J. Fish. Biol. 40:381–394; 1992.

Buhler, D.R.; Rasmusson, M.E.; Shanks, W.E. Chronic oral DDT toxicity in juvenile coho and chinook salmon. Toxicol. App. Pharmacol. 14:535–555; 1969.

Buijse, A.D.; Houthuijzen, R.P. Piscivory, growth, and size-selective mortality of age 0 pikeperch (*Stizostedion lucioperca*). Can. J. Fish. Aquat. Sci. 49:894–902; 1992.

Chevalier, J.R. Cannibalism as a factor in first year survival of walleye in Oneida lake. Trans. Am. Fish. Soc. 102:739–744; 1973.

Coutant, C.C. Poor reproductive success of striped bass from a reservoir with reduced summer habitat. Trans. Am. Fish. Soc. 116:154–160; 1987.

Coutant, C.C.; Benson, D.L. Summer habitat suitability for striped bass in Chesapeake Bay: reflections on a population decline. Trans. Am. Fish. Soc. 119:757–778; 1990.

Cowey, C.B.; Sargent, J.R. Fish nutrition. Adv. Mar. Biol. 10:383–492; 1972.

Cowey, C.B.; Mackie, A.M.; Bell, J.G., eds. Nutrition and Feeding in Fish. Orlando, FL: Academic Press; 1985.

Cowey, C.B.; Sargent, J.R. Nutrition. In: Hoar, W.S.; Randall, D.J.; Brett, J.R., eds. Fish

Physiology. vol. VIII. Bioenergetics and Growth. New York: Academic Press; 1979:p. 1–70.

Cunjak, R.A. Physiological consequences of overwintering in streams: the cost of acclimatization? Can. J. Fish. Aquat. Sci. 45:443–452; 1988.

Cunningham, G.L.; Shuter, B.J. Interaction of low pH and starvation on body weight and composition of young-of-the-year smallmouth bass (*Micropterus dolomieui*). Can. J. Fish. Aquat. Sci. 43:869–876; 1986.

Davis, K.B.; Simco, B.A. Salinity effects on plasma electrolytes of channel catfish, *Ictalurus punctatus*. J. Fish. Res. Bd. Can. 33:741–746; 1976.

Dawson, A.S.; Grimm, A.S. Quantitative seasonal changes in the protein, lipid and energy content of the carcass, ovaries and liver of adult female plaice, *Pleuronectes platessa* L. J. Fish. Biol. 16:493–504; 1980.

Dehn, P.F.; Schirf, V.R. Energy metabolism in largemouth bass (*Micropterus floridanus salmoides*) from stressed and non-stressed environments: adaptations in the secondary stress response. Comp. Biochem. Physiol. 84A:523–528; 1986.

DeVlaming, V. The effects of food deprivation and salinity changes on reproductive function in the estuarine gobiid fish, *Gillichthys mirabilis*. Biol. Bull. 141:458–471; 1971.

Docker, M.F.; Medland, T.E.; Beamish, F.W.H. Energy requirements and survival in embryo mottled sculpin (*Cottus bairdi*). Can. J. Zool. 64:1104–1109; 1986.

Dygert, P.H. Seasonal changes in energy content and proximate composition associated with somatic growth and reproduction in a representative age-class of female English sole. Trans. Am. Fish. Soc. 119:791–801; 1990.

Eure, H.E.; Esch, G.W. Effects of thermal effluent on the population dynamics of helminth parasites in largemouth bass. In: Gibbons, J.W.; Sharitz, R.R., eds. Thermal Ecology. CONF-730505. Springfield, VA: Natl. Technical Infor. Center; 1974:p. 207–215.

Fabacher, D.L.; Chambers, H. A possible mechanism of insecticide resistance in mosquitofish. Bull. Environ. Contam. Toxicol. 6:372–376; 1971.

Fisher, R.A. The Genetical Theory of Natural Selection, 2nd ed. New York: Dover; 1958.

Fleming, I.A.; Gross, M.R. Latitudinal clines: a trade-off between egg number and size in Pacific salmon. Ecology 71:1–11; 1990.

Foltz, J.W.; Norden, C.R. Seasonal changes in food consumption and energy content of smelt (*Osmerus mordax*) in Lake Michigan. Trans. Am. Fish. Soc. 106:230–234; 1977.

Freeman, H.C.; Idler, D.R. The effect of polychlorinated biphenyl on steroidogenesis and reproduction in the brook trout *Salvelinus fontinalis*. Can. J. Biochem. 53:666–670; 1975.

Friedman, K.J.; Easton, D.M.; Nash, M. Temperature-induced changes in fatty acid composition of myelinated and non-myelinated axon phospholipids. Comp. Biochem. Physiol. 83B:313–319; 1986.

Gall, G.A.E. Influence of size of eggs and age of female on hatchability and growth in rainbow trout. Calif. Fish Game 60:26–35; 1974.

Geyer, H.J.; Scheunert, I.; Bruggemann, R.; Matthies, M.; Christian, E.; Steinberg, W.; Zitko, V.; Kettrup, A.; Garrison, W. The relevance of aquatic organisms' lipid content to the toxicity of lipophilic chemicals: toxicity of lindane to different fish species. Ecotoxicol. Environ. Saf. 28:53–70; 1994.

Ghittino, P. Nutrition and fish diseases. In: Halver, J.E., ed. Fish Nutrition. San Diego, CA: Academic Press; 1989:p. 681–713.

Gibbons, J.W.; Bennett, D.H.; Esch, G.W.; Hazen, T.C. Effects of thermal effluent on body condition of largemouth bass. Nature 274:470–471; 1978.

Glebe, B.D.; Leggett, W.C. Temporal, intra-population differences in energy allocation and use by american shad (*Alosa sapidissima*) during the spawning migration. Can. J. Fish. Aquat. Sci. 38:795–805; 1981.

Graham, T.P. Chronic malnutrition in four species of sunfish in a thermally loaded impoundment. In: Gibbons, J.W.; Sharitz, R.R., eds. Thermal Ecology. AEC symposium series, CONF-730505. Springfield, VA: Natl. Technical Infor. Center; 1974:p. 151–157.

* Grimes, D.V. Vitellolipid and vitelloprotein profiles of environmentally stressed and non-stressed populations of striped bass. Trans. Am. Fish. Soc. 122:636–641; 1993.

Gross, M.R. Sneakers, satellites and parentals: polymorphic mating strategies in North American sunfishes. Z. Tierpsychol. 60:1–26; 1982.

Gurr, M.I. The role of lipids in the regulation of the immune system. Prog. Lipid Res. 22:257–287; 1983.

Halver, J.E., ed. Fish Nutrition, 2nd ed. San Diego, CA: Academic Press; 1989.

Headrick, M.R.; Carline, R.F. Restricted summer habitat and growth of northern pike in two southern Ohio impoundments. Trans. Am. Fish. Soc. 122:228–236; 1993.

Heming, T.A.; Buddington, R.K. Yolk absorption in embryonic and larval fishes. In: Hoar, W.S.; Randall, D.J., eds. Fish Physiology. vol. XIA. The Physiology of Developing Fish. San Diego, CA: Academic Press; 1988:p. 407–446.

Henderson, B.A.; Nepszy, S.J. Reproductive tactics of walleye (*Stizostedion vitreum*) in Lake Erie. Can. J. Fish. Aquat. Sci. 51:986–997; 1994.

Henderson, P.A.; Wong, J.L.; Nepszy, S.J. Reproduction of walleye in Lake Erie: allocation of energy. Can. J. Fish. Aquat. Sci. 53:127–133; 1996.

Henderson, P.A.; Holmes, R.H.A.; Bamber, R.N. Size-selective overwintering mortality in the sand smelt, *Atherina boyeri* Risso, and its role in population regulation. J. Fish. Biol. 33:221–233; 1988.

Henderson, R.J.; Tocher, D.R. The lipid composition and biochemistry of freshwater fish. Prog. Lipid Res. 26:281–347; 1987.

Herting, G.E.; Witt, A. The role of physical fitness of forage fishes in relation to their vulnerability to predation by bowfin (*Amia calva*). Trans. Am. Fish. Soc. 96:427–430; 1967.

Hester, F.J. Effects of food supply on fecundity in the female guppy, *Lebistes reticulatus* (Peters). J. Fish. Res. Bd. Can. 21:757–764; 1964.

Hoar, W.S. General and Comparative Physiology, 3rd ed. Englewood Cliffs, NJ: Prentice-Hall; 1983.

Hodson, P.V.; McWhirter, M.; Ralph, K.; Gray, D.; Thivierge, D.; Carey, J.; Van der Kraak, G.; Whittle, D.; Levesque, M-C. Effects of bleached kraft mill effluent on fish in the St. Maurice River, Quebec. Environ. Toxicol. Chem. 11:1635–1651; 1992.

* Hutchings, J.A. Age-and size-specific costs of reproduction within populations of brook trout, *Salvelinus fontinalis*. Oikos 70:12–20; 1994.

Isley, J.J. Effects of water temperature and energy reserves on overwinter mortality in young-of-the-year largemouth bass (*Micropterus salmoides*). MS thesis, Southern Illinois Univ., Carbondale, IL; 1981.

Jezierska, B.; Hazel, J.R.; Gerking, S.D. Lipid mobilization during starvation in the rainbow trout, *Salmo gairdneri* (Richardson), with attention to fatty acids. J. Fish. Biol. 21:681–692; 1982.

Johnson, T.B.; Evans, D.O. Temperature constraints on overwinter survival of age-0 white perch. Trans. Am. Fish. Soc. 125:466–471; 1996.

Kamler, E. Early life history of fish: an energetics approach. New York: Chapman & Hall; 1992.

Katti, S.R.; Sathyanesan, A.G. Changes in tissue lipid and cholesterol content in the catfish *Clarias batrachus* (L.) exposed to cadmium chloride. Bull. Environ. Contam. Toxicol. 32:486–490; 1984.

⁕Keast, A.; Eadie, J. Growth in the first summer of life: a comparison of nine co-occurring fish species. Can. J. Zool. 62:1242–1250; 1984.

King, J.R.; Murphy, M.E. Periods of nutritional stress in the annual cycles of endotherms—fact or fiction? Am. Zool. 25:955–964; 1985.

Klaunig, J.E.; Lipsky, M.M.; Trump, B.F. Biochemical and ultrastructural changes in teleost liver following subacute exposure to PCB. J. Environ. Pathol. Toxicol. 2:953–963; 1979.

Kleiber, M. The Fire of Life. An Introduction to Animal Energetics. New York: John Wiley; 1961.

Knutsen, G.M.; Tilseth, S. Growth, development, and feeding success of atlantic cod larvae *Gadus morhua* related to egg size. Trans. Am. Fish. Soc. 114:507–511; 1985.

Kramer, R.H.; Smith, L.L. Formation of year classes in largemouth bass. Trans. Am. Fish. Soc. 91:29–41; 1962.

Kwain, W.; McCauley, R.W.; MacLean, J.A. Susceptibility of starved juvenile smallmouth bass, *Micropterus dolomieui* (Lacepede) to low pH. J. Fish. Biol. 25:501–504; 1984.

⁕LaL, B.; Singh, T.P. Impact of pesticides on lipid metabolism in the freshwater catfish, *Clarias batrachus,* during the vitellogenic phase of its annual reproductive cycle. Ecotoxicol. Environ. Saf. 13:13–23; 1987.

Lamon, M.S.; Ward, J.A. Measurements of reproductive effort from successive reproductive cycles for the Asian cichlid *Etroplus maculatus.* Environ. Biol. Fish. 8:311–320; 1983.

Larson, R.J. Seasonal cycles of reserves in relation to reproduction in *Sebastes.* Environ. Biol. Fish. 30:57–70; 1991.

Leatherland, J.F.; Sonstegard, R.A. Pathobiological responses of feral teleosts to environmental stressors: interlake studies of the physiology of Great Lakes salmon. In: Cairns, V.W.; Hodson, P.V.; Nriagu, J.O., eds. Contaminant Effects on Fisheries. New York: John Wiley; 1984:p. 115–149.

Leray, C.; Nonnotte, G.; Nonnotte, L. The effect of dietary lipids on the trout erythrocyte membrane. Fish Physiol. Biochem. 1:27–35; 1986.

Lipsky, M.M.; Klaunig, J.E.; Hinton, D.E. Comparison of acute response to PCB in liver of rat and channel catfish: a biochemical and morphological study. J. Toxicol. Environ. Health 4:107–121; 1978.

Love, R.C. The Chemical Biology of Fishes. vol. 2: Advances 1968–1977. New York: Academic Press; 1980.

Maceina, M.J.; Isley, J.J. Factors affecting growth of an initial largemouth bass year class in a new Texas reservoir. J. Freshwat. Ecol. 3:485–492; 1986.

MacFarlane, R.B.; Norton, E.C.; Bowers, M.J. Lipid dynamics in relation to the annual reproductive cycle in yellowtail rockfish (*Sebastes flavidus*). Can. J. Fish. Aquat. Sci. 50:391–401; 1993.

McKee, M.J.; Knowles, C.O.; Buckler, D.R. Effects of aluminum on the biochemical composition of Atlantic salmon. Arch. Environ. Contam. Toxicol. 18:243–248; 1989.

MacKinnon, J.C. Summer storage of energy and its use for winter metabolism and gonad maturation in American plaice (*Hippoglossoides platessoides*). J. Fish. Res. Bd. Can. 29:1749–1759; 1972.

McMaster, M.E.; van der Kraak, G.J.; Portt, C.B.; Munkittrick, K.R.; Sibley, P.K.; Smith, I.R.; Dixon, D.G. Changes in hepatic mixed-function oxygenase (MFO) activity, plasma steroid levels and age at maturity of a white sucker (*Catostomus commersoni*) population exposed to bleached kraft pulp mill effluent. Aquat. Toxicol. 21:199–218; 1991.

Mazeaud, M.M.; Mazeaud, F. Adreneric responses to stress in fish. In: Pickering, A.D., ed. Stress and Fish. New York: Academic Press; 1981:p. 49–75.

✦Medford, B.A.; Mackay, W.C. Protein and lipid content of gonads, liver and muscle of northern pike (*Esox lucius*) in relation to gonad growth. J. Fish. Res. Bd. Can. 35:213–219; 1978.

Miller, T.J.; Crowder, L.B.; Rice, J.A.; Marschall, E.A. Larval size and recruitment mechanisms in fishes: toward a conceptual framework. Can. J. Fish. Aquat. Sci. 45:1657–1670; 1988.

Minton, J.W.; McLean, R.B. Measurements of growth and consumption of sauger (*Stizostedion canadense*): implication for fish energetics studies. Can. J. Fish. Aquat. Sci. 39:1396–1404; 1982.

Miranda, L.E.; Hubbard, W.D. Winter survival of age-0 largemouth bass relative to size, predators, and shelter. North Am. J. Fish. Manage. 14:790–796; 1994.

Morris, R.; Bull, J.M. Studies on freshwater osmoregulation in the ammocoete of *Lampetra planeri* (Block). II. The effect of deionized water and temperature on sodium balance. J. Exp. Biol. 48:597–609; 1968.

Munkittrick, K.R.; Dixon, D.G. Growth, fecundity, and energy stores of white sucker (*Catostomus commersoni*) from lakes containing elevated levels of copper and zinc. Can. J. Fish. Aquat. Sci. 45:1355–1365; 1988.

Munkittrick, K.R.; McCarty, L.S. An integrated approach to aquatic ecosystem health: top-down, bottom-up or middle-out? J. Aquat. Ecosystem Health 4:77–90; 1995.

Munro, A.D. General idntroduction. In: Munro, A.D.; Scott, A.P.; Lam, T.J., eds. Reproductive Seasonality in Teleosts: Environmental Influences. Boca Raton, FL: CRC Press, Inc.; 1990:p. 1–11.

Murty, A.S.; Devi, A.P. The effect of endosulfan and its isomers on tissue protein, glycogen, and lipids in the fish *Channa punctata*. Pesticide Biochem. Physiol. 17:280–286; 1982.

Myers, R.A. Demographic consequences of precocious maturation of Atlantic salmon (*Salmo salar*). Can. J. Fish. Aquat. Sci. 41:1349–1353; 1984.

✦Newsome, G.E.; Leduc, G. Seasonal changes of fat content in the yellow perch (*Perch flavescens*) of two Laurentian lakes. J. Fish. Res. Bd. Can. 32:2214–2221; 1975.

Nordlie, F.G.; Leffler, C.W. Ionic regulation and the energetics of osmoregulation in *Mugil cephalus* Linnaeus. Comp. Biochem. Physiol. 51A:125–131; 1975.

Oliver, J.D.; Holeton, G.F.; Chua, K.E. Overwinter mortality of fingerling smallmouth bass in relation to size, relative energy stores, and environmental temperature. Trans. Am. Fish. Soc. 108:130–136; 1979.

Pandey, A.K.; Pandey, K. Biochemical estimation of lipid in liver and muscle of some fresh water fishes. Environ. Ecol. 12:880–883; 1994.

Pepin, P.; Myers, R.A. Significance of egg and larval size to recruitment variability of temperate marine fish. Can. J. Fish. Aquat. Sci. 48:1820–1828; 1991.

✦Post, J.R.; Evans, D.O. Size-dependent overwinter mortality of young-of-the-year yellow perch (*Perca flavescens*): laboratory, in situ enclosure, and field experiments. Can. J. Fish. Aquat. Sci. 46:1958–1968; 1989.

Prosser, C.L. Comparative Animal Physiology, 3rd ed. Toronto: WB Saunders; 1973.

✦Rajan, M.R. Sublethal effects of textile mill effluent on protein, carbohydrate and lipid content of different tissues of the fish *Cyprinus carpio*. Environ. Ecol. 8:54–58; 1990.

Ram, R.N.; Sathyanesan, A.G. Mercuric chloride induced changes in the protein, lipid and cholesterol levels of the liver and ovary of the fish *Channa punctatus*. Environ. Ecol. 2:113–117; 1984.

Rao, K.S.P.; Rao, K.V.R. Changes in the tissue lipid profiles of fish (*Oreochromis mossambicus*) during methyl parathion toxicity—a time course study. Toxicol. Lett. 21:147–153; 1984.

Reznick, D.N.; Braun, B. Fat cycling in the mosquitofish (*Gambusia affinis*): fat storage as a reproductive adaptation. Oecologia 73:401–413; 1987.

Rice, C.D.; Kergosien, D.H.; Adams, S.M. Innate immune function as a bioindicator of pollution stress in fish. Ecotoxicol. Environ. Saf. 33:186–192; 1996.

Roche, H.; Jouanneteau, J.; Peres, G. Effects of adaptation to different salinities on the lipids of various tissues in sea dace (*Dicentrarchus labrax* Pisces). Comp. Biochem. Physiol. 74B:325–330; 1983.

Roff, D.A. Reproductive strategies in flatfish: a first synthesis. Can. J. Fish. Aquat. Sci. 39:1686–1698; 1982.

Roff, D.A. On being the right size. Am. Nat. 118:405–422; 1981.

Roff, D.A. Optimizing development time in a seasonal environment: the "ups and downs" of clinal variation. Oecologia 45:202–208; 1980.

Rowe, D.K.; Thorpe, J.E.; Shanks, A.M. Role of fat stores in the maturation of male Atlantic salmon (*Salmo salar*) parr. Can, J. Fish. Aquat. Sci. 48:405–413; 1991.

Runnells, R.A.; Monlux, W.S.; Monlux, A.W. Principles of Veterinary Pathology, 7th ed. Ames, IA: Iowa State University Press; 1965.

Sandstrom, O.; Neuman, E.; Karas, P. Effects of a bleached pulp mill effluent on growth and gonad function in Baltic coastal fish. Wat. Sci. Tech. 20:107–118; 1988.

Sargent, R.C.; Taylor, P.D.; Gross, M.R. Parental care and the evolution of egg size in fishes. Am. Nat. 129:32–46; 1987.

Savitz, J. Effects of starvation on body protein utilization of bluegill sunfish (*Lepomis macrochirus* Rafinesque) with a calculation of caloric requirements. Trans. Am. Fish. Soc. 100:18–21; 1971.

Schmidt-Nielsen, K. Scaling: why is animal size so important? Cambridge, MA: Cambridge Univ. Press; 1984.

Schultz, E.T.; Clifton, L.M.; Warner, R.R. Energetic constraints and size-based tactics: the adaptive significance of breeding-schedule variation in a marine fish (Embiotocidae: *Micrometrus minimus*). Am. Nat. 138:1408–1430; 1991.

Scott, D.P. Effect of food quantity on fecundity of rainbow trout *Salmo gairdneri*. J. Fish. Res. Bd. Can. 19:715–731; 1962.

Shelton, W.L.; Davies, W.D.; King, T.A.; Timmons, T.J. Variation in the growth of the initial year class of largemouth bass in West Point Reservoir, Alabama and Georgia. Trans. Am. Fish. Soc. 108:142–149; 1979.

Sheridan, M.A.; Allen, W.V.; Kerstetter, T.H. Seasonal variations in the lipid composition of the steelhead trout, *Salmo gairdneri* Richardson, associated with the parr-smolt transformation. J. Fish. Biol. 23:125–134; 1983.

Shul'man, G.E. Life Cycles of Fish. New York: John Wiley; 1974.

Shuter, B.J.; Post, J.R. Climate, population viability, and the zoogeography of temperate fishes. Trans. Am. Fish. Soc. 119:314–336; 1990.

Shuter, B.J.; MacLean, J.A.; Fry, F.E.J.; Regier, H.A. Stochastic simulation of temperature effects on first year survival of smallmouth bass. Trans. Am. Fish Soc. 109:1–34; 1980.

Slobodkin, L.B. Energy in animal ecology. Adv. Ecol. Res. 1:69–101; 1962.

＊ Smit, G.L.; Hattingh, J.; Ferreira, J.T. The physiological responses of blood during thermal adaptation in three freshwater fish species. J. Fish. Biol. 19:147–160; 1981.

Somanath, B. Effect of acute sublethal concentration of tannic acid on the protein, carbohydrate and lipid levels in the tissues of the fish *Labeo rohita*. J. Environ. Biol. 12:107–112; 1991.

Stanley, J.G.; Colby, P.J. Effects of temperature on electrolyte balance and osmoregulation in the alewife (*Alosa pseudoharengus*) in fresh and sea water. Trans. Am. Fish. Soc. 100:624–638; 1971.

Stearns, S.C. The Evolution of Life Histories. Toronto: Oxford University Press; 1992.

Thompson, J.M.; Bergersen, E.P.; Carlson, C.A.; Kaeding, L.R. Role of size, condition, and lipid content in the overwinter survival of age-0 Colorado squawfish. Trans. Am. Fish. Soc. 120:346–353; 1991.

Thorpe, J.E.; Miles, M.S.; Keay, D.S. Developmental rate, fecundity and egg size in Atlantic salmon, *Salmo salar* L. Aquaculture 43:289–305; 1984.

Toneys, M.L.; Coble, D.W. Mortality, hematocrit, osmolality, electrolyte regulation, and fat depletion of young-of-the-year freshwater fishes under simulated winter conditions. Can. J. Fish. Aquat. Sci. 37:225–232; 1980.

Toneys, M.L.; Coble, D.W. Size-related first winter mortality of freshwater fishes. Trans. Am. Fish. Soc. 108:415–419; 1979.

＊ Tulasi, S.J.; Reddy, P.U.M.; Rao, J.V.R. Accumulation of lead and effects on total lipids and lipid derivatives in the freshwater fish *Anabas testudineus* (Block). Ecotoxicol. Environ. Saf. 23:33–38; 1992.

＊ Tyler, A.V.; Dunn, R.S. Ration, growth, and measures of somatic and organ condition in relation to meal frequency in winter flounder, *Pseudopleuronectes americanus*, with hypothesis regarding population homeostasis. J. Fish. Res. Bd. Can. 33:63–75; 1976.

Walkey, M.; Meakins, R.H. An attempt to balance the energy budget of a host-parasite system. J. Fish. Biol. 2:361–372; 1970.

Ware, D.M. Relation between egg size, growth and natural mortality of larval fish. J. Fish. Res. Bd. Can. 32:2503–2512; 1975.

Watanabe, T. Importance of the study of broodstock nutrition for further development of aquaculture. In: Cowery, C.B.; Mackie, A.M.; Bell, J.G., eds. Nutrition and Feeding in Fish. London: Academic Press; 1985:p. 395–414.

Wedemeyer, G.A.; McLeay, D.J.; Goodyear, C.P. Assessing the tolerance of fish and fish populations to environmental stress: the problems and methods of monitoring. In: Cairns, V.W.; Hodson, P.V.; Nriagu, J.O., eds. Contaminant Effects on Fisheries. New York: John Wiley; 1984:p. 164–195.

Werner, E.E. Niche partitioning by food size in fish communities. In: Clepper, H., ed. Predator-Prey Systems in Fisheries Management. Washington, DC: Sport Fishing Institute; 1979.

Wicker, A.M.; Johnson, W.E. Relationships among fat content, condition factor, and first-year survival of Florida largemouth bass. Trans. Am. Fish. Soc. 116:264–271; 1987.

Wilkens, N.P. Starvation of the herring, *Clupea harengus* L.: survival and some gross biochemical changes. Comp. Biochem. Physiol. 23:503–518; 1967.

Wootton, R.J. Effect of food limitation during the breeding season on the size, body components and egg production of female sticklebacks (*Gasterosteus aculeatus*). J. Anim. Ecol. 46:823–834; 1977.

8

Lipids and Essential Fatty Acids in Aquatic Food Webs: What Can Freshwater Ecologists Learn from Mariculture?

Yngvar Olsen

8.1. Introduction

Only a few decades have passed since the major discovery that certain fatty acids, or more precisely, the family of fatty acids denoted n-6 or ω6 polyunsaturated fatty acid (PUFA), are essential for humans (Holman et al., 1964; Hansen et al., 1963). These fatty acids are characterized by having their first double bond between carbon number 6 and 7 from the methyl end. In the late 1970s, it was shown that also the family of n-3 or ω3 fatty acids, characterized by having the first double bond between carbon number 3 and 4 from the methyl end, are essential for normal growth and function of the human brain (Leaf, 1993; Bjerve, 1991; Neuringer et al., 1988; Crawford et al.,1981, 1976). The impact of dietary ω3 fatty acids on the incidences of vascular and coronary diseases was first described by Danish scientists (Dyerberg et al., 1977; 1975) who compared the food habits and health of Greenland Inuits with people from Denmark (representative for the Western industrialized world). This classic work brought a major change in thinking about how the current role of lipids in metabolism is viewed; from being a metabolic fuel compound usually associated with obesity in the Western world to an essential compound in a wide range of metabolic functions. Since then, it has been thoroughly demonstrated that ω3 fatty acids affect human health by reducing blood lipid levels, by reducing the rate of atherosclerosis and thrombus formation and by altering inflammatory and immune responses (see review by Leaf, 1993).

The interest in cultivating marine fish species is old (Rollefsen, 1934), but the early pioneers in this century usually failed in their attempts to make marine larvae survive the early feeding stages. The Japanese were pioneers in using the brackish water rotifer *Brachionus plicatilis* as live feed for cultured marine larvae in the early 1960s (see review by Nagata and Hirata, 1986; Ito, 1960), but they did not achieve major success until they realized that food quality was important, especially in respect to essential fatty acids (EFA) (Fukusho, 1977, Kitajima and Koda, 1976). Marine fish larvae turn out to have high dietary requirements for ω3 fatty acids (Watanabe et al., 1978). The Japanese and Danish findings were published at approximately the same time, and in the time that followed, it has been confirmed that many cultured marine animals have high ω3 requirements.

The above Japanese findings and the mass rearing techniques simultaneously developed for *B. plicatilis* (Hirata, 1980) have been very important for the later developments in marine larviculture and in mariculture in general. *B. plicatilis* is used worldwide as a live feed for marine larvae, and considerable efforts have been made worldwide during the past decade to improve and adapt the Japanese methods to species with commercial potential. These studies have shown that marine fish larvae, in particular those living in relatively cold water or with high growth potential, have high and rather specific requirements for highly unsaturated ω3 fatty acids (ω3 HUFA). It has also become evident that all groups of organisms, including crustaceans and shellfish, exhibit enhanced survival, growth, and viability when their EFA requirements are met during the juvenile stage. Aspects related to marine larval ω3 HUFA requirements and the techniques to manipulate lipids and fatty acid composition of zooplankton organisms used as live feed have been dealt with in numerous of conferences during the past decade, and ω3 HUFA requirements are probably still the most frequent nutritional question brought up in larval nutrition studies.

The research on lipids and EFA in marine larvae and food zooplankton (mainly in *B. plicatilis* and brine shrimp *Artemia* sp.) have provided a fundamental knowledge base that is useful in the field of food-quality research related to natural food webs, including freshwater food webs. The existing knowledge is not always directly applicable to freshwater systems, but many of the principal mechanisms and conclusions, as well as the more applied perspectives, may be useful for freshwater biologists.

The main goal here is to summarize knowledge on lipids and EFAs established in mariculture that may have relevance also for freshwater food webs. The chapter is not a literature review but a presentation of selected studies. I address and speculate on interesting observations without directly critically evaluating findings and conclusions made by others.

8.2. Results and Discussion

8.2.1. Some Important Lipids and Fatty Acids

Many textbooks as well as several chapters in this book (e.g., Arts, Cavaletto and Gardner, Landrum and Fisher, Parrish, and Wainman et al.) describe the common types of lipids normally found in freshwater organisms. I only briefly mention the lipid and fatty acid groups that are important to understand the principal issues treated.

8.2.1.1. Essential Fatty Acids

The fatty acids found in biological matter are grouped according to their chemical or biological properties. From the perspective of animal nutrition, fatty acids may be grouped into those that cannot be synthesized de novo, although being essential for animal growth and development (EFA), and those that can be synthesized (nonessential). The EFAs must be supplied in the food. Both animal and plant cells can catabolize, elongate, and desaturate (dehydrogenate) fatty acids through successive steps. Animal cells cannot, however, desaturate fatty acids at points less than between carbon 9 and 10 from the methyl end (n-9 or ω9), even though they need such fatty acids to function adequately and to grow. This finding, which may not be strictly true for all terrestrial animals, implies that all ω6 and ω3 fatty acid bonds of marine origin have been formed by organisms of other kingdoms, among which marine algae are believed to be most important. The importance of bacteria in this regard is still an open question.

Many animals can elongate and desaturate the shorter C18-precursors of ω6 and ω3 fatty acids, linoleic acid and linolenic acid, respectively. Many carnivorous animals have, however, little or no ability to modify C18 fatty acids through anabolic reactions, whereas they are able to modify C22 HUFA through catabolic chain-shortening reactions (Sargent et al., 1993b). It is nevertheless still useful to group EFA of marine organisms in two groups, the linoleic acid (ω6) family and the linolenic acid family (ω3) (Fig. 8.1).

High contents of long-chain ω3 HUFA (e.g., eicosapentaenoic acid [EPA] or docosahexaenoic acid [DHA]) are found only in marine plants, whereas the shorter linoleic acid (18:2ω6) and to a lesser extent also linolenic acid (18:3ω3)

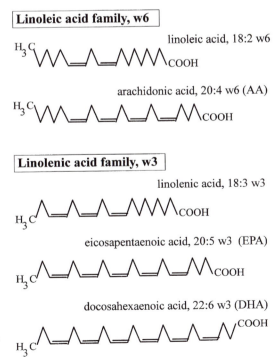

FIGURE 8.1. Families of essential fatty acids (EFA) and chemical structures of some common fatty acids within the two families.

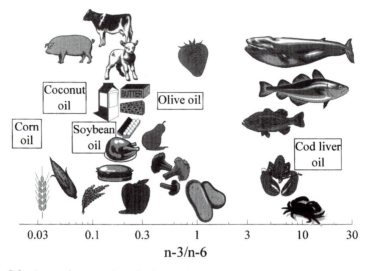

FIGURE 8.2. Approximate ratios of ω3 to ω6 essential fatty acids of some terrestrial and marine products and organisms. (Data from Skjervold, 1992.)

are more abundant in terrestrial plants (Fig. 8.1). This difference is found also in animals of marine and terrestrial systems, because the fatty acid distribution of animals is primarily determined by the composition of their dietary fatty acids. Generally, ω3 fatty acids dominate over ω6 fatty acids by a factor 5–20 in marine food webs (Fig. 8.2). In terrestrial food webs, the opposite is true, and few agricultural plants store high amounts of linolenic acid (e.g., olives), whereas many store linoleic acid (e.g., soybeans). Thus, the human diets of the Western world are logically enough strongly dominated by ω6 fatty acids. In contrast to this, it is believed that the diet of ancient humans contained more balanced quantities of ω6 and ω3 fatty acids (Leaf, 1993; Weber, 1989).

Freshwater and marine fish contain high levels of HUFA but with a slightly higher level of ω6 fatty acids compared with ω3 fatty acids in freshwater versus marine fish (Ahlgren et al., 1994; Henderson and Tocher, 1987). This is probably a result of the greater importance of food of terrestrial origin in most freshwater food webs. It is also notable that freshwater fish (e.g., rainbow trout) seem to have higher capacity than marine fish to elongate and desaturate C18 PUFA into DHA and EPA (Sargent et al., 1993b; Owen et al., 1975).

8.2.1.2. Lipid Classes

Lipids are, depending on their polarity, grouped as neutral and polar lipids (Fig. 8.3). Triacylglycerides (TAG) and wax esters (WE) are neutral lipids and abundant storage lipids that provide metabolic energy through oxidative catabolism. The TAG molecule is a dominant energy and carbon storage product in animals and is ubiquitous in terrestrial animals. Many marine animals in the lower food

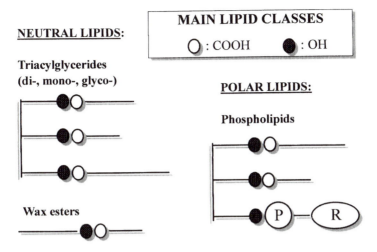

FIGURE 8.3. Main lipid classes of marine animals and their schematic chemical structures.

web, however, store most of their energy as WEs, which are simple esters constituting a long-chain fatty alcohol moiety bound to a fatty acid moiety. WEs, which have a lower melting point and higher energy content per weight than TAGs, are most abundant in cold-blooded animals (mainly invertebrates) living in cold waters, be it in the deep parts of the oceans or at high latitudes (Sargent and Henderson, 1986). Some animals that store WEs may exhibit very high contents of these compounds toward the end of the growth season (>50% of the dry weight) (Sargent and Henderson, 1986).

The basic unit of phospholipids (PL), present in all kingdoms, is a diacylglyceride molecule with a phosphate group bound to an organic group in the second terminal position of the glycerol (R, Fig. 8.3). Common organic groups of PLs are ethanolamine, choline, and inositol, and the respective PLs are denoted phospatidylethanolamine, phospatidylcholine, and phospatidylinositol. Aquatic cold-blooded animals normally have PLs characterized by a high fraction of PUFA. PLs are, together with cholesterol and sphingolipids, ubiquitous constituents of cell membranes and are therefore both structurally and functionally important. PL molecules, or their ancient chemical relatives, were undoubtedly critical constituents during the formation of primitive cells and organisms on earth.

8.2.2. Methodological Considerations

A methodological constraint experienced by most researchers of marine larval nutrition and live feed production was the large sample size normally needed for lipid and fatty acid analysis. The scale of typical laboratory feeding experiments with very small fish larvae and zooplankton was restricted by both biological and economical constraints, and even relatively large experimental pilot systems could not produce the amount of material (>1 g) normally used. This called for efforts to scale down the analytical methods for determination of lipid and fatty acids.

Most laboratories now use quantitative methods in their determination of fatty acids (see Parrish, this volume). By adding an internal fatty acid standard before extraction of the sample (normally 19:0 or 21:0), it is possible to derive quantitative values for individual fatty acids after careful calibration, establishment of response curves, and the use of on-column injection of the material (see Rainuzzo, 1993). The total weight of fatty acids in the sample can then be estimated as the sum of the individual fatty acids, which is the basis for the estimation of percentage fatty acid distribution. Absolute fatty acid content (e.g., milligrams fatty acid per gram dry material) and relative fatty acid distribution (i.e., percentage of a given fatty acid of total fatty acids) are complementary assessment criteria for nutritional value. The quantitative content of given EFA is obviously important during evaluations of the EFA requirements of given species. The percentage distribution is important because different EFAs tend to compete with each other in many enzymatic reactions. The relative proportions of the competing EFAs of the diet or the ratio between specific EFAs may then be more important than their absolute quantitative contents.

Quantitative determination of fatty acids was not common some 10–15 years ago. Most values reported were in terms of percentage content of the total fatty acids, primarily because the quantitative method was little known. Unfortunately, the relative values in the older literature cannot easily be converted to quantitative values, because the absolute sum of fatty acids, which forms the 100% value, is normally not known. It is important to realize that the quantitative sum of fatty acids is not equivalent to the quantitative content of lipids, which was indeed frequently measured in earlier studies. Approximate estimates of the absolute fatty acid contents based on percentage fatty acids and quantitative lipid content, thus assuming that lipids constitute the 100% level for fatty acids, will overestimate fatty acid values to some degree. The fraction of fatty acids to total lipids is variable, dependent on the material analyzed.

Algae show normally relatively low fractions of fatty acids to lipids (e.g., 32–53%) (Reitan et al., 1994b). Animals, however, tend to exhibit higher fractions, in particular fat animals with high content of TAGs. For example, the fatty acid fraction of total lipids found for relatively fat Atlantic salmon (S. Salar) was higher than that for the leaner B. plicatilis (Fig. 8.4). This apparent discrepancy is, however, part of the same general pattern of variation between lipid content and the fractioin of fatty acids, suggesting that this is representative for animals that store TAGs, including freshwater animals (Fig. 8.4).

The international cooperation in aquaculture research that developed gradually through the 1980s exposed the need for intercalibration of methods for lipid and fatty acid determination. This was necessary for comparison of results derived from different laboratories and therefore also for the progress of the research. The most recent calibration exercise has revealed that our methodological capabilities have indeed improved during the past decade. This exercise, which was run by the International Council for the Exploration of the Sea, involved a standardized

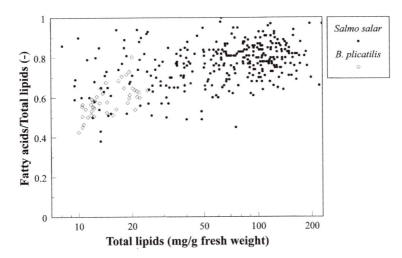

FIGURE 8.4. Relative contribution of fatty acids to the total lipids of farmed Atlantic salmon and the rotifer *Brachionus plicatilis* as a function of the lipid content of the animal.

predefined method for biological and chemical treatment of the samples (described by Coutteau and Sorgeloos, 1995). Most of the laboratories that took part in the intercalibration demonstrated acceptable variation in replicates within and between the participating laboratories, both for fatty acid and lipid analysis. However, some laboratories still did not provide absolute values for fatty acid content, and there were still methodological questions (e.g., whether the internal standard fatty acids should be added before or after extraction).

The experience with the intercalibration exercise has shown that both methods used and values of fatty acids and lipids should be treated critically. Another complicating issue is that the literature does not always specifically express the units of the presented values. For example, fatty acid contents are frequently presented as "percentage" without information on whether this percentage value refers to dry matter, lipids, or the sum of fatty acids. Experience in mariculture clearly shows that concern for methods and intercalibration measures, as well as rigorously detailing the units used to express fatty acid values, are important for the scientific progress. These issues should be treated seriously by both new and well-established researchers of EFAs and food quality.

8.2.3. Physiological Requirements of Marine Animals

Cold-blooded aquatic animals living at higher latitudes require high proportions of PUFA in their cell membranes to maintain efficient intercellular transport processes and cell metabolism in general. With this simple principle in mind, there is admittedly still major confusion regarding the specific requirements of EFAs of marine animals (Sargent et al., 1993b). Considerable efforts have been made during the past decade, but experimental as well as empirical determination of

EFA requirements have turned out to be complicated, especially for the early stages of marine fish larvae. I suggest that the main reasons for this are that

- It took some time to realize that DHA was the most critical component in larviculture, in part because of high larval DHA requirements and the inability in providing sufficient analytic material.
- The ability of chain elongation and desaturation of short C18 ω3 fatty acids to DHA is inefficient, and perhaps not even possible, in many cultured marine species. This implies that individual ω3 fatty acids must be treated separately in studies of marine larval EFA requirements.
- The live feed organisms used worldwide, and in particular the crustacean *Artemia* sp., are characterized by rapid DHA catabolism compared with that of the common natural zooplankton, resulting in severe losses of DHA in a short time. This instability in DHA, and the fact that early stages cannot be fed formulated feed, have made research on larval EFA requirements a challenge.
- The early experience with raising salmonid juveniles did not contribute to our understanding of EFA requirements for larval stages of fish. Salmonids accept formulated feed made from fish meal from the very beginning.

Although our knowledge on EFA requirements of cultured marine fish larvae is still limited, we have learned substantial lessons through research efforts on larval EFA metabolism and live feed production. The understanding of the structural and metabolic functions of fatty acids has been improved, and the finding that EFA must be treated separately and not as families is important. It is also apparent that DHA is the most important component of the ω3 family (Watanabe, 1993; Watanabe et al., 1989). The requirements of ω6 fatty acids in marine larvae are still unknown, but arachidonic acid (AA; 20:4ω6), which is present in low proportions in marine animals, is likely to have an important metabolic role (Sargent et al., 1993b).

8.2.3.1. General Evaluation of EFA Requirements

The following considerations of EFA requirements are mainly restricted to ω3 fatty acids and marine fish larvae. EFAs supplied in the diet and also from the larval yolk sac in the very early larval stages play an important role in the following physiological key processes of marine animals:

- Anabolic processes and growth—general requirements related to de novo formation of biomass, morphological development, and tissue differentiation
- Membrane transport and metabolism—inter- and intracellular membrane transports, respiration, and enzyme activity in general
- Regulation of metabolism—prostaglandins and their hormonal functions

8.2.3.2. Anabolic Processes and Growth

Most of the common marine fish species have a spawning strategy that is characteristic of r-strategists. The number of spawned eggs is high, the eggs are small and hatch early, the mother fish invests little in a single egg, and the larvae are relatively immature at the time of hatching. The newly hatched larvae have to rely on its yolk sac material before they start feeding (i.e., normally for a few days to a month, depending on the species). The composition of the maternal yolk sac material is dependent on the nutritional state of the mother. Tissue differentiation, including eye, brain, and neural development in general, takes place just after hatching. Neural tissues, including eyes, contain very high amounts of DHA (Sargent et al., 1993a; Mourente et al., 1991), implying that this fatty acid is very important for very young stages of marine fish larvae. The expected high DHA requirements during the early stages, which may be deduced based on the rapid growth and differentiation of neural tissues of brain and eyes, are supported by the finding that eggs of marine species tend to be very rich in DHA (Rainuzzo et al., 1992; Sargent et al., 1993b). From an evolutionary point of view, we may assume that essential components that are found in high quantities in the eggs are important for normal growth and development during the early stages of life. It is easy to imagine that inadequate brain function and vision, demonstrated for DHA-deficient herring larvae (Sargent et al., 1993a), will be fatal for marine larvae, which are visual feeders.

The EFA requirements of marine fishes are also substantial in the later juvenile stages, when the rate of increase in body biomass is still high. The contents of DHA, and also shorter ω3 fatty acids such as EPA, are high in juveniles as well as in adult stages of many marine species (see below). The specific growth rates of fish larvae and older stages of juveniles are, however, normally much higher than that of the adult fish. Larvae and juveniles of cold water species, for example, cod (*Gadus morhua*) and Atlantic halibut (*Hippoglossus hippoglossus*), will double their biomass within 1–2 weeks, and some fast-growing species such as turbot (*Scophthalmus maximus*) may have a biomass turnover time of <3 d (Reitan et al., 1993). Their rapid growth, along with their inefficient digestion of food in the very early stages (Øie et al., 1997), suggests that the EFA requirements for body growth are relatively high. The requirements will gradually become reduced with age, because older fish have a lower specific growth rate. Even adult stages of marine fish exhibit high EFA requirements compared with, for example, common species in agriculture.

There are pronounced differences in dietary EFA requirements of marine fish species (Reitan et al., 1994a; Koven et al., 1993, 1990; Mourente et al., 1993; Watanabe, 1993; Izquierdo et al., 1989; Watanabe et al., 1989). Temperature is normally believed to be decisive. More specifically, it is assumed that fish living at high latitudes and species that reside permanently in deep water will normally have higher EFA requirements than fish residing in warmer water. Associated with these trends is the relative ability of the different species to elongate and desaturate shorter EFA to long-chain PUFA such as DHA. High metabolic flexibility to modify EFA within the essential families will probably make the species more flexible with regard to the composition of the dietary lipids. High metabolic flexibility is apparently characteristic for many salmonids, including rainbow trout (Sargent et al., 1993b; Owen et al., 1975).

The ability of marine animals to synthesize AA from shorter ω6 precursors is not known, but this trait is presumably not very important because AA seems to be present in low, although significant, amounts in most biological material and therefore also in all types of food (see below).

It is commonly believed that carnivorous fish show less ability to modify EFAs through anabolic reactions than planktivorous and omnivorous species (Sargent et al., 1993a). Such differences can be understood from an evolutionary point of view. It has not been very important for carnivorous fish to retain the ability of fatty acid elongation and desaturation during their evolution as compared with groups feeding on more suboptimal diets; this is simply because carnivorous fish to a large extent feed on an overall well-balanced diet relative to their own requirements, with respect to both EFA and other essential compounds. Most of the marine species of interest for aquaculture, except for the salmonids, which are not strictly marine, are carnivorous species with high EFA requirements.

8.2.3.3. Membrane Transport and Metabolism

The physiological justification for the relatively high EFA requirements of aquatic animals in general, and for marine fish larvae in particular, is their high need for PUFA used in PL synthesis (Fig. 8.5). Most enzymes, whether they control the cellular exchange of ions or metabolic processes, are associated with membranes. The PLs of the cell membranes are synthesized by enzyme systems that exhibit higher affinity for PUFA than for other fatty acids. Under conditions of excess supply, we know that PUFAs are preferentially esterified to the sn2 position (C2 position) of both TAGs and PLs, whereas saturated and monounsaturated fatty acids are esterified to sn1 and sn3 positions (terminal positions). This discrimination has its origin in the enzymes involved and may explain why one out of two fatty acids belonging to PLs (i.e., 50%) is expected to be polyunsaturated, whereas

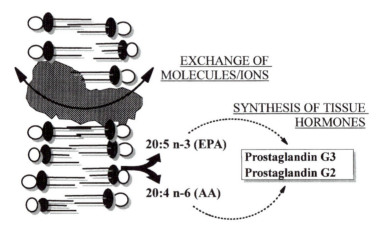

FIGURE 8.5. Schematic illustration of a cell membrane and important metabolic and regulatory processes taking place across the membrane.

one out of three will tend to be polyunsaturated in TAGs (Sargent et al., 1993b). This enzyme mechanism, along with other differences in enzyme specificity, will secure some degree of genetic control of the composition of the membranes, which is critical from a functional point of view. The fatty acid composition of the membrane affects the activity of the enzymes, and a certain fraction of PUFA, or more specifically DHA, is needed to ensure optimum activity of the enzymes associated with the cell membranes (Dratz and Holte, 1992; Stubbs, 1992).

Enzymes of PL synthesis seem to exhibit low specificity for individual PUFA (Mourente et al., 1991). Even though the enzymes show higher affinity to PUFA than to monounsaturated and saturated fatty acids, the enzymes may still incorporate enhanced levels of non-EFA (polyunsaturated, monounsaturated, and saturated fatty acids) in the membrane PLs under conditions of EFA deficiency, resulting in a reduced membrane fraction of PUFA (Sargent et al., 1993b). We may anticipate that a reduced fraction of PUFA, which is species-specific, in the membranes will, at some point, result in gradually reduced activity of all membrane-bound enzymes. The full impact of a suboptimal membrane composition can hardly be imagined, but the physiological capacity or the general health of deficient specimens will surely be reduced. This explains why inadequate dietary EFA supplies may affect animal health in a general fashion.

8.2.3.4. Regulation of Metabolism

From lipid studies of humans and other warm-blooded animals, we have learned that the C20 fatty acids (EPA and AA) of the membrane PLs are precursors in prostaglandin synthesis (Fig. 8.5). These prostaglandins are themselves precursors for a number of compounds known as tissue hormones. Older literature describing the formation and interactions of prostaglandins may be confusing, but a useful description is given by Weber (1989). He states that prostaglandin G_3, which is synthesized from EPA, is a regulatory antagonist to prostaglandin G_2, which is synthesized from AA. Both AA and EPA are derived from the membrane PLs (see further speculations in Sargent et al., 1993b). The G_3/G_2 ratio, which is believed to affect or control many cellular processes (Weber, 1989), will then depend on the ratio of EPA to AA in the membranes and finally on the ratio of the specific $\omega3$ to $\omega6$ in the diet. Marine fish may be less able to elongate EFA than humans, but we can most likely apply our knowledge from human research to fish and other animals. Keep in mind, though, that the values of typical and critical ratios of fatty acids EPA/AA and prostaglandins G_2/G_3 are most likely different in fish and humans.

The relatively poor enzymatic control (i.e., genetic control) of PL composition (i.e., EPA/AA or $\omega3/\omega6$ ratios) and of the successive prostaglandin synthesis (G_3/G_2 ratio) implies that the dietary fatty acid composition of animals will more or less directly affect their regulatory hormonal processes (review by Bell et al., 1991; Sargent et al., 1993b; Weber, 1989). The current recommended dietary ratio $\omega3/\omega6$ for humans is roughly in the range of 0.1–1, but this range is still partly based on an educated guess. The human diet during the Stone Age period is believed to have had a ratio ω3 to ω6 in the upper part of this range (0.5–1) (Skjervold, 1992; Weber, 1989), whereas the typical Western diet of today contains a much higher fraction of ω6 fatty acids (<0.1, see Fig. 8.2). The change in dietary lipid composition for humans is a consequence of the increasing population and our industrial agricultural practices, and this change has been rapid on an evolutionary time scale (Skjervold, 1992). Some have suggested that this change in diet is a primary cause of typical Western life-style diseases, be it cardiovascular diseases or inflammatory and allergic disorders (Leaf, 1993; Weber, 1989).

The proposed mechanism of prostaglandin synthesis and the potential metabolic impacts may easily explain why many marine animals, be it marine larvae or Atlantic salmon, cannot grow efficiently on food rich in ω6 fatty acids or if the ω3/ω6 ratio is low. There are no obvious reasons why the long-chain polyunsaturated AA, or its 22C analogue (22:5ω6), which may be common in membrane phospholipids (Stubbs, 1992), should not work satisfactorily as a membrane component. It is likely that food too rich in ω6 fatty acids may yield an unacceptable, even fatal, prostaglandin composition for marine species.

8.2.3.5. General Considerations and Concluding Remarks

The general considerations on EFA requirements of marine species are probably also valid for freshwater species, but there may be major systematic differences at the species level. Freshwater fish have during their evolution consumed more EFAs originating from terrestrial sources, with a higher fraction of ω6 fatty acids than the food of marine species at the same trophic level. This is probably reflected both in the normal fatty acid composition of freshwater fish and their actual requirements for specific EFA. Literature indeed suggests that AA is more abundant in freshwater than in marine food webs, whereas EPA is less abundant (Ahlgren et al., 1994; Henderson and Tocher, 1987). DHA seems to be equally important in both systems. There is no reason to believe that the quantitative physiological requirements of PUFA of freshwater fish species adapted to cold water are different from those of cold-water marine fish species. The metabolic flexibility to elongate and desaturate short EFA may, however, be systematically higher in freshwater fish. If so, freshwater species may be less dependent on high levels of DHA in the diet, because DHA may partly be synthesized from 18:3ω3.

Experience in mariculture reveals a fundamental difference in DHA metabolism among species of zooplankton and fish. Whereas the metabolism of fatty acids tends to vary more or less similarly in all species, there are two distinct patterns of DHA metabolism. Some species exhibit a DHA-conservative metabolism, characterized by a lower catabolism of DHA than of all other fatty acids during starvation, resulting in increasing percentage DHA of total fatty acids. Other species tend to catabolize DHA much faster than all the other fatty acids, yielding a reduced percentage of DHA. It is well known that DHA is the most important fatty acid in the metabolic, short-term adaptation to low temperature for many species living in a fluctuating environment (Williams and Hazel, 1992; Olsen and Skjervold, 1991; Farkas et al., 1980). I suggest that the different patterns in DHA metabolism specified above characterize two groups of animals with different evolutionary histories. The DHA-conservative species are presumably evolutionarily adapted to low temperatures (e.g., salmon, halibut, turbot, cod, red feed, and other coastal copepods at high latitudes), whereas the DHA-catabolizing species are adapted to high temperatures (e.g., *B. plicatilis, Artemia* sp.).

8.2.4. Methods for Evaluation of EFA Requirements

The requirements of essential components of animals can theoretically be determined by feeding the animal food with variable contents of the critical component while all other factors are kept constant. A typical result of a successful experiment is that the animal growth rate is positively related to the concentration of the critical component below a certain critical concentration and constant at concentrations above this level (i.e., saturation kinetics). The inflection point can be interpreted as the actual requirement for the specific essential component. This method is not always suitable for estimation of EFA requirements of larval stages for the following reasons:

- It is not straightforward, or perhaps even possible, to establish a gradient of a single EFA in live feed, and synergetic effects of other fatty acids within the essential family cannot be excluded.
- The composition of the live feed organism varies with time (i.e., lipids, fatty acid composition, and protein contents), and specific countermeasures must be implemented to reduce the effect of this (Reitan, 1994).
- Pure fatty acid extracts are not easily available, and in any event, they are expensive.
- Metabolic conversion of dietary EFA takes place in animals, and the conversion rates presumably depend on the composition of the dietary lipids.
- The exposure time of the test experiments affects the magnitude of the response, because some time will always pass before the ultimate state variables (e.g., variation in larval survival and growth rate) are manifested. General and unified criteria for early assessment of malnutrition in larvae are not well established.
- Sudden mortality caused by non-nutritional boundary conditions, such as inadequate physical and microbial cultivation conditions, is still common in larval cultures. This makes reproducibility of replicate experiments low, and small amounts of material still often restrict our analytical capability.
- Boundary conditions may affect the critical level of the component under study (i.e., temperature, bacteria associated with the larvae, or maternal supply to early larval stages).

Alternatives of the above approach have been applied in several studies of EFA requirements with some success (Koven et al., 1993, 1990; Mourente et al., 1993; Watanabe, 1993; Izquierdo et al., 1989; Watanabe et al., 1989). The typical range for marine fish larvae/juveniles is 5–40 mg ω3 HUFA per gram dry matter of food. Another conclusion is that DHA seems to be more important than EPA

(Watanabe, 1993). It is, however, again emphasized that the biochemical instability of the live feed makes values derived for larvae only approximate (Reitan, 1994).

An alternative, but admittedly less quantitative, way to assess EFA requirements for larval growth and survival is to compare the fatty acid composition of cultured larvae with that of naturally occurring larvae of the same species raised on natural food. Comparisons of fatty acid profiles of larvae of Atlantic halibut (*H. hippoglossus*), which were cultured with rotifers and *Artemia,* with the profile of larvae feeding on copepods in natural systems, have clearly demonstrated the problem of providing sufficient amounts of DHA for this species through rotifers and *Artemia,* both typical warm water–adapted zooplankton species. Halibut larvae cultured on a diet of rotifers and *Artemia* typically contain ≈10% DHA of their total fatty acids, whereas larvae cultured with copepods contain ≈40% during the early feeding stages (J.O. Evjemo, A. Olsen, and others, unpublished results). This may indicate DHA deficiency of the former larvae, and it certainly demonstrates that the DHA supplied with cultured live feed is much lower than in nature. The above approach cannot unequivocally prove DHA deficiency, but we should strive to enhance the DHA level of cultured live feed. It is also important to recognize that when DHA values are equal in both larval groups, any hypothesis of DHA deficiency is likely to be rejected.

Analysis of the percentage fatty acid distribution of fish larvae in the late part of the larval stage and its live food has been useful in evaluating larval EFA requirements. This method is quick compared with most other methods, and it is also reliable provided that a comprehensive database for the EFA-sufficient conditions is available as a reference for comparison.

The hypothesis that DHA is a critical component for many species during the larval and later stages has been independently supported by starvation experiments, which represent a third method for evaluation of EFA requirements. Most marine fish larvae and copepods, and an uncommon strain of *Artemia* as well (*Artemia sinica;* Evjemo et al., 1997), retain DHA with higher efficiency than all other fatty acids during starvation (Rainuzzo, 1993). Some other strains of *Artemia* and *B. plicatilis* tend to catabolize DHA more efficiently than other fatty acids (Evjemo et al., 1997; Olsen et al., 1993a). When an animal tends to retain a specific EFA level during starvation, this is interpreted as a high requirement of that specific EFA, and the percentage value of total fatty acids may yield an estimate of the quantitative requirements.

Efficient DHA retention, as shown by an enhanced percentage DHA of total fatty acids during starvation, is typical for starving, wild Atlantic salmon that migrate upstream for spawning. Their DHA retention efficiency during starvation is found to be higher at low than at high temperatures, whereas most other fatty acids are only moderately affected by temperature (Olsen and Skjervold, 1991). Figure 8.6 shows percentage DHA in the salmon flesh as a function of the latitude of the river outlet in the sea. The regression analysis (see legend of Fig. 8.6) showed that the percentage DHA of total fatty acids increases significantly ($P < .05$) by on average 0.76% per latitude. The water temperature of the northern rivers is generally 3–6°C lower than that of the southern rivers, but mountains and

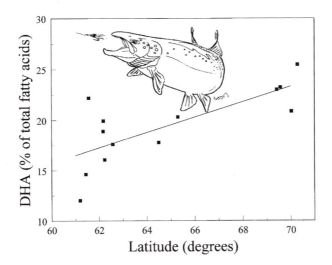

FIGURE 8.6. Average percentage DHA of total fatty acids in starved male spawners of wild Atlantic salmon caught in rivers during the autumn (5–18 fish per river, total of 134 fish), as a function of the latitude of the river outlet in the sea along the Norwegian coast. (Data from Olsen and Skjervold, 1991.) The curve shows the regression line expressed by the equation: %DHA = (0.76 ± 0.21) latitude − (29.8 ± 2.6) (r^2 = 0.55, P <.05 for slope <0, t-test).

glaciers disturb, to some extent, this general relationship in the south. The results illustrate the importance of DHA in temperature acclimation of fish.

8.2.5. Symptoms of EFA Deficiency

It is important to distinguish between quantitative deficiency in the EFA supply and the effects of inadequate composition of the EFA actually supplied (i.e., the ω3/ω6 and DHA/EPA ratios). A satisfactory quantitative supply of both EFA families may still be nutritionally inadequate if the EFA composition within the families is inadequate or if the families themselves are supplied in inappropriate ratios. A marine fish fed saturating amounts of terrestrial lipids rich in ω6 fatty acids is most likely not to survive.

It is also important to recognize that questions of quantitative supply and composition always must be evaluated relative to the actual requirements of each species. This is because species requirements may be very different, depending on their metabolic flexibility to elongate and desaturate shorter EFA precursors, in short their evolutionary dietary history. Finally, EFA deficiency may in some cases cause mortality and in other cases merely reduced viability and health (Watanabe, 1993). These symptoms are species-dependent and are a function of developmental stage and the realized degree of EFA deficiency.

Fish, and probably also zooplankton, are presumably very sensitive to EFA deficiency during the egg and early larval stages, but organisms that can elongate short precursors of EFA efficiently will presumably be better able to tolerate inadequate fatty acid composition in the larval stage than organisms that do not.

The mother deposits high amounts of EFA during egg formation, be it DHA for marine fish larvae (Sargent et al., 1993b) or other EFAs for other taxa. We may anticipate that egg quality, fecundity, and the number of viable offspring are the factors to be affected first by deficiency in EFA. The consequence for the premature organism will, in most cases, be ultimately fatal.

Poor outcomes can also be expected for EFA deficiency during the stages when neural tissues, including eyes and brain, are developing. A brain-damaged, blind fish cannot survive in nature or in culture, and the ultimate symptoms or effects of deficiency will be reduced growth, enhanced mortality, and inadequate behavior (Sargent et al., 1993a). An early symptom of DHA deficiency known from many cultures of marine larvae and juveniles is reduced viability to all types of environmental stress (Watanabe, 1993). Malpigmentation is linked to inadequate dietary EFA consumption during early stages of feeding (Fig. 8.7) (Reitan et al., 1994a).

The effect of EFA deficiency on older stages of fish or zooplankton may be more diverse and presumably less drastic than for younger stages. Considering the metabolic function of highly unsaturated EFA in membrane PLs and prostaglandin synthesis, we may anticipate a gradual effect of inadequate EFA nutrition on growth and survival rather than a threshold-type response. Metabolism and animal health are likely to become gradually reduced as the EFA supply becomes insufficient or as the composition of the available EFA (i.e., ratios ω3/ω6 and DHA/EPA) becomes suboptimal relative to the species requirements. The chance that adult marine zooplankton may ever experience quantitative EFA deficiency is probably low, but inadequate composition of their dietary EFA is possible because phytoplankton exhibits both variable and unbalanced fatty acid composition (see Wainman et al., this volume, and below). The shape of the response curve to inadequate EFA nutrition is unknown, but an optimum-type response curve is most likely. This occurs because there are ultimate requirements for both quantita-

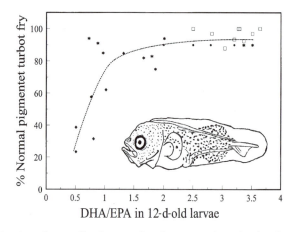

FIGURE 8.7. Fraction of normally pigmented and metamorphosed turbot fry (cultivation day 23–30) as a function of the larval tissue DHA/EPA ratio at day 12 during first feeding. (Data from Reitan et al., 1993 [●], 1994a [◆]; Øie et al., 1997 [□]; and unpublished data [*].)

tive supply and composition of EFA and because highly polyunsaturated EFAs (e.g., DHA) are potentially toxic in high concentrations. As a result of these underlying mechanisms, the optimum is most likely wide for most organisms, in particular for species with high metabolic flexibility or low specific requirements. The optimum range may turn out to be more narrow with respect to reproduction and in the very early larval stages. These speculations are further elaborated below.

8.2.6. Fatty Acid Transport and Metabolism in Food Webs

The literature provides a rapidly accumulating database for lipid and fatty acid metabolism in aquatic organisms and the patterns of fatty acid transport in the lower parts of aquatic food webs. I present some results from marine algae, the rotifer *B. plicatilis,* which is used as live feed for marine larvae worldwide, and farmed Atlantic salmon (*S. salar*). The results are mostly previously unpublished.

8.2.6.1. Algae

Many species of plankton algae store energy as carbohydrates, but some species use lipids or more precisely, TAGs, for storage (Reitan et al., 1994b; Olsen, 1989). This obviously has a great impact on the absolute ω3 fatty acid content of algae. Lipid-storing species are probably found within all taxonomic groups of algae, and most algal species used in marine larviculture seem to be lipid-storing species. Algae are treated in Wainman et al. (this volume), and only some general features of marine species are covered in this section.

8.2.6.1.1. Lipids of Algae

The limiting factor for algal growth is crucial for the absolute lipid content of lipid-storing species, whereas this is of less importance for species that store carbohydrates. Nutrient limitation tends to depress the rate of cell division more extensively than it does of carbon accumulation through photosynthesis, and nutrient limitation should therefore result in algal cells that are rich in carbon storage products, be it lipids or carbohydrates (Reitan et al., 1994b; Olsen, 1989; Shifrin and Chisholm, 1981). Energy (i.e., light) and carbon limitation should instead result in a lower level of storage products (i.e., lipids) in the algal cells, because energy or carbon compounds are limiting cell division.

Lipids in algae normally range from 5 to 70% of dry weight, with 15–30% and 8–13% as the most common ranges for species that tend to store lipids and carbohydrates, respectively. The highest values are representative for severely nutrient-limited lipid-storing species. The fraction of fatty acids is normally in the range of 30–50% of total lipids in algae (Reitan et al., 1994b). The above generalization is only valid for rough comparisons with other organisms or products.

8.2.6.1.2. Essential Fatty Acids of Algae

De novo synthesis of high amounts of long-chain ω3 PUFA, like EPA and DHA, occurs only in algae, in particular marine benthic and planktonic species

(Pohl, 1982). Although some marine animals are able to elongate shorter ω3 and ω6 fatty acids, we know that all ω3 and ω6 double bonds of fatty acids present in marine systems have been synthesized de novo in marine plants or bacteria.

The content of ω3 fatty acids in marine algae is related to the species (genetic) and environmental growth conditions (see also Napolitano, this volume). Values

TABLE 8.1. Content of ω3 fatty acids (percentage of total fatty acids) in species and groups of marine phytoplankton.[a]

	EPS	DHA	Sum ω3
Dinoflagellates (n = 22)			
Mean	6.9	23	45
SD	5.6	6	12
Range	0–20	12–34	26–68
Selected species			
Amphidinium cartiri	20	24	66
Cryptecodinium cohnii (heterotrophic)	0	30	31
Gymnodinium sp.	14	32	68
Diatoms (n = 12)			
Mean	14	1	17
SD	7.6	0.8	7.4
Range	7–30	0–4	10–32
Selected species			
Lauderia borealis	30	0	32
Skeletomema costatum	14	2	18
Chaetoceros sp.	18	4	25
Phaeodactylumt ricornutum	24	4	32
Chrysophytes (n = 13) (covers many taxonomic groups)			
Mean	15	4	30
SD	9.8	4.5	11
Range	0–28	0–17	12–47
Selected species			
Syracosphaera carterae	4	8.6	45
Coccolithus huxley	17	0	20
Isochrysis galbana	0.5	17	35
Pavlova lutheri	26	1	47
Green algae (n = 14) (covers many taxonomic groups)			
Mean	9	0.3	29
SD	12	1	13
Range	0–40	0–4	3–48
Selected species			
Chlorella minutissima	40	0	41
Tetraselmis sp.	10	0.2	28
Nannochloris atomus	15	0.3	22

[a]n, Number of observations included. Data from review by Olsen (1989) and references therein and Reitan et al. (1994b).

compiled from the literature (Table 8.1, references in legend) show that there are pronounced differences in the fatty acid profiles of the various species but also characteristic features for the taxonomic groups. All dinoflagellates contain high levels of DHA, a moderate although variable content of EPA, and a high sum of ω3 fatty acids. The heterotrophic dinoflagellate *Cryptecodinium cohnii* contains a high level of DHA, whereas no other ω3 fatty acid is present in significant amounts (<1%). Among the C18 fatty acids, 18:4ω3 appears to be most common. The diatoms are, by contrast, characterized by a high level of EPA and a low content of other ω3 fatty acids. Some diatom species contain minor quantities of 18:4ω3 and DHA, but EPA is generally prominent. Some chrysophytes exhibit a moderate or high DHA and EPA content, whereas others contain relatively high amounts of short-chain ω3 fatty acids (18:3ω3, 18:4ω3). The variability in ω3 fatty acid content is also pronounced in the green algae. Only small amounts of DHA are found, but some species contain relatively high contents of EPA while having little of shorter ω3 fatty acids. Some green algae exhibit relatively low levels of ω3 fatty acids.

The fatty acid profiles of marine algae are somewhat dependent on the culture conditions. In both marine and freshwater algae, the percentage ω3 fatty acids of total fatty acids tends to be slightly higher in nutrient-sufficient than in nutrient-deficient cells, whereas deficient cells contain a higher fraction of saturated and monounsaturated fatty acids (e.g., 16:0 and 18:1) (Reitan et al., 1994b; Siron et al., 1989). The effect of culture conditions on the fatty acid composition of algae is usually less pronounced than the broad range of fatty acid compositions found among species.

A high percentage ω3 fatty acids in nutrient-sufficient algae is not necessarily equivalent to a higher quantitative content of these fatty acids, at least not in lipid-storing species. This is because of the antagonistic effect of enhanced quantitative ω3 content under nutrient limitation, which tends to counteract the effect of reduced percentage content.

8.2.6.1.3. Conclusion

A general conclusion regarding fatty acid profiles of algae is that dinoflagellates are rich in DHA; diatoms are rich in EPA; chrysophytes show variable ω3 profiles with either 18:3, 18:4, EPA, or DHA as the dominant ω3 fatty acid; and the diverse group of green algae shows even more variable profiles, with 18:3, 18:4, or EPA as the dominant ω3 fatty acids. Some species exhibit a high content of a single ω3 fatty acid in combination with relatively low levels of the others. The high content and purity of DHA in many dinoflagellates and in some chrysophytes are noteworthy. Oils derived from marine animals are generally more complex and balanced.

Changes in the growth conditions (e.g., light intensity, nutrient availability) will most likely affect the level of total lipids more strongly than the fatty acid distribution in lipid-storing algae, whereas the same manipulations will affect the lipid content of carbohydrate-storing species only moderately. Generally, it may be concluded that differences in percentage ω3 fatty acid contents of species reported in the literature reflect genetic differences (or differences resulting from methodological problems, see above) rather than environmental or experimental conditions.

8.2.6.2. Zooplankton

Fatty acid and lipid accumulation by marine zooplankton depends on the type of storage lipid (Fig. 8.3); WEs or TAGs. Zooplankton used as live feed, and normally also fish, store TAGs.

8.2.6.2.1. Lipid Content and Lipid Composition

The relative amounts of PL and TAG in species that store TAGs (TAG-zooplankton) depend mainly on the lipid content, or the nutritional state, of the zooplankton (general scheme in Fig. 8.8). The contents of TAG are highly variable, whereas the quantitative content of PL for given species is believed to be relatively less variable and independent of the nutritional state of the zooplankton (Rainuzzo et al., 1994). The percentage content of both TAG and PL of total lipids will accordingly vary strongly (Fig. 8.8).

Young stages of WE-zooplankton show the same characteristic features as TAG-zooplankton, and very high WE contents in young stages are most unlikely in nature. Older stages may, however, be very rich in WE. This was thoroughly demonstrated by Sargent and coworkers through their classical work on marine copepods in the food web (Sargent, 1989; Sargent and Henderson, 1986). The fact that many important herbivorous zooplankton (e.g., *Calanus finmarchicus*) synthesize WEs has a great impact on the oils that are extracted from planktivorous fish such as capelin, herring, and salmon. The fatty alcohol moiety of WEs in herbivorous zooplankton are normally long-chained and monounsaturated (22:1, 20:1) and are synthesized by the zooplankton themselves. Carnivorous zooplankton may incorporate larger fractions of short-chain saturated fatty alcohol moieties than herbivorous zooplankton.

Fish that feed on WE-zooplankton will oxidize the monounsaturated fatty alcohols, which are then transformed into monounsaturated fatty acids of equal chain length. Long-chain monounsaturated fatty acids are major components of marine oils derived from planktivorous fish. These characteristic fatty acids, which originate from herbivorous WE-zooplankton, may be useful as tracers in food web studies. These fatty acids are also selectively used as catabolic fuel (Olsen and Skjervold, 1991; Sargent and Henderson, 1986). It is noteworthy that the fatty acid metabolism of the zooplankton, and consequently the fatty acid composition at higher levels in the food web, are partly controlled by zooplankton metabolism or genetics.

The lipid content of marine zooplankton is partly dependent on the lipid content of their food. *B. plicatilis* is a typical TAG species that shows optimum growth at relatively high temperature (Olsen et al., 1993a). The impact of the food lipid content on the lipid contents of the rotifer is low in the lower and intermediate

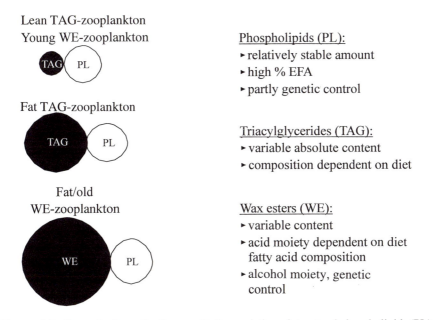

Lean TAG-zooplankton
Young WE-zooplankton

Phospholipids (PL):
- relatively stable amount
- high % EFA
- partly genetic control

Fat TAG-zooplankton

Triacylglycerides (TAG):
- variable absolute content
- composition dependent on diet

Fat/old
WE-zooplankton

Wax esters (WE):
- variable content
- acid moiety dependent on diet fatty acid composition
- alcohol moiety, genetic control

FIGURE 8.8. General scheme for the quantitative variation of structural phospholipids (PL), triacylglycerides (TAG), and wax esters (WE) and general properties of the different lipid components. The circles express the quantitative relationship between the lipid compartments. TAG-zooplankton, zooplankton that use TAGs as energy store; WE-zooplankton, zooplankton that primarily use WE as energy store.

range of food lipid contents (30–200 mg $\cdot$ g^{-1} DW), but the rotifers' lipid level becomes more rapidly enhanced for very fat food (>250 mg $\cdot$ g^{-1} DW; Fig. 8.9).

 B. plicatilis tends to metabolize fat more efficiently under conditions that promote rapid growth (>0.3 $\cdot$ d^{-1}) than under conditions of more severe food limitation (Øie and Olsen, 1997), resulting in a lipid content as low as 70–80 mg $\cdot$ g^{-1} DW at high growth rates. By comparison, low growth rates yield lipid contents of 130–150 mg $\cdot$ g^{-1} DW (Fig. 8.9). Starvation experiments have moreover shown that 50–60 mg lipids $\cdot$ g^{-1} DW appears to be the lowest viable level in our *B. plicatilis* strain.

Starvation experiments have demonstrated that the quantitative PL content of the rotifer is relatively constant and independent of food composition and feeding conditions (shaded area in Fig. 8.9, from Rainuzzo et al., 1994). Fast-growing rotifers may then contain as low as 20–30 mg TAG $\cdot$ g^{-1}DW, whereas slow-growing rotifers exhibit higher levels (Fig. 8.9). This confirms that the general schemes for lean and fat TAG-zooplankton in Figure 8.8 fit *B. plicatilis* well.

8.2.6.2.2. Fatty Acids of TAG-Zooplankton

B. plicatilis is believed to have relatively low requirements for EFA. This is indicated by its ability to maintain rapid growth despite a very low EFA content in

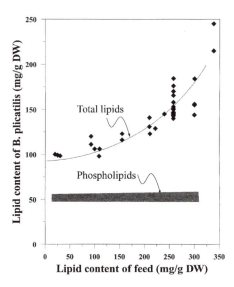

FIGURE 8.9. Lipid contents of *B. plicatilis* as a function of the lipid content of its food. The approximate content of phospholipids in the rotifer has been indicated (shaded area, based on Rainuzzo et al., 1994). Values are valid for rotifer populations growing at relatively low growth rates $(0.05–0.1 \cdot \text{day}^{-1})$ and fixed food composition.

both its food and its tissues. It is crucial that the rotifer contains a satisfactory amount of EFA when used as live food for marine fish larvae. The effort to produce rotifers with high and stable EFA content has been a main challenge in establishing a feasible live food technology for marine fish larvae. An adequate EFA can presently be achieved through appropriate manipulation of the dietary lipids, a method that is commonly denoted the fatty acid or lipid enrichment technique (Dhert et al., 1993; Olsen et al., 1993b; Watanabe et al., 1983). In this regard, the goal is to undertake the EFA enrichment to make rotifers a satisfactory food organism for fish larvae, not to enhance rotifer growth rates.

The fatty acid metabolism of the rotifer is most likely equal to that of other species of TAG-zooplankton. However, *B. plicatilis* has low EFA requirements in contrast to many zooplankton species living in cold waters. This difference is expressed by the pattern of catabolic degradation of EFA that has been assimilated in its tissues, and in particular, that of DHA. It is well documented that *B. plicatilis,* as well as *Artemia franciscana,* selectively catabolize DHA during starvation (Evjemo et al., 1997; Dhert et al., 1993; Olsen et al., 1993a). Many other zooplankton species (J.O. Evjemo, unpublished results), as well as larvae (Rainuzzo, 1993) and adult fish (Olsen and Skjervold, 1991) showing high DHA requirements, will selectively retain DHA during starvation.

The fatty acid distribution of TAG-zooplankton tissues is a dynamic variable that is highly dependent of the immediate consumption of dietary EFA. Synthesis of fatty alcohols in WE-zooplankton will, to some extent, disturb the relationship for this group. The kinetics of EPA and DHA accumulation in rotifers following changes in dietary lipid composition are illustrated in Figure 8.10. The rotifers used in the experiment were cultured for more than five generations with Baker's yeast, which had a very low EFA level (Y-rotifers). Their initial ω3 fatty acid content was 1.5% of total fatty acids, with undetectable quantities of EPA and

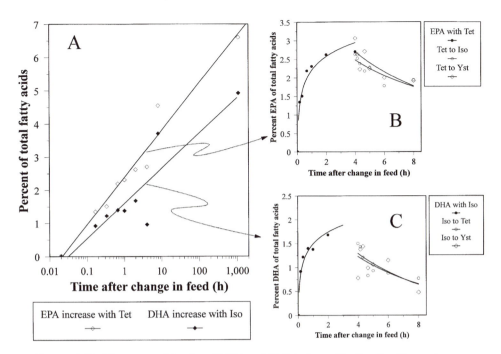

FIGURE 8.10. Time course kinetics of DHA and EPA in the rotifer *B. plicatilis* on changes in dietary lipids. (A) Increase in DHA and EPA with time after replacing Baker's yeast (Yst) with *I. galbana* or *Tetraselmis* sp., respectively, as food. (B) Restoration of EPA after changing back from *Tetraselmis* sp. (4-h value from A) to Baker's yeast and *I. galbana*. (C) Restoration of DHA after changing back from *I. galbana* (4-h value from A) to Baker's yeast or *Tetraselmis* sp.

DHA. When these cultures were given the algae *Isochrysis galbana* or *Tetraselmis* sp. instead of Baker's yeast, the fatty acid distribution of the rotifers showed rapid changes (Fig. 8.10).

This strain of *I. galbana* contained 5–6% DHA (of total fatty acids) and undetectable quantities of EPA. The Y-rotifers fed this algae showed a logarithmic increase in DHA content for more than 40 successive days during sustained feeding (Fig. 8.10A), whereas no change in EPA could be detected during the first day (EPA not shown). More than 60% of the change in percentage DHA of total fatty acids took place during the first day, and the rate of increase decreased gradually after the first half-hour of incubation.

The present strain of *Tetraselmis* sp. contained 8–10% EPA and no detectable quantities of DHA. Y-rotifers offered this algae showed a logarithmic increase in percentage EPA for more than 40 d, whereas no changes in DHA took place during the first day (Fig. 8.10A, DHA not shown). Also in this case, more than 60% of the changes in percentage EPA took place during day 1.

The steeper slope of the EPA curve compared with the DHA curve (Fig. 8.10A) is noteworthy, and it probably reflects the fact that the percentage EPA in *Tetra-*

selmis sp. is higher than the percentage DHA in *I. galbana*. In fact, absolute DHA contents of *I. galbana* were two to three times higher than EPA contents of *Tetraselmis* sp. This clearly shows that the relative fatty acid distribution of the food algae has a greater impact on the rate of change in fatty acid distribution than the absolute fatty acid content (Olsen et al., 1993b). The catabolic rate of DHA degradation in the rotifers may regrettably also have been slightly higher than the catabolic rate of EPA (see above), but this is believed to be less important.

Rotifers fed *I. galbana* for 4 h (lower curve, Fig. 8.10A) showed a logarithmic reduction in the percentage DHA content when given Baker's yeast or *Tetraselmis* sp. (Fig. 8.10C). By contrast, rotifers given *Tetraselmis* sp. for 4 h (upper curve, Fig. 8.10A) showed a logarithmic reduction in percentage EPA when fed with Baker's yeast or *I. galbana* for another 4 h (Fig. 8.10B). The two types of feed used during the second stage differed with respect to EFA contents but gave the same time course for EPA or DHA reduction. This illustrates how dietary EPA and DHA contents, rather than the type or taxonomic class of the food, are decisive in determining the fatty acid profile of rotifers. This implies that any metabolic conversion of DHA and EPA, which are believed to be among the most labile fatty acids of the rotifer tissues (Olsen et al., 1993a), is insignificant compared with the net changes in fatty acid distribution taking place when the food is changed.

The above dynamic response of the fatty acid composition of rotifer tissues following changes in fatty acid composition of the diet is probably typical for TAG-zooplankton, including freshwater species. The change in different fatty acids, expressed in terms of percentage of total fatty acids, can be interpreted as a continuous dilution of the fatty acids contained in zooplankton tissues due to the constant consumption of new dietary fatty acids.

If no selective catabolism or anabolism of fatty acids takes place in the rotifer tissues, we can anticipate that their fatty acid composition will become very close to that of their feed, provided that the rotifers are cultivated for a long time on a fixed diet, exceeding the lifetime of the individual rotifer (i.e., more than five doublings of biomass or 15 d). The fatty acid compositions of dietary lipids and rotifers, grown for more than five generations on a fixed diet, are indeed very closely related, but the fatty acid profiles are not identical (Fig. 8.11). The DHA content is lower in rotifers than in the food, whereas the content of docosapentaenoic acid (DPA) (22:5ω3) is slightly higher. The differences found in shorter ω3 fatty acids and ω6 fatty acids are relatively minor. The distribution of monounsaturated fatty acids suggests reduced levels of the long-chain fatty acids 20:1 and 22:1 and increased level of the short-chain 16:1 and 18:1 fatty acids in rotifers compared with their food. Saturated fatty acids were generally found in slightly higher proportions in the rotifers than in the diet. The data in Figure 8.11 therefore suggest that *B. plicatilis* selectively catabolizes DHA into DPA (and EPA?). It also catabolizes the long-chained monounsaturated fatty acids 20:1 and 22:1, considered to be important fuel in the marine food web (Sargent and Henderson, 1986), at a slightly higher rate than most other fatty acids. No indications of chain elongation have been found for this strain *of B. plicatilis,* which is not surprising considering the EFA-rich diets used to cultivate the rotifers.

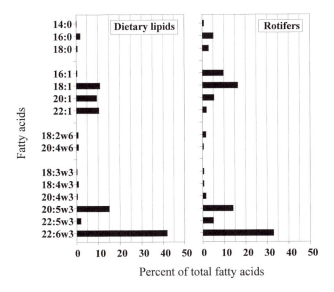

FIGURE 8.11. Fatty acid profiles (% of total fatty acids) of *B. plicatilis* and its food. The rotifers grew for more than five biomass doublings on fixed food (i.e., dietary lipids), and equilibrium in their fatty acid profiles was then assumed to occur.

The above relationship between dietary and tissue fatty acid composition of *B. plicatilis* is valid over a wide range of EFA food levels, as illustrated by total ω3 and ω6 contents depicted in Figure 8.12. The percentage ω3 of rotifers grown to equilibrium with a fixed food supply (see above) increases linearly with the percentage content in the food up to about 60% ω3 fatty acids (Fig. 8.12A). For even higher dietary ω3 levels, the content in the rotifers remained nearly constant. In agreement with Figure 8.11, the ω3 value of the rotifers was slightly lower than that of the food, which is caused primarily by their faster DHA catabolism (Olsen et al., 1993a,b). The highest levels of n-3 fatty acids shown in Figure 8.12A are hardly found in nature.

The pattern of variation is somewhat different for the ω6 fatty acids. They seem to accumulate in slightly higher fractions in the rotifers than in the feed (Fig. 8.12B). Most values in the figure are above unity, and some values are well above. In any event, the positive relationship between dietary and tissues levels is apparent also for ω6 fatty acids. The principal pattern of variation (Fig. 8.12) is similar for all individual fatty acids of *B. plicatilis*. There is a positive relationship between the dietary fatty acids and fatty acid contents of the rotifers, but each fatty acid has its own characteristic response (i.e., slope of curve).

I suggest that EFA kinetics of TAG-zooplankton is generally comparable with that of *B. plicatilis*. Differences in EFA requirements will, however, most probably affect the responses of the individual EFA (Figs. 8.10–8.12). Species exhibiting, for example, high DHA requirements will probably tend to retain DHA when the dietary supply of DHA is low. We may also expect that species capable of

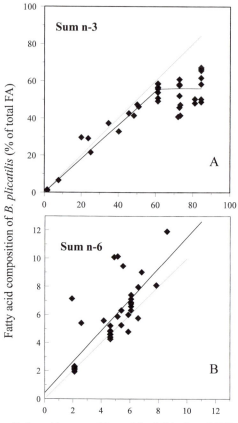

FIGURE 8.12. Percentage ω3 (A) and ω6 (B) of total fatty acids of *B. plicatilis* as a function of the respective percentage ω3 and ω6 in the rotifer feed (rotifers grown at equilibrium). The dotted line is the 1:1 line, whereas the solid curves are regression lines. Equations of regression lines are (A) $\%\omega3_{rotifer} = 0.91 \cdot \%\omega3_{feed}$ (range, 0–62%ω3; $r^2 = 0.93$); (B) $\%\omega6_{rotifer} = 1.11 \cdot \%\omega6_{feed} + 0.4$ (range, 0–8%ω6; $r^2 = 0.63$).

elongating EFA will respond to deficiency by chain elongation and desaturation of short-chain EFA. Both selective retention and chain elongation will, to some extent, affect the kinetic patterns of the individual EFAs shown in Figures 8.10–8.12. The patterns may therefore be slightly different for TAG species characterized by higher EFA requirements than for *B. plicatilis*.

8.2.6.3. Fish

Lipids and fatty acid kinetics of marine fishes are exemplified through data for Atlantic salmon, an andronomous species. Most fish species store TAGs either in the flesh or in the liver, and a strong relationship between animal lipid composition and dietary lipids is clearly evident (see below). The main difference between salmon, which is adapted to cold water, and *B. plicatilis,* which is genetically adapted to warm waters, is the higher level of PUFA in salmon PLs than in rotifer PLs. My working hypothesis is that the fatty acids of salmon will vary with dietary and environmental conditions basically in the same way as the fatty acids in TAG-zooplankton adapted to cold water.

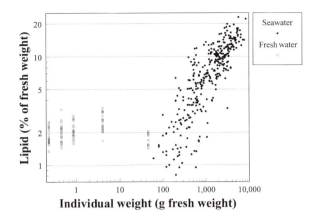

FIGURE 8.13. Lipid content of farmed Atlantic salmon as a function of the age or the weight of individual fish. Values pertain to newly hatched larvae in fresh water, through the stages from transfer from fresh water to seawater and production and slaughtering sizes. Food was maintained at a constant composition throughout the life stage of the salmon.

8.2.6.3.1. Lipid

Figure 8.13 illustrates how lipids in Atlantic salmon, which store the TAGs in its flesh, may vary during its lifetime, from hatching in fresh water through transfer to seawater cages until slaughtering at a weight of 3–7 kg. The lipid content of juvenile salmon increases slightly after hatching in April, reaching a level of 2–3% of fresh weight by late summer. From late summer to January, juveniles show a reduction in lipid contents and reach a lipid level by the early winter that seems to be representative for that of a fish transferred to seawater in late spring.

The lipid content of small individuals just transferred to seawater was variable. This is not surprising in the light of the fact that the juvenile fish originated from several producers of juvenile salmon, sampled from nine fish farms along the Norwegian coastline (Olsen and Skjervold, 1995). Fish were given food with identical composition from the same manufacturer all through the production cycle. The feed was based on fish meal and fish oil mainly derived from typical planktivorous fish species, most likely capelin and herring. The lipid content of the feed was 22% of dry weight or 5–6% of fresh weight (assuming dry weight to be 25% of fresh weight). This lipid level is lower than that normally found in marine fishes, which store TAGs in the flesh (Table 8.2).

The lipid content of the farmed salmon flesh increased steadily through their 2-year cycle in seawater, with the highest rate of increase occurring in the first year when salmon weigh about 1 kg. The present growth of farmed salmon is faster than that in 1988–1989 when the experiments were carried out. The lipid content reaches a typical average of 13% of fresh weight at the time of slaughtering, with pronounced variations among individuals. The individual variation in salmon lipid contents is most likely coupled to its life cycle, considering that the feed has been supplied in excess through the production cycle. Although domesticated, we may anticipate that farmed Atlantic salmon do prepare for their migration up rivers to

TABLE 8.2. Lipid content and percentage EFA of total fatty acids in marine species.[a]

	Lipid, % of fresh weight	%EPA	%DHA	%AA	%Sum ω3	%Sum ω6	ω3/ω6
Atlantic salmon (farmed) (*Salmo salar*)	14–23	5.6	9.0	0.4	20.4	5.2	3.9
Trout (farmed) (*Oncorhynchus mykiss*)	10	4.6	12.8	0.4	22.4	4.6	4.9
Mackerel (*Scomber scombus*)	3–30	5.7	13.7	0.5	25.0	2.5	10.0
Herring (*Clupea harengus*)	14	9.4	9.9	0.3	24.8	1.7	14.6
Sprat (*Sprattus sprattus*)	18	6.1	12.9	0.2	24.2	1.6	15.1
Atlantic halibut (*Hippoglossus hippoglossus*)	10	2.7	2.9	0.4	7.7	2.1	3.7
Greenland halibut (*Reinhardtius hippoglossoides*)	13	3.0	4.2	0.3	9.5	1.6	5.9
Turbot (*Scophthalmus maximus*)	2.4	8.1	18.6	1.3	35.9	6.2	5.8
Plaice (*Pleuronectes platessa*)	1.4	17.0	10.4	6.6	34.3	8.2	4.2
Wolf-fish (*Anarhichas* sp.)	2.5	10.8	10.1	4.4	26.7	6.5	4.1
Redfish (*Sebastes* sp.)	2.8	8.3	11.3	0.5	24.2	2.3	10.5
Anglerfish (*Lophius piscatorius*)	0.1	6.9	34.2	4.3	43.0	5.6	7.7
Cod (*Gadus morhua* L)	0.3	14.5	36.8	1.8	54.3	3.1	17.5
Saithe (*Pollachius virens*)	0.3	10.7	29.0	1.9	43.6	3.6	12.1
Ling (*Molva molva*)	0.2	8.3	41.6	3.3	53.0	4.1	12.9
Pollack (*Pollachius pollachius*)	0.2	11.6	45.0	2.3	59.0	3.2	18.4
Haddock (*Melanogrammus aeglefinus*)	0.2	16.1	31.4	4.6	50.3	5.7	8.8
Tusk (*Brosme brosme*)	0.2	6.3	41.6	2.4	50.3	3.5	14.4
Blue ling (*Molva byrkelange*)	0.1	8.5	30.8	2.4	42.5	3.6	11.8
Sole (*Solea solea*)	0.5	5.9	16.2	3.0	29.2	4.3	6.8
Tuna (*Thunnus thunnus*)	0.2	3.6	26.9	5.8	32.9	8.2	4.0
Picked dogfish (*Squalus acanthias*)	6.4	8.0	18.3	2.8	31.4	4.8	6.5
Skate (*Raja batis*)	0.2	5.2	32.7	5.1	44.0	6.5	6.8
Crab (*Cancer pagurus*)	1.8	15.6	16.5	2.1	38.5	4.0	9.6
Deep-water prawn (*Pandalus borealis*)	0.8	17.4	16.0	1.2	35.3	2.4	14.7
Lobster (*Homarus vulgaris*)	0.6	19.1	12.8	4.0	34.0	6.1	5.6
Scallops (*Pecten maximus*)	0.4	20.4	21.8	1.3	50.5	2.4	21.0

[a]Data from *Facts about Fish*, information folder prepared by Directorate of Fisheries, P.O. Box 185, N-5002 Bergen, Norway, and the Norwegian Seafood Export Council, Skippergt. 35/39, N-9005 Tromsø, Norway.

spawn. Migration and spawning are highly energy-requiring activities, and it is known that salmon hardly feed during migrations and that their lipid level when caught in the rivers is considerably lower than that given in Figure 8.13 (3–4% of fresh weight) (Olsen and Skjervold, 1991). The lipid level of adult salmon in the sea is most likely higher than in most other species. The pattern of changes in salmon lipids with age, including the reduction before spawning (Fig. 8.6), may be general and valid also for freshwater fish, which store lipids in the flesh.

The absolute structural lipid content of salmon flesh, which mainly consists of membrane PLs and other membrane constituents, is believed to be rather constant, as was also shown for the rotifer (Fig. 8.9). Figure 8.13 suggests that the lower threshold for total flesh lipids is in the range of 1–1.5% of fresh weight. This level is of the same magnitude as the quantitative level of PLs found in salmon flesh, some 1% of fresh weight in adult individuals (Ø. Lie, personal communication). This implies that the salmon juveniles transferred to seawater (see Fig. 8.13) contained very low, although variable, contents of TAGs. Thus the energy stores left after the energy-requiring processes of smoltification and adaptation to seawater were very low. The other implication, further elaborated below, is that the fatty acid distribution in the salmon at the time of transfer to seawater reflected the composition of the salmon membrane PLs. Tissue TAGs, however, were reflected in older salmon in which the PLs constituted only a minor fraction of total lipids.

8.2.6.3.2. Fatty Acid Composition

When salmon are given the same food and dietary lipids throughout their life, as much as 80% of the variation in their ω3 fatty acid contents can be explained by the variation in only two factors: the lipid content of the flesh and the body weight of the fish, which relates to the age (Olsen and Skjervold, 1995). The effects of light levels and seawater temperature on ω3 fatty acid contents, as represented by the latitude survey of salmon along the Norwegian coast, explained <1% of the variation in ω3 fatty acids. Although not further dealt with here, it should be noted that salmon respond to changes in dietary lipid composition in a fashion similar to that of *B. plicatilis*.

The pattern of variation found for the quantitative content of the dominant EFAs (EPA and DHA) through the salmon life cycle are, as expected, similar to the pattern of variation found for total lipids (Fig. 8.14A). DHA is most dominant during the early life stages in fresh water and during the first part of the seawater stage. The average quantitative contents of EPA and DHA at the time of slaughtering (weight >3 kg) are about 7 and 12 mg $\cdot$ g^{-1} fresh weight, respectively. The corresponding total ω3 and ω6 fatty acid contents are 22 and 5.5 mg $\cdot$ g^{-1} fresh weight, respectively, with linoleic acid as the dominant ω6 fatty acid. Thus, DHA and EPA presumably make up 80–90% of the ω3 fatty acids in salmon flesh, and ω3 fatty acids dominate over ω6 fatty acids by a factor of 4.

The percentage content of EPA and DHA in salmon flesh during the life cycle clearly demonstrates the strong dominance of DHA through the early developmental stages (Fig. 8.14B). In fact, DHA makes up 20–50% of total fatty acids all through the juvenile stage in fresh water and through the first few months of

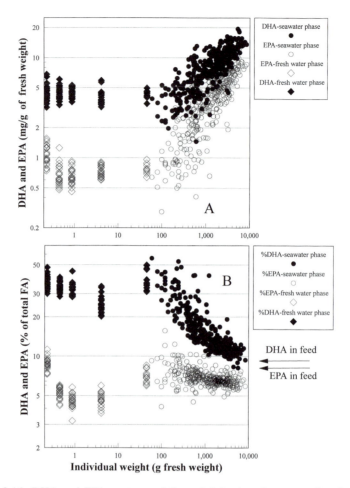

FIGURE 8.14. DHA and EPA contents of farmed Atlantic salmon as a function of the weight of individual fish. (A) Absolute fatty acid contents expressed as milligrams fatty acids per gram of fresh weight. (B) Relative fatty acid contents expressed as percentage fatty acids of total fatty acids (for details, see legend of Fig. 8.13).

growth in seawater. The major conclusion is that the percentage DHA in the salmon flesh is gradually approaching the value of the feed as the salmon grows in seawater, but that the average DHA content remains about 3% (of total fatty acids) higher in the salmon than in the feed (arrow, Fig. 8.14B). The difference between *B. plicatilis* and its food DHA is of the same magnitude but inverse of that for salmon (Fig. 8.11). This, I presume, is a typical metabolic difference between cold-blooded animals, which are genetically adapted to cold (salmon) or to warm (rotifer) conditions. The rotifer will catabolize DHA selectively even at low temperatures, whereas salmon will probably not do this at any temperature.

In contrast to DHA, the percentage EPA of salmon flesh is stabilized at a level slightly below the percentage level in the food. In fact, it becomes gradually

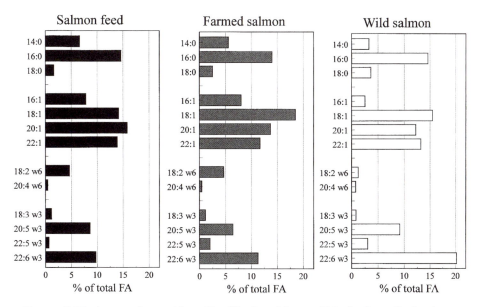

FIGURE 8.15. Average fatty acid profiles (% of total fatty acids) of salmon feed, salmon individuals >3 kg (marketable), and wild, starved Atlantic salmon males caught in Norwegian rivers during autumn.

reduced as the salmon grows in seawater. This reduction, corresponding to about a 2% (of total fatty acids) difference between big and small salmon, may originate from selective EPA catabolism or chain elongation and desaturation of EPA to DHA, eventually through shorter ω3 fatty acids (Sargent et al., 1993b). This explanation is, however, speculative.

The average fatty acid profile of adult salmon (>3 kg), its feed, and for curiosity, wild salmon caught in Norwegian rivers can be compared with respect to EFA (Fig. 8.15). The relationship between monounsaturated fatty acids in salmon and their feed exhibits exactly the same pattern of variation as in the rotifers; the percentage long-chain monounsaturated 20:1 and 22:1 of total fatty acids is lower in salmon than in the feed, 18:1 is more abundant in the salmon, and 16:1 is found in identical fractions (see Fig. 8.11 for comparison). The percentage content of individual saturated fatty acids and ω6 fatty acids is almost identical in food and fish. With regard to ω3 fatty acids, not mentioned above, both salmon and rotifers exhibit slightly enhanced percentage 22:5 ω3 (DPA) of total fatty acids relative to the food.

Wild salmon, which were much leaner than farmed salmon (3–4% lipids of fresh weight), exhibit similar fatty acid distributions. The DHA level was, however, significantly higher in wild versus farmed salmon, whereas the level of linoleic acid (18:2ω6) was lower (Fig. 8.15). This is to be expected, because the percentage DHA of a lean farmed salmon, containing 3–4% lipids (of fresh weight), is also expected to be some 20% of total fatty acids (see below). The enhanced level of linoleic acid in the farmed salmon, however, originates from plant oils added to the salmon feed in trace amounts (Fig. 8.15). This affects the ω3/ω6 ratio of the salmon relatively strongly, and it demonstrates the effect that a minor fraction of dietary lipids of terrestrial origin, common in food of freshwater fishes, can have on the fatty acid composition of fish (Ahlgren et al., 1994; Henderson and Tocher, 1987).

Fish lipid content was the second factor that contributed to the variation in EFA in salmon (Olsen and Skjervold, 1995). It is not surprising that the absolute contents of EPA and DHA are both positively related to the lipid content of the

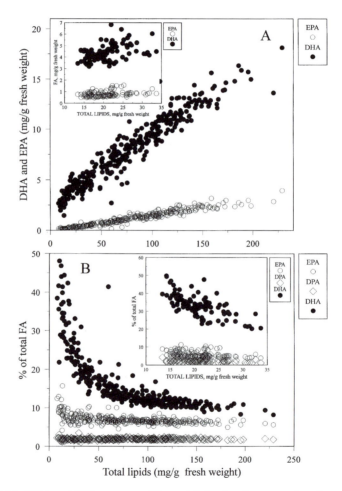

FIGURE 8.16. DHA and EPA contents of farmed Atlantic salmon as a function of the lipid content of the individual fishes. The insets show data for juveniles in fresh water. (A) Absolute fatty acid contents expressed as milligrams fatty acids per gram of fresh weight. (B) Relative fatty acid contents expressed as percentage fatty acids of total fatty acids (DPA also shown, see details in legend of Fig. 8.13).

salmon (Fig. 8.16A). Values obtained for juvenile salmon in the freshwater stage (see inset, Fig. 8.16A) fit well with values obtained for older stages in seawater. The inverse relationship between percentage ω3 EFA and lipid content is also quite evident (Fig. 8.16B). The percentage content of 22:5ω3 (DPA) is almost constant and independent of the lipid content of the salmon, the percentage EPA shows a minor increase in very lean salmon (less than two times), and relative DHA abundance increases more than four times in a pronounced and nonlinear fashion as the lipid content of the salmon decreases, particularly when the lipid content is <50 mg · g^{-1} fresh weight. Again, values for juveniles fit well into the general relationship for older stages (see inset, Fig. 8.16B).

The pattern of variation shown in Figure 8.16B, characterized by an inverse relationship between percentage EFA and lipid content (cf. Ahlgren et al., 1994), can be understood on the basis of the compartment scheme for TAG-zooplankton (Fig. 8.8). TAGs and PLs are the main lipid compartments in salmon flesh. The PLs contain higher fractions of PUFA than the TAGs, and their composition is, as discussed above, to some extent genetically controlled. Assuming that salmon PLs constitute 10 mg · g^{-1} fresh weight, it may be deduced that PLs will be the dominant lipids in salmon with very low total lipid content (Fig. 8.13; see also lean TAG-zooplankton case, Fig. 8.8). The fatty acid composition obtained for very lean salmon (Fig. 8.16B) will then primarily reflect the fatty acid composition of its PLs. This again implies that DHA is the dominant PUFA of the salmon membranes and that 30–40% of the individual fatty acids of the cell membranes are DHA. This also explains why percentage DHA tends to increase in starving marine fish, fish larvae, and zooplankton (i.e., in animals with high EFA requirements).

In contrast to lean salmon, fat salmon may be characterized by a dominance of TAGs (Fig. 8.13; see also fat TAG-zooplankton case, Fig. 8.8), and the fatty acid composition of fat salmon will therefore mainly reflect the fatty acid composition of the salmon TAGs. The impact of PLs on total flesh fatty acid composition will be 4–8% in salmon >3 kg. The slightly enhanced DHA levels in large salmon, compared with the percentage content in the food (Figs. 8.14 and 8.15), can therefore be explained solely by the DHA contribution of the cell membrane PLs. This implies that the percentage DHA of salmon TAGs becomes almost identical to that of the feed.

The flesh lipid contents of marine fish differ strongly, primarily as a result of how the fish store fat. Species that accumulate lipid in the muscle, such as salmon, may show very high contents of flesh lipids after periods of active food consumption. However, species that store lipid in the liver exhibit flesh lipid levels of the same magnitude as the typical PL levels of their muscles. Table 8.2 reviews lipid and EFA contents of some marine fish species. The ranges of variation in lipids as well as in percentage fatty acids are quite pronounced. If the percentage EFA content of the diverse fish species presented in Table 8.2, among which not all are recognized as cold-water species, is plotted as a function of their lipid content (Fig. 8.17), we obtain more or less the same relationship as that for salmon (Fig. 8.16).

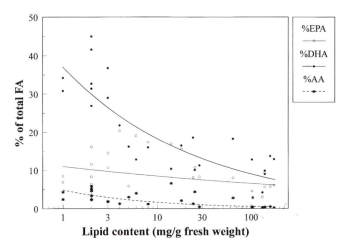

FIGURE 8.17. Percentage DHA, EPA, and AA as a function of flesh lipid content in selected marine fish species, including species that store lipids in the liver and species that store lipids in the flesh (see Table 8.2; data obtained from the Norwegian Seafood Export Council).

8.2.6.4. General Conclusions

Studies of EFA dynamics in zooplankton and fish must be based on the understanding that lipids are a diverse group of molecules with very different properties and metabolic functions. TAGs serve primarily as a source of metabolic energy generating ATP for respiration and anabolic reactions. PLs, which have the highest content of PUFA, are primarily structural and fundamentally important for transport across membranes and enzymatic reactions in the cell. They also supply fatty acid precursors for prostaglandin synthesis. The fish and zooplankton TAGs are characterized by dominance of dietary fatty acids, whereas the composition of PLs is subjected to some, but not complete, genetic control by the fact that the enzymes that form PLs select for PUFA. The absolute content of PL in flesh tissues is relatively constant, albeit species-dependent. Very lean specimens may exhibit a fatty acid composition in their flesh which primarily reflects their PLs. The fatty acid composition of fat individuals will, however, to a greater extent reflect the composition of their tissue TAGs and in turn their diet. General compartment models, sensitivity to considerations of the species-specific evolutionary adaptations to temperature, and the strategy of lipid deposition are important to consider to understand lipid and EFA dynamics of fish and zooplankton, be they freshwater or marine species.

8.2.7. Relevance of Mariculture Research

This section sums up some of the evaluations and conclusions made above. The experience in mariculture has clearly shown that major concern for analytical methods and intercalibration measures is important for the scientific progress. Moreover, rigorously detailing of the exact units used to express published fatty acid values is very important to avoid misunderstandings.

Research in mariculture has contributed significantly to the physiological understanding of EFA deficiency in aquatic animals, including the general symptoms of deficiency, and to the establishment of methods to assess EFA deficiency. If used with some care, these methods and the general base of knowledge are relevant for freshwater organisms as well.

EFAs are needed for growth, efficient membrane transport, and synthesis of prostaglandins that regulate metabolism. Fecundity and the number of viable offspring are especially sensitive to EFA deficiency. This may be because EFA deficiency is critical during stages when neural tissues, including eyes and brain, are developing. Other ultimate symptoms or effects of EFA deficiency are reduced growth rate, enhanced mortality, reduced viability to environmental stress, and inadequate behavior.

Identification of the important physiological role of DHA is another contribution from mariculture that is important for freshwater studies. Marine species show two distinct patterns of DHA metabolism; some species exhibit lower catabolism of DHA than of all other fatty acids during starvation, whereas other species tend to catabolize DHA much faster than all other fatty acids. It is postulated that these different patterns in DHA metabolism characterize species evolutionarily adapted to low and high temperatures, respectively. This hypothesis should be further elaborated both for marine and freshwater animals.

The species metabolic flexibility to elongate and desaturate short EFA is decisive for their dietary EFA requirements. The dietary EFA requirements of a species will, for example, be lower if it is capable to synthesize DHA from 18:3ω3. It is an interesting question if there is a systematic difference in the ability to elongate and desaturate short EFA moieties between freshwater and marine species. Otherwise, there is no reason to believe that EFA requirements of freshwater fish are different from that of marine fish species.

The relationship between dietary fatty acid composition and animal fatty acid composition has been very well described in mariculture research through use of model organisms of algae, zooplankton, and fish. It is important that studies of EFA dynamics in aquatic animals must be based on the understanding that TAG and WE primarily serve as sources of metabolic energy, whereas PLs, rich in PUFA, are structural lipids important for functions such as cellular transports, enzymatic reactions, and prostaglandin synthesis. The fatty acid composition of TAG is mainly determined by the dietary fatty acids, whereas the composition of PL is controlled by enzymes that select more strongly for PUFA (i.e., partly genetic control). It is in this regard important to recognize that very lean specimens will exhibit total fatty acid composition, which reflects the composition of their PL, whereas fat individuals will exhibit compositions close to their TG or WE. This is frequently overlooked in ecological studies. General compartment models with structural and storage lipids included, sensitivity to considerations of the species-specific evolutionary adaptations to temperature, and the strategy of lipid deposition are important factors to consider to understand lipid and EFA dynamics of both freshwater or marine species.

8.2.8. Evaluation of Ecological Effects of Essential Fatty Acids

Without doubt, aquatic animals must be supplied with specific EFAs of both the $\omega3$ and the $\omega6$ families to reproduce and grow. It also appears that the fatty acids with longest chain length and number of double bonds, such as AA, DHA, and EPA, are more essential than the C18 moieties and that most animals are able to catabolize the longer moieties within both families into the shorter C18 moieties. Our quantitative understanding of animals' EFA requirements and metabolism, including their flexibility to modify EFAs through anabolic reactions, is insufficient. This lack of knowledge also applies to the general range of variation in physiological requirements between species of zooplankton and fish, both with regard to the absolute EFA amounts and the relative composition of EFAs.

I suggest that the actual and potential ecological impacts of EFA deficiency must be evaluated on the basis of a sound knowledge of the physiological function of EFA in animals and the general physiological symptoms of EFA deficiency mentioned above. One should not evaluate EFA deficiency on the basis of the general idea that EFAs are just other types of elemental nutrients such as carbon (or energy), phosphorus, or nitrogen. This is because animals may suffer from EFA deficiency, which is expected to result in a reduced rate of reproduction, even though phosphorus or nitrogen strictly limit the biomass of the population. In other words, EFA may affect the intrinsic growth and reproduction rates of animal populations but never the carrying capacity of the system, which must be controlled by a biologically conservative factor (e.g., phosphorus cannot be produced by organisms). The most obvious effect of EFA deficiency on zooplankton and fish communities is therefore likely to be an alteration of community structure (i.e., species composition).

The relevance of EFA deficiency in natural communities will, however, depend on

• Whether the EFA supplied in the food is clearly in excess relative to carbon, energy, and other elemental supplies
• Whether there are pronounced differences in EFA requirements between species of zooplankton and fish

Regarding the first question, we know that phytoplankton species are characterized by having few dominant fatty acids in relatively high concentrations compared with animals (Table 8.1, Napolitano, this volume). We also know that algae generally contain high levels of EFA, although there are some green algae and cyanobacteria that appear to have low contents (Ahlgren et al., 1992; Olsen, 1989; Pohl, 1982). This implies that situations in which one phytoplankton species is entirely dominant for long periods of time represent a potential threat of EFA deficiency for herbivorous zooplankton, especially if the phytoplankton

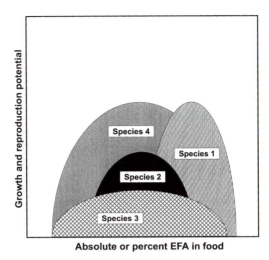

FIGURE 8.18. Growth and reproduction potential of hypothetical animal species as a function of their EFA availability, absolute values of EFA, or relative distribution.

species is characterized by low EFA levels or unbalanced composition of individual EFAs. Similar general evaluations can hardly be made for animals on higher trophic levels.

The second question is probably the most crucial. Based on a knowledge of the physiological functions of EFA and the known effects of EFA deficiency, it may be suggested that the biological response curve to EFA deficiency is smooth and probably not a threshold curve but rather a typical optimum curve. In Figure 8.18, a situation with four model animal species, either zooplankton or fish, is illustrated. Relatively strong effects might occur if the food of these species were to exhibit highly variable EFA levels. The X-axis expresses either the absolute amount of EFA or the relative EFA composition, and the Y-axis shows specific growth and reproduction rates, which are strongly affected by suboptimal EFA supply. The more general effect parameters, represented by the term *state of health* or *physiological state,* should in this regard be understood as equivalent to growth and reproduction rates.

Figure 8.18 shows that model species 1 and 2 have relatively high requirements and low metabolic flexibility of EFA elongation. These species will stay healthy and grow fast when EFA is abundant. This is, in fact, crucial for these species becoming dominant components of the community. Species 1 has a high growth potential, but this will become expressed only during very high EFA supply in the food. Species 3 and 4 are characterized by low EFA requirements, including high metabolic flexibility. These species will stay healthy and grow relatively fast under low EFA supplies, and they will also compete well under conditions of high EFA supply. Species 4 has a high growth potential, and this will become expressed for a wide range of EFA supplies. Species 3 is the typically slow-growing omnivorous species with low EFA requirements. It is ubiquitous in nature, although not very abundant.

The situation in Figure 8.18 is purely speculative. It illustrates, however, that the ecological impact of EFA deficiency may be pronounced under the conditions assumed above. The assumption that EFA deficiency affects reproduction and survival of early developmental stages is robust. It is also obvious that species that show reduced reproduction and growth will be less abundant compared with other species. What is then left for further elaboration and speculation is the extent to which that EFA requirements vary between animal species.

8.3. Concluding Remarks

Research on EFA requirements of cultivated marine fish has been extensive during the past decade, and the present chapter has highlighted some lessons learned concerning fatty acid metabolism and trophic transport and metabolism. There are major similarities between zooplankton and fish that store TAGs, and species-specific evolutionary adaptation to temperature is suggested to be important for their EFA requirements. Two different strategies are evident in the way that animals metabolize DHA, which is by far the most important fatty acid during physiological adaptation to low temperature. Many marine zooplankton species store WEs, and this will strongly affect both metabolism and transport of fatty acids and lipids. These species are very important food for harvested commercial species but not in mariculture. They have therefore not been treated here.

Another lesson learned is that studies of EFA dynamics of zooplankton and fish must be based on the understanding that PLs of cell membranes have important structural and metabolic functions, whereas TAGs and WEs are primarily energy depots. Sophisticated metabolic studies must involve fatty acid composition of individual PLs of the membranes. The general compartment model, constituting pools of metabolic active lipids and metabolic passive lipids, is important for interpreting EFA dynamics and contents of all animals.

Data for total contents of ω3 and ω6 fatty acids, which represent the two families of EFAs, cannot adequately express food quality, at least not for marine cold-blooded animals. It is therefore necessary to consider individual EFAs within the two families. This is another important lesson learned in mariculture. It will probably be adequate to include the C20 and C22 EFAs in studies of marine organisms, but one can still not take for granted that C18 moieties are unimportant for freshwater organisms with high metabolic flexibility to elongate short EFA.

Acknowledgments. The late Professor Harald Skjervold challenged me, in our joint research on ω3 fatty acids, to explore the space between human health and mariculture. I will always be grateful for his great inspiration. I also thank my colleagues from the university and SINTEF who have contributed to making this chapter possible.

References

Ahlgren, G.; Blomqvist, P.; Boberg, M.; Gustavsson, I.B. Fatty acid content of the dorsal muscle—an indicator of fat quality in freshwater fish. J. Fish. Biol. 45:131–157; 1994.

Ahlgren, G.; Gustavsson, I.B.; Boberg, M. Fatty acid content and chemical composition of freshwater algae. J. Phycol. 28:37–50; 1992.

Bell, J.G.; McVicar, A.H.; Park, M.T.; Sargent, J.R. Effects of high dietary linoleic acid on fatty acid compositions of individual phospholipids from tissues of Atlantic salmon (*Salmo salar*): association with stress susceptibility and cardiac lesion. J. Nutr. 121: 1163–1172; 1991.

Bjerve, K.S. Requirements for adults and elderly. In: Simopoulos, A.P.; Kifer, R.R.; Barlow, S.M.; Martin, R.E., eds. Health Effects of Omega3 Polyunsaturated Fatty Acids in Seafoods. Basel: Karger; 1991:p. 26–30.

Coutteau, P.; Sorgeloos, P. Intercalibration exercise on the qualitative and quantitative analysis of fatty acids in *Artemia* and marine samples used in mariculture. In: Howell, B.; Olsen, Y.; Iglesias, J., eds. Cooperative Research Report 211. Copenhagen: ICES; 1995.

Crawford, M.A.; Hassam, A.G.; Stevens, P.A. Essential fatty acid requirements in pregnancy and lactation with special reference to brain development. Prog. Lipid Res. 20:30–40; 1981.

Crawford, M.A.; Hassam, A.G.; Williams, G.; Whitehouse, W.L. Essential fatty acids and fetal brain growth. Lancet. 1:452–453; 1976.

Dhert, P.; Sorgeloos, P.; Devresse, B. Contributions towards a specific DHA enrichment in the live food *Brachionus plicatilis* and *Artemia* sp. In: Reinertsen, H.; Dahle, L.A.; Jørgensen, L.; Tvinnereim, K., eds. Proceedings from the International Conference on Fish Farming Technology, Trondheim, Norway, August 9–12, 1993. Rotterdam, The Netherlands: Balkema; 1993:p. 109–115.

Dratz, E.A.; Holte, L.L. The molecular spring model for the function of docosahexaenoic acid (22:6ω3) in biological membranes. In: Sinclair, A.; Gibson, R., eds. Essential Fatty Acids and Eicosanoids. The Third International Congress on Essential Fatty Acids and Eicosanoids. Adelaide, Australia, March 1–5, 1992. Champaign, IL: American Oil Chemists' Society; 1992:p. 122–127.

Dyerberg, J.; Bang, H.O.; Hjørne, N. Plasma cholesterol concentration in Caucasian Danes and Greenland west coast Eskimos. Dan. Med. Bull. 24:52–55; 1977.

Dyerberg, J.; Bang, H.O.; Hjørne, N. Fatty acid composition of the plasma lipids in Greenland Eskimos. Am. J. Clin. Nutr. 28:958–966; 1975.

Evjemo, J.O.; Coutteau, P.; Olsen, Y.; Sorgeloos, P. The stability of docosahexaenoic acid (DHA) in two *Artemia* species following enrichment and subsequent starvation. Aquaculture 155:135–148; 1997.

Farkas, T.; Csengeri, I.; Majoros, F.; Oláh, J. Metabolism of fatty acids in fish III. Combined effect of environmental temperature and diet on formation and deposition of fatty acids in the carp, *Cyprinus carpio* Linnaeus 1758. Aquaculture 20:29–40; 1980.

Fukusho, K. Nutritional effects of the rotifer, *Brachionus plicatilis,* raised by baking yeast on larval fish of *Oplegnathus fasciatus,* by enrichment with *Chlorella* sp. before feeding. Bull. Nagasaki Pref. Inst. Fish. 3:152–154; 1977 (in Japanese).

Hansen, A.E.; Wiese, H.F.; Boelsche, A.N. Role of linoleic acid in infant nutrition. Pediatrics 31 (suppl):171–192; 1963.

Henderson, R.J.; Tocher, D.R. The lipid composition and biochemistry of freshwater fish. Prog. Lipid Res. 26:281–347; 1987.

Hirata, H. Culture methods of the marine rotifer, *Brachionus plicatilis*. Min. Rev. Data File Fish. Res. 1:27–46; 1980.

Holman, R.T.; Caster, W.O.; Wiese, H.F. The essential fatty acid requirement of infants and the assessment of their dietary intake of linoleate by serum fatty acid analysis. Am. J. Clin. Nutr. 14:70–75; 1964.

Ito, T. On the culture of mixohaline rotifer *Brachionus plicatilis* O.F. Muller. Rep. Fac. Fish. Mie Pref. Univ. 3:708–740; 1960 (in Japanese).

Izquierdo, M.S.; Watanabe, T.; Takeuchi, T.; Arakawa, T.; Kitajima, C. Requirements of red seabream *Pagrus major* for essential fatty acids. Nippon Suisan Gakkaishi 55(5):859–867; 1989.

Kitajima, C.; Koda, T. Lethal effects of the rotifer cultured with baking yeast on the larval sea bream, *Pagrus major,* and the increase rate using the rotifer recultured with *Chlorella* sp. Bull. Nagasaki Pref. Inst. Fish. 2:113–116; 1976 (in Japanese).

Koven, W.M.; Tandler, A.; Sklan, D.; Kissel, G.W. The association of eicosapentaenoic and docosahexaenoic acids in the main phospholipids of different-age *Sparus aurata* larvae with growth. Aquaculture 116:71–82; 1993.

Koven, W.M.; Tandler, A.; Kissel, G.W.; Sklan, D.; Friezlander, O.; Harel, M. The effect of dietary (ω3) polyunsaturated fatty acids on growth, survival and swim bladder development in *Sparus aurata* larvae. Aquaculture 91:131–141; 1990.

Leaf, A. Omega-3 PUFA, an update: 1986–1993. Omega-3 News 1/93. VIII:1–4; 1993.

Mourente, G.; Rodriguez, A.; Tocher, D.R.; Sargent, J.R. Effects of dietary docosahexaenoic acid [DHA; 22:6ω3] on lipid and fatty acid composition and growth in gilthead sea bream (*Sparus aurata* L.) larvae during first feeding. Aquaculture 112:79–98; 1993.

Mourente, G.; Tocher, D.R.; Sargent, J.R. Specific accumulation of docosahexaenoic acid [22:6ω3] in brain lipids during development of juvenile turbot *Scophthalmus maximus* (L). Lipids 26:871–877; 1991.

Nagata, W.D.; Hirata, H. Mariculture in Japan: past, present, and future prospectives. Mini Rev. Data File Fish. Res. 4:1–38; 1986.

Neuringer, M.; Andersdon, G.J.; Connor, W.E. The essentiality of omega 3 fatty acids for the development and function of the retina and the brain. Annu. Rev. Nutr. 8:517–541; 1988.

Øie, G.; Makridis, P.; Reitan, K.I.; Olsen, Y. Protein and carbon utilization of rotifers (*Brachionus plicatilis*) in first feeding of turbot larvae (*Scophthalmus maximus* L.). Aquaculture 153:103–122; 1997.

Øie, G.; Olsen, Y. Protein and lipid content of the rotifer *Brachionus plicatilis* during variable growth and feeding conditions. Hydrobiologia 358:251–258; 1997.

Olsen, Y. Cultivated algae as a source of omega-3 fatty acids. In: Fish, Fat and Your Health. Proceedings of the International Conference on Fish Lipids and Their Influence on Human Health. Svanøy, Norway: Svanøy Foundation; 1989:p. 50–61.

Olsen, Y.; Skjervold, H. Variation in content of ω3 fatty acids in Atlantic salmon produced along the Norwegian coast, with special emphasize on the impact of non-dietary factors. Aquaculture Int. 3:22–35; 1995.

Olsen, Y.; Skjervold, H. Impact of latitude on ω3 fatty acids in wild Atlantic salmon. Omega 3 News VI:1–4; 1991.

Olsen, Y.; Reitan, K.I.; Vadstein, O. Dependence of temperature on loss rates of rotifers, lipids, and ω3 fatty acids in starved *Brachionus plicatilis* cultures. Hydrobiologia 255/256:13–20; 1993a.

Olsen, Y.; Rainuzzo, J.R.; Reitan, K.I.; Vadstein, O. Manipulation of lipids and ω3 fatty acids in *Brachionus plicatilis*. In: Reinertsen, H.; Dahle, L.A.; Jørgensen, L.; Tvinnereim, K., eds. Proceedings from the International Conference on Fish Farming Technology, Trondheim, Norway, August 9–12, 1993. Rotterdam, The Netherlands: Balkema; 1993b:p. 101–108.

Owen, J.M.; Adron, J.W.; Middleton, C.; Cowey, C.B. Elongation and desaturation of dietary fatty acids in turbot *Scophthalmus maximus* and rainbow trout *Salmo gairdneri*. Lipids 10:528–531; 1975.

Pohl, P. Lipids and fatty acids in algae. In: Zaborsky, O.R., ed. Handbook of Biosolar Resources. Boca Raton, FL: CRC Press; 1982:p. 383–404.

Rainuzzo, J.R. Lipids in early stages of marine fish. PhD thesis, University of Trondheim, Norway; 1993.

Rainuzzo, J.R.; Reitan, K.I.; Olsen, Y. Effect of short- and-long term lipid enrichment on total lipids, lipid class and fatty acid composition in rotifers. Aquaculture Int. 2:19–32; 1994.

Rainuzzo, J.R.; Reitan, K.I; Jørgensen, L. Comparative study on the fatty acids and lipid composition of four marine fish larvae. Comp. Biochem. Physiol. 103B(1):21–26; 1992.

Reitan, K.I. Nutritional effects of algae in first-feeding of marine fish larvae. PhD thesis, University of Trondheim, Norway; 1994.

Reitan, K.I.; Rainuzzo, J.R.; Olsen, Y. Influence of lipid composition of live feed on growth, survival and pigmentation of turbot larvae, *Scophthalmus maximus* L. Aquaculture Int. 2:33–48; 1994a.

Reitan, K.I.; Rainuzzo, J.R.; Olsen, Y. Effect of nutrient limitation on fatty acid and lipid content of marine algae. J. Phycol. 30:972–979; 1994b.

Reitan, K.I.; Rainuzzo, J.R.; Øie, G.; Olsen, Y. Nutritional effects of algal addition in first feeding of turbot (*Scophthalmus maximus* L.) larvae. Aquaculture 118:257–275; 1993.

Rollefsen, G. The eggs and the larvae of the halibut (*Hippoglossus hippoglossus*). K. Norske Vidensk. Selsk. Skr. 7:20–23; 1934.

Sargent, J.R. Wax esters, long chain monoenoic fatty acids and polyunsaturated fatty acids in marine oils: resource and nutritional implications. In: Fish, Fat and Your Health. Proceedings of the International Conference on Fish Lipids and Their Influence on Human Health. Svanøy, Norway: Svanøy Foundation; 1989:p. 43–49.

Sargent, J.R.; Bell, M.V.; Tocher, D.R. Docosahexaenoic acid and the development of brain and retina in marine fish. In: Drevon, C.A.; Baksaas, I.; Krokan, H.E., eds. Omega-3 Fatty Acids: Metabolism and Biological Effects. Basel, Switzerland: Birkhäuser Verlag; 1993a:p. 139–149.

Sargent, J.R.; Bell, J.G.; Bell, M.V.; Henderson, R.J.; Tocher, D.R. The metabolism of phospholipids and polyunsaturated fatty acids in fish. In: Lahlou, B.; Vitiello, P., eds. Aquaculture: Fundamental and Applied Research. Coastal and Estuarine Studies, Vol. 43. Washington, D.C.: American Geophysical Union; 1993b:p. 103–124.

Sargent, J.R.; Henderson, R.J. Lipids. In: Corner, E.D.S.; O'Hara, S.C.M., eds. The Biological Chemistry of Marine Copepods. Oxford: Clarendon Press; 1986:p. 59–108.

Shifrin, N.S.; Chisholm, S.W. Phytoplankton lipids: Interspecific differences and effects of nitrate, silicate and light-dark cycles. J. Phycol. 17:374–384; 1981.

Siron, R.; Giusti, G.; Berland, B. Changes in the fatty acid composition of *Phaeodactylum tricornutum* and *Dunaliella tertiolecta* during growth and under phosphorus deficiency. Mar. Ecol. Prog. Ser. 55:95–100; 1989.

Skjervold, H. Lifestyle diseases and the human diet. How should the new discoveries influence future food production? Collection of articles printed in The Journal of Dairy Industry of Norway. Norway: Ås-Trykk; 1992:p. 48.

Stubbs, C.D. The structure and function of docosahexaenoic acid in membranes. In: Sinclair, A.; Gibson, R., eds. Essential Fatty Acids and Eicosanoids. The Third International Congress on Essential Fatty Acids and Eicosanoids, Adelaide, Australia, March 1–5, 1992. Champaign, IL: American Oil Chemists' Society; 1992:p. 116–121.

Watanabe, T. Importance of docosahexaenoic acid in marine larval fish. J. World Aquaculture Soc. 24(2):152–161; 1993.

Watanabe, T.; Izquierdo, M.S.; Takeuchi, T.; Satoh, S.; Kitajima, C. Comparison between eicosapentaenoic and docosahexaenoic acid in terms of essential fatty acid efficiency in larval red seabream. Nippon Suisan Gakkaishi 55(9):1635–1640; 1989.

Watanabe, T.; Kitajima, C.; Fujita, S. Nutritional values of live organisms used in Japan for mass propagation of fish: a review. Aquaculture 34:115–143; 1983.

Watanabe, T.; Kitajima, C.; Arakawa, T.; Fukusho, K.; Fujita, S. Nutritional quality of rotifer, *Brachionus plicatilis,* as a living feed from the viewpoint of essential fatty acids for fish. Bull. Jpn. Soc. Sci. Fish. 44:1109–1114: 1978 (in Japanese).

Weber, P.C. Are we what we eat? Fatty acids in nutrition and in cell membranes: cell functions and disorders induced by dietary conditions. In: Fish, Fat and Your Health. Proceedings of the International Conference on Fish Lipids and Their Influence on Human Health. Svanøy, Norway: Svanøy Foundation; 1989:p. 9–18.

Williams, E.E.; Hazel, J.R. The role of docosahexaenoic acid-containing molecular species of phospholipid in the thermal adaptation of biological membranes. In: Sinclair, A.; Gibson, R., eds. Essential Fatty Acids and Eicosanoids. The Third International Congress on Essential Fatty Acids and Eicosanoids, Adelaide, Australia, March 1–5, 1992. Champaign, IL: American Oil Chemists' Society; 1992:p. 128–133.

9

Influence of Lipids on the Bioaccumulation and Trophic Transfer of Organic Contaminants in Aquatic Organisms

Peter F. Landrum and Susan W. Fisher

9.1. Introduction

9.1.1. Sources of Contaminant Gain and Loss in Aquatic Systems

In aqueous systems, organisms are exposed to contaminants via multiple routes (Fig. 9.1). The extent of contaminant accumulation ultimately depends on the extent and mode of interaction with diverse contaminated media. The influence of lipids on contaminant uptake likewise varies according to the route by which the exposure takes place and the lipophilic character of the contaminant. Thus, it is necessary to clarify the environmental sources of contaminants for accumulation. The means by which contaminants, once accumulated, can be eliminated from an organism can also depend on organism lipid content. This elimination can be modified by the route, contaminant lipophilicity, and extent of contamination of the environmental compartment into which elimination occurs.

A few definitions will help guide the discussion. When contaminants are freely dissolved in water, uptake by aquatic organisms occurs when the contaminant is absorbed across a respiratory surface, the gills (Nichols et al., 1991, 1990), or less commonly, across the cuticular epidermis or exoskeleton (Lien et al., 1994; Lien and McKim, 1993). If uptake exceeds elimination, then *bioconcentration* has occurred. If the ultimate source of the contaminant is not water (i.e., if the contaminant is bound to some environmental medium such as sediment or is present in an organism's food), *bioaccumulation* from solid media can occur. Because some fraction of the contaminant may detach from solid media and enter the dissolved phase, some uptake of the contaminant directly from water may occur (Bruner et al., 1994a). However, it is difficult to separate the fraction of the contaminant that originates from solid media and the fraction that is directly absorbed from water; bioaccumulation, thus, is understood to include both routes of entry into an organism. Once a contaminant moves from abiotic media such as water or sediment into living organisms, the contaminants can subsequently move through the food chain when, for example, contaminated prey are ingested. *Trophic transfer* is the term that describes the transfer of contaminants between

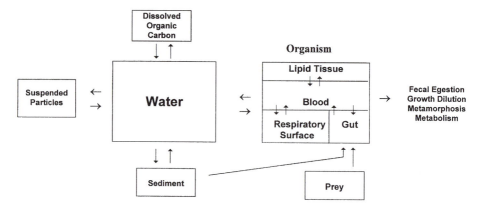

FIGURE 9.1. Sources of contaminant gain and loss for aquatic organisms. Contaminants in aquatic organisms are accumulated from both water and food. The contaminant distributes within the aquatic environment in accordance with its characteristics and those of the ecosystem between water, sediment, suspended material including particles, and dissolved organic matter. This distribution will dictate the bioavailability of the contaminant and depends on the characteristics of the interaction between the organism and phases into which the contaminant is distributed. The accumulation of a contaminant in an organism depends on the rate of exposure, the rate of distribution within the organism, and the characteristics (e.g., lipid content) of the tissues and is balanced by biotransformation and elimination processes to yield the net accumulation.

trophic levels via ingestion. A special instance of trophic transfer, known as *biomagnification,* occurs when contaminant loads increase significantly with each successive trophic level. In both cases, the primary route of exposure is thought to be from contaminated food.

Accumulated contaminants can also be eliminated from an organism. Because most contaminants are generally accumulated through passive partitioning and because the partitioning is an equilibrium process, contaminants can be eliminated from an organism's body by diffusion. The process depends on the relative affinity of the contaminant for the organism and the compartment into which the contaminant is being eliminated. Several internal physiological processes dictate the rates at that contaminants will be transported internally, deposited in lipid-rich storage, or removed via some excretory process such as fecal egestion. Contaminants that are susceptible to metabolism can be eliminated via chemical alteration. In addition, contaminant concentrations can be diluted through tissue growth even though the contaminants are still present. Such growth dilution may reduce the biological effects associated with contamination. Finally, contaminants can be lost by offloading to neonates through reproduction or, in the case of animals that undergo metamorphosis, through ecdysis. If an animal is actively bioaccumulating, then uptake must exceed loss. When the rates of uptake and loss are equal, the organism is at steady state.

9.1.2. Organism Adiposity and Internal Distribution of Contaminants

Although contaminants can be accumulated from several environmental compartments, once the contaminants are bioaccumulated, organism lipids become the dominant force controlling the dynamics of distribution and the manifestation of toxic effects. Lipids in organisms make up a significant portion of cell membranes, are particularly concentrated around neurons, and are the fundamental energy stores for organisms. Thus, lipids serve as membrane barriers to accumulation and distribution of contaminants, the sites of toxic action where changes in membrane function are disturbed by contaminant interaction, and the ultimate storage site for contaminants that have high lipid solubility. The interaction of contaminants with lipids depends, in large part, on the ability of the contaminant to dissolve in the lipid, and contaminants that have low lipid solubility will not cross membranes except through facilitated or active transport sites (Hunn and Allen, 1974). By contrast, highly hydrophobic contaminants will be found more strongly associated with lipid membranes and storage lipids of the organism.

Initial efforts to understand the interactions of organic contaminants and lipids come from the pharmacological literature and are focused on understanding the distribution of drugs within an organism (Bickel, 1984). Contaminants that are freely dissolved in the blood are available for transport among tissues whereas those that are sequestered by blood proteins and lipids are less mobile. Further, contaminants that are highly lipophilic eventually end up in organism lipid stores away from their receptor sites for toxic action However, lipid mobilization by an organism under stress can release contaminants back to the general circulation. In addition, contaminants of differing lipophilicity differ in their ability to be assimilated because crossing the cell membranes into an organism requires the contaminant to pass through lipid membranes. Thus, differing lipid solubilities along with other molecular features such as molecular size produce differences in the rates and extent of bioaccumulation. Because of the interest in developing better drugs with more improved transfer efficiencies and rates, the partitioning of contaminants between water and representatives of lipid materials has been studied. The best small molecule found to be representative of lipid materials was n-octane-1-ol (usually referenced as octanol). The partitioning behavior between octanol and water provides a standardized reference source for partitioning that has proved very useful for evaluating the lipophilicity of contaminants (reviewed in Leo et al., 1971).

The kinetics and efficiency of transport into organisms are determined by the rate of presentation of contaminants to the membrane, the chemical activity (chemical potential) difference between the source compartment and the organism tissue, and the contaminant's resistance for crossing membranes. Because membranes are composed primarily of lipid bilayers, the ability of contaminants to dissolve into, and move through, membranes dictates part of the kinetics of transport (Hunn and Allen, 1974). The rate of exit from the membrane into the circulating fluid and transport to the site of action or storage can also significantly dictate the rate of accumulation of lipophilic contaminants. The same factors affect the transfer of the contaminants out of a tissue. In summary, when the lipid solubility of the contaminant is high relative to its solubility in the source compartment, the contaminant will tend to pass into and accumulate in the organism.

The impact of lipids on these kinetic processes is most easily observed in their effect on the rate of elimination. Organisms with higher lipid contents exhibit elimination rates that are substantially slower than observed for leaner organisms (Van den Huevel et al., 1991; Landrum, 1988). Differences in the overall elimination rates among organisms reflect the capacity of the organisms relative to the source compartment, and decreases in elimination rate are reflected in increasing bioaccumulation. In some cases, the rate of accumulation can increase with increasing lipid content as was observed for the zebra mussel (Bruner et al., 1994a). The mechanism for this increase is tied to maintenance of the concentration gradient between the source compartment and the site of uptake. This presumes that the rate processes involved in the distribution within the organism are not rate-limiting. If distribution limitations become a significant portion of the rate process, then distribution to the storage lipid will become disconnected from the uptake rate process and lipid content will no longer be tightly coupled to uptake rate. This was clearly demonstrated with the accumulation of trifluralin in rainbow trout, in which the uptake clearance rates declined in proportion to the intercompartmental transfer rates with increasing organism size despite the increased lipid content for large fish (Schultz and Hayton, 1994).

9.1.3. Nonlipid Factors Affecting Internal Distributions

In addition to the lipophilicity of the contaminant, the size of the molecule may preclude its dissolution into the membrane and, therefore, accumulation by the organism. In the extreme case of polymers, it is clear that even hydrophobic (highly lipophilic) molecules such as polymers of polydimethylsiloxane are not accumulated and are only found on the surface of organisms (Kukkonen and Landrum, 1995; Opperhuizen et al., 1987). The failure of molecules to accumulate can also be due to binding to extracellular materials such as dissolved organic carbon (Bruggeman et al., 1984) or decreased permeability of the membrane resulting from molecular size limitations, >9.5 Å (Saito et al., 1990; Opperhuizen et al., 1985; Zitko, 1980). Both mechanisms result in reduced bioavailabilities. The effective molecular size range for interaction with lipids not only has an upper limit, at which the lipophilic membranes act as essentially impermeable barriers, but there is also a lower molecular size cutoff. At the lower end, the membranes are permeable to small un-ionized molecules. This size cutoff is <50 a.m.u. (Walter and Gutknecht, 1986). The extra permeability is not related to the lipophilicity of the contaminant but rather inversely related to the molecular size. The molecular volume dependence was attributed to the membrane properties, with the lipid behaving more like a polymer than a liquid hydrocarbon (Walter and Gutknecht, 1986).

Within the effective molecular size range, this relative solubility between the source compartment and the organism's lipid is a useful predictor of the potential extent of accumulation. For contaminants that contain ionizable functional groups, the pK of the contaminant will influence the final storage site in the organism (Hunn and Allen, 1974). For instance, pentachlorophenol (PCP) ionizes and has a pKa of 4.74 (Westhall, 1985); its penetration into the organism requires, in general, that the un-ionized molecule penetrate the membrane (Stehly and Hayton, 1990), which is reflected by the apparent lipophilicity of the contaminant as reflected by apparent changes in the octanol/water partition coefficient with pH (Kaiser and Valdmanis, 1982). In the Great Lakes, where the pH is about 8, PCP would be highly ionized (approximately 0.05% in the un-ionized form) so penetration of the lipophilic membrane to enter the animal will be limited. In the case of *Diporeia,* the uptake clearance for PCP (log K_{ow} 5.01; Westhall, 1985) from Lake Michigan water was 3.74 ml $\cdot$ g^{-1} $\cdot$ h^{-1} (Landrum and Dupuis, 1990), whereas the uptake clearance for pyrene, a contaminant of similar log K_{ow} (5.2), was 131 ml $\cdot$ g^{-1} $\cdot$ h^{-1} (Landrum, 1988). Similarly, the distribution within the organism may become limited because of ionization in the circulatory fluid of the organism, which limits distribution within and, in some cases, elimination from the organism. The impact of ionization is equally important for contaminants having basic as well as acidic functional groups. The relative state of ionization will dictate the degree of penetration into the lipophilic environment of the membrane due to changes in lipid solubility of the contaminant.

9.2. Prediction of Bioconcentration and Bioaccumulation

9.2.1. Bioconcentration

Once the contaminant enters the organism and is distributed to the final storage site, the relative solubility in lipid will permit predictability. As with the distribution between the circulating fluid and the tissue, the partitioning characteristics of the contaminant between 1-octanol and the source can help predict the accumulation potential of the contaminant. This approach works as well for terrestrial organisms as it does for aquatic organisms (Kenega, 1980). For aquatic organisms exposed in contaminated water, nonpolar contaminants are accumulated in proportion to the octanol/water partition coefficient (K_{ow}) of contaminants. This was first demonstrated by a prediction of the bioconcentration of contaminants by fish. The log of the bioconcentration factor (log BCF is defined as log of the ratio of the contaminant concentration in the organism to the contaminant concentration in the water) was linearly correlated with the log K_{ow} (Neely et al., 1974). The use of this approach was further demonstrated through the work of Veith et al., (1980, 1979) and Mackay (1982) on fish. Subsequently, the bioconcentration in both invertebrates and macrophytes was examined with respect to log K_{ow} (Gobas et al., 1991; Connell, 1988; Hawker and Connell, 1986). In all cases, there was evidence of nonlinearity for large, very hydrophobic contaminants. Some of this nonlinearity was accounted for by molecular size as stated above. Steric proper-ties (e.g., molecular size, surface area, or configuration) of even relatively small molecules such as polychlorinated biphenyls (PCBs) can influence the relative bioaccumulation of contaminants compared with traditional log BCF − log K_{ow} relationship (Shaw and Connell, 1984). In addition, there is evidence that 1-oc-tanol becomes a less ideal solvent for larger molecules. Thus, accounting for the relative solubility in octanol removes some of the observed nonlinearity in BCF prediction (Banerjee and Baughman, 1991).

Because the contaminant lipophilicity as measured by K_{ow} and extent of con-taminant accumulation are so well related, a convention developed to normalize the contaminant concentrations to the lipid contents of organisms. This technique reduced the variability between organism species and resulted in improved pre-dictions for accumulation from water (Barron, 1990; Connell, 1988). However, cases remain that demonstrate the limitations of lipid normalization to totally account for the variation in contaminant accumulation among species. For in-stance, in Lake Baikal lipid-normalized BCF-K_{ow} relationships had different slopes for two fish species (Kucklick et al., 1994). Similarly, the BCFs for lake trout and white fish from Siskiwit Lake on Isle Royle exhibit significant vari-ability even with lipid normalization; the regression with log K_{ow} was weak for pesticides, and the correlation for PCBs was even more variable (Swackhamer and Hites, 1988). In some cases, the absence of improved relationships despite lipid normalization may be due to inclusion of multiple contaminant classes in the regression (Axelman et al., 1995; Connell, 1988). Where contaminant characteris-tics change, the interaction with lipids also changes. Thus, even in a single species, the slopes of the relationships between lipid-normalized BCF and log K_{ow} are different for different contaminant classes (Axelman et al., 1995). Thus, predictability will depend on both the composition of the lipids and the charac-teristics of the contaminant. Both characteristics will contribute to an interaction that will determine the relative contaminant solubility in the organism's lipids and the ability of log K_{ow} to predict that solubility interaction. This predictability will generally be good within a species and class of contaminants (Axelman et al., 1995; Connell, 1988). However, the variance in the predicted BCF may be sub-stantial if attempts are made to predict across species and contaminant classes (Connell, 1988). For instance, the intercepts for regressions of log K_{ow} against the lipid-normalized BCF range over an order of magnitude among organisms whereas the slopes vary from 0.844 to 1.0. (Connell, 1988). Despite the above-mentioned caveats, lipid normalization of contaminant concentrations remains the most viable and useful method of predicting BCFs and serves an important role in screening new contaminants for potential BCFs.

9.2.2. Bioaccumulation

The variance in the predicting contaminant accumulation, using such predictors as log K_{ow}, increases substantially compared with predictions from aqueous ex-posures when attempting to predict the thermodynamic limits for contaminated sediment exposures. The increased variance results from the organic matrix of the sediment competing with the organism lipids for the relative solubility of the contaminant. There is a hypothesis that if a correction is applied for the amount of organic matter in the sediment, generally expressed as organic carbon, then the thermodynamic limit for bioaccumulation becomes relative to the freely dissolved contaminant concentration in the interstitial water, which is representative of the chemical activity (DiToro et al., 1991; McFarland and Clark, 1989; McFarland, 1984). According to this model, the derived biota sediment accumulation factor (BSAF), in which the concentrations in the sediment are carbon-normalized and those in the organism are lipid-normalized, should be invariant with log K_{ow}. The theoretical thermodynamic partitioning relationship between the sediment organic matter and the organism lipid leads to an estimated constant value of 1.7 (McFarland and Clark, 1989). This value represents the relative capacity of organism lipid and sediment organic matter on a unit mass basis and should not vary with log K_{ow}. In studies that have measured BSAF among organisms and sediments, the variance can exceed 100-fold between the lowest and highest values even for the same contaminants (Brannon et al., 1993; Lee, 1992; Lake et al., 1990). The magnitude of this range was suggested to result from an absence of true equilibrium/steady state for some of the data.

The difficulty appears to be in accounting for the bioavailability of the organic contaminants in the sediment matrix. If the uptake clearance from the sediment (the amount of sediment cleared of contaminant per mass of organism per hour) is used as a measure of bioavailability, simple normalization to the amount of organic matter in the sediment is inadequate to explain all the observed variance by a factor of 10 (Landrum and Faust, 1994). Further, bioavailability is not even the same for contaminants of similar hydrophobicity from a single sediment (Harkey et al., 1994a; Landrum and Faust, 1991). This change in bioavailability appears to be driven in part by differences in the distribution of contaminants among the varying particles within the sediment matrix (Kukkonen and Landrum, 1996; Harkey et al., 1994b). Further, these distributions do not always correlate with the amount of organic carbon in a particular sediment fraction (Kukkonen and Landrum, 1996; Harkey et al., 1994b). Thus, it is not surprising that the relationship between sediment concentration and bioaccumulation is not regular even when normalized for organic carbon and lipid. Prediction among sediments will likely require additional factors to better account for the interaction of the contaminant with the sediment.

9.3. Factors Affecting Prediction

9.3.1. Methods for Measuring Lipid Content

Among the factors that affect the ability to predict either bioconcentration from water or bioaccumulation from sediments is the method used to measure the lipid content of the organism. Lipids have been extracted through the use of several organic solvent combinations. Most frequently, the solvent chosen for the lipid extraction is determined by that which most efficiently extracts the contaminant of interest from the various matrices. This often leads to incomplete extraction, sometimes in excess of 50%, of the total lipids and is generally a failure to extract the polar lipids (Ewald, 1996). Further, the solvents specifically used for lipid extraction sometime fail to completely extract the contaminants, particularly in leaner organisms with a difference of nearly 40% in some cases (Ewald, 1996). There has been little effort to intercalibrate the relative extractability of the various solvent systems with a more conventional system, chloroform:methanol, to extract lipids (Bligh and Dyer, 1959; Folch et al., 1957). These extraction differences may lead to substantial differences in comparing the lipid-normalized bioconcentration across studies and among species of varying lipid composition. We suggest that the substantial variability in lipid extractability among data sets could be eliminated by standardizing to the Bligh-Dyer extraction scheme, which uses a chloroform:methanol extracting solvent and is known to extract both polar and nonpolar lipids (Randall et al., 1991). It would not be necessary for all studies to use a single extraction technique, but there should be an intercalibration between the extraction solvent of choice and the Bligh-Dyer scheme. This would permit calculation to common units. This scheme can even be adapted for the measurement of lipid content for small sample masses (Gardner et al., 1985).

9.3.2. Lipid Composition and Bioaccumulation

The relative lipid compositions among species also contributes to the observed variation in bioaccumulation potential. The generally held view is that neutral storage lipids are the most important class for the bioaccumulation of nonpolar contaminants. Thus, species with higher neutral lipid fractions would presumably accumulate higher contaminant concentrations on a lipid-normalized basis. However, the evidence for the role of lipid composition is limited. Several studies have focused on the relative extractability of the contaminant of interest and the corresponding lipid to suggest the role of lipid composition on bioaccumulation (Ewald, 1996; de Boer, 1988; Schneider, 1981). Another approach examining the role of lipid composition to compare across species and contaminant characteristics used different portions of the lipids as normalizing factors. When this is done, prediction of bioaccumulation was better related to total lipid content than to any one lipid fraction (Kucklick et al., 1996; Stange and Swackhamer, 1994).

The relative distribution of the moderately lipophilic herbicide triallate was investigated by using autoradiography (Arts et al., 1995) and mass spectrometric techniques (Arts et al., 1996) and showed that the contaminant resides primarily in the storage lipids and in the highly lipophilic nervous system tissues. Although the triallate was found in storage lipids as expected, the bioaccumulation did not always correlate with the triglycerol content of the amphipods (Arts et al., 1995). This suggests a role for other lipids or other storage sites in some species that were not detected by the above techniques.

Some experiments have examined this issue more directly. One such study extracted lipids from phospholipid-rich organisms and from triglyceride-rich organisms. The extracted lipids were coated onto filters, and the relative partitioning onto the filters was determined. The data indicated that the filters coated with triglyceride-rich lipids accumulated nearly twice the amount of tetrachlorobiphenyl compared with filters containing lipids with 20% or more phospholipids (Ewald and Larsen, 1994). From this more direct study, the bioaccumulation by biota was deduced to vary with triglyceride composition of the species.

In another study, the two lipid types, buoyant nonpolar and polar membrane-associated lipids, were separated by physical means, and the corresponding accumulation of contaminant in each phase along with the corresponding lipid composition of each phase were examined. On a lipid-normalized basis, both phases contained essentially the same concentration of the nonmetabolized contaminants, suggesting that the polarity of the lipid when it exists in the organism does not dramatically affect the accumulation of nonpolar contaminants (Gardner et al., 1990).

The ability to accurately determine the influence of differing lipid compositions contaminant bioavailability is not completely resolved, and new approaches to study the issue are required. An additional interesting finding of the Gardner et al. (1990) study was a dynamic imbalance (differences in normalized lipid concentrations) among the lipid pools that continued for a relatively long time period (>24 h) in mysids. Thus, before comparing among lipid pools, it is critical that adequate exposure time be allowed to ensure that the pools are dynamically in balance (i.e., have equal chemical activities [lipid-normalized contaminant concentrations] among the lipid pools). If this is not the case, the analyses may artificially indicate that there are differences in the contaminant distribution relative to lipid content when, in fact, the organism is out of steady state relative to the distribution among the various pools. The absence of balance among the lipid pools is more likely to occur for larger animals in which the distribution processes require longer times before pools attain equal chemical activity.

9.4. Mimicking Bioconcentration with Semipermeable Membrane Devices

With the recognition that the lipid content of organisms drives the bioaccumulation of nonpolar contaminants, attempts have been made to develop nonbiological surrogates. These surrogates would directly sample the chemical activity of systems without the difficulty of maintaining live organisms or accounting for processes such as biotransformation that complicate the interpretation. Early efforts used solvent-filled dialysis membranes or hexane "droplets" as surrogates for accumulation of contaminants (Södergren and Okla, 1988; Södergren, 1987). This early work produced encouraging comparisons between the accumulation by nonbiological surrogates and organisms. However, there were difficulties with the hexane-filled bags and with the dialysis tubing. The hexane tended to dialyze out of the bags, and the dialysis membranes degraded over time. More recent approaches have used thin polymeric film membrane bags with a lipid material for filler (Huckins et al., 1990a) and an improved method for contaminant recovery from these samplers (Huckins et al., 1990b). Because these semipermeable membrane devices (SPMDs) are filled with lipid materials, they should more accurately reflect the contaminant capacity of lipids in organisms. Further, unlike solvents, the large molecular size of the lipids precludes their diffusion out of the sampler into the water.

The properties of these samplers have been well characterized by deploying them alongside caged organisms including both fish and bivalves (Ellis et al., 1995; Prest et al., 1992). In both cases, the SPMDs performed well in sampling the environment for nonpolar organic contaminants. In one study, the results of the SPMDs were compared with another method that measured concentrations in ultrafilter permeates (Ellis et al., 1995). The SPMDs gave equivalent results with the ultrafilter permeates in indicating the amount of contaminants bioavailable for aqueous exposures. However, caged (channel catfish) and feral fish (carp and sauger) residues exhibited fewer contaminants. The differences between caged and feral fish may be due to mobility of the feral population reflecting differing sources, differences in species, and enhanced metabolism in the feral organisms. In general, the SPMDs had higher concentrations of contaminants compared with the fish. This may reflect more rapid accumulation, the absence of biotransformation, and/or rapid elimination of the contaminants from fish (Ellis et al., 1995). When contaminant concentrations in the samplers were compared with caged bivalves (*Corbicula fluminea*), there were again differences between the mussels and the SPMDs. In general, the SPMDs sequestered a wider range of contaminants, and there were differences in the distribution between the clams and the SPMDs. Further, the clams sequestered greater concentrations than the SPMDs, which may be the result of greater uptake rates and additional pathways such as accumulation from food (Prest et al., 1992). It is clear that SPMDs can take a long time to come to true equilibrium, and this can be further modified by biofouling in aquatic systems. However, these devices hold the potential to provide a good baseline for aqueous exposures. Because they are passive storage devices, they will never reflect the active processes that govern the uptake and loss of contaminants by aquatic organisms. This also serves to suggest that although bioconcentration and bioaccumulation, even when evaluated at steady state, may be proportional to the storage capacity (lipid content) of the organism, neither may reflect the chemical activity of a given contaminant in the system. Both bioconcentration and bioaccumulation depend on the toxicokinetics that are dictated by chemical and biological processes affecting contaminant exposure and loss and the available contaminant sources.

9.5. Toxicity and the Role of Lipid

9.5.1. Release of Sequestered Contaminant During Metabolism

Because persistent organic contaminants accumulate in lipid material, alteration of the lipid pool can alter the toxicity of contaminants. For example, when lipid-rich organisms such as salmon are starved to reduce the total lipid, stored toxins are also released, which can produce deleterious results (Ewald, 1996; Bickel, 1984). Further, organisms that are rich in lipid can sequester toxic contaminants in storage sites, thereby removing the toxin from the site of toxic action (Geyer et al., 1994, 1993; van den Huevel et al., 1991; Bickel, 1984). Thus, the relative storage capacity of organisms needs to be considered not only for bioaccumulation but also for toxicity. Normalization to the amount of lipid in the organism is suggested to reduce this variation when comparing toxicity values for neutral organic contaminants among species (Geyer et al., 1994). Normalized contaminant concentration data lead to the idea of the survival of the fattest. This concept has been incorporated into a model of exposure and toxic response for narcotic contaminants. Not only will the fattest organisms have additional storage capacity, the higher fat content will impart additional energy reserves to help the organism weather stresses (Lassiter and Hallam, 1990).

The significance of lipid content and composition will increase as attempts are made to establish residue effects concentration in organisms. Residue effects concentrations are the measured concentrations of contaminant, on a whole-body basis, that are equated with a toxic response. Residue effects concentrations are under development in an attempt to move aquatic toxicology from the use of the external environment as a measure of exposure to the internal dose as the exposure measurement (McCarty and Mackay, 1993). There have been several studies demonstrating the reduction in overall range of doses required to produce mortality, particularly for nonpolar narcotics (e.g., polycyclic aromatic hydrocarbons [PAHs], chlorinated benzenes, PCBs), compared with the use of external concentrations (e.g., for fish, as reviewed by McCarty and Mackay, 1993; for amphipods, Landrum et al., 1994, 1991, and Landrum and Dupuis, 1990; for daphnids, Pawlisz and Peters, 1993a,b). The remaining variance can be attributed to three potential causes: the relative capacity for a contaminant of one species or group of organisms over another (i.e., differences in lipid content), the inherent difference in the sensitivity of one species or portion of the population versus another for the contaminant, and the relative biotransformation capability of different organisms. For contaminants that act as narcotics and are not readily biotransformed, differences in the sensitivity among and within species can be largely accounted by adjusting for the lipid composition and content of the organisms. For example, the variation in the lethal body burden (residue effects concentration) in fathead minnows exposed to chlorinated hydrocarbon contaminants was reduced 50% by accounting for the minnows' lipid content (van Wezel et al., 1995). Specific comparisons among species and the role of lipid content on residue effects concentrations have yet to be fully explored. However, based on the intraspecies role of lipid and the general understanding of the storage of toxins and their impact on the toxic response (see above), it is clear that variation in the storage lipid capacity among organisms will help explain the variance among species when comparing residue-based effects.

9.5.2. Lipids and Membrane Narcosis

The direct interactions of contaminants with membrane lipids can result in narcosis. Two reviews on this subject (van Wezel and Opperhuizen, 1995; Mullins, 1954) suggest that the function of the lipid membrane becomes impaired when a sufficient molar volume of contaminant becomes dissolved in the membrane. The in vivo membrane burden of toxicant that produces narcosis is 40–160 mmol $\cdot$ kg^{-1} lipid, and this range corresponds to approximately 3 ml $\cdot$ kg^{-1} on a volume basis (Mullins, 1954). The change in membrane function is thought to occur because the ion permeability of the membranes increases due to an increase in fluidity of the lipids with the solubilization of contaminants (van Wezel and Opperhuizen, 1995). Such changes in membrane fluidity have been demonstrated in the presence of the narcotic benzyl alcohol through the use of nuclear magnetic resonance (NMR) (Ma et al., 1992). These NMR studies demonstrated that the thermodynamic character of lipid membranes changes with the introduction of a narcotic.

9.5.3. Effect of Toxins on Lipid Metabolism and Function

The incorporation of modified fatty acid molecules, such as chlorinated fatty acids, into the lipid biomolecules and subsequently into membrane structures is a newly recognized problem. The presence of chlorine in the fatty acid moiety causes a bend in the molecule similar to that produced by a double bond. Changes in chemical character and molecular conformation may or may not be recognized by the biochemistry of the organism. When these chlorinated fatty acids are incorporated into lipid bilayers, the membrane character may be altered. Further, incorporation of lipids containing chlorinated fatty acids into storage lipids should not initially affect the organism, but when these are mobilized for energy, there may be less energy available (Ewald, 1996). In highly polluted areas, up to 1% of the total fatty acids in an organism may be chlorinated (Håkansson et al., 1991).

The exposure to contaminants can also directly affect the lipid content, either through effects on lipid metabolism or by increasing the overall stress, thus increasing catabolism with resultant reductions in lipid. However, the exposure to chemicals need not always result in reductions in lipids. Rainbow trout (*Oncorhynchus mykiss*) exposed to heptadecane not only sequester the contaminant in the lipid stores but metabolize the heptadecane to fatty acids that become incorporated into the neutral and phospholipid fractions of the organism (Cravedi and Tulliez, 1986).

More often, however, exposure to contaminants results in a reduction in lipids. For instance, when rainbow trout were exposed to PCP, the surviving organisms at the high dose showed significant reductions in total lipids (van den Huevel et al., 1991). Other similar examples exist for exposures to both organic and inorganic contaminants. For example, the accumulation of lead in fish yielded reductions in total lipids, phospholipids, and cholesterol. This was accompanied by an increase in lipase and free fatty acids (Tulasi et al., 1992). Both of these effects were observed during the preparation for reproduction. Organic pesticides, γ-BHC and malathion, were observed to affect lipid metabolism during the vitellogenic phase of the annual reproductive cycle in the catfish *Clarias batrachus* (Lal and Singh, 1987). These contaminants affected both nonpolar lipid and phospholipid metabolism. In particular, these pesticides inhibited the esterification of free fatty acids to acyl glycerides and also affected their mobilization from liver to gonads (Lal and Singh, 1987). Thus, the impact of contaminants on lipid metabolism, particularly in preparation for reproduction, provides some insight into mechanisms for reproductive impairment of fish and presumably to other aquatic organisms exposed to environmental contaminants.

9.6. Relevance of Food Chain Transfer to Bioaccumulation

9.6.1. Relevance of Trophic Transfer to Bioaccumulation

Previous sections have dealt primarily with the uptake of contaminants directly from water and the role of lipids in determining bioconcentration. However, living organisms can also accumulate contaminants via consumption of contaminated food. Dietary transfer of contaminants has been a controversial subject for several decades. A large number of investigators have argued that uptake of contaminants from dissolved form in water eclipses accumulation from any other source and, thus, can be considered the primary source of contaminant exposure in the aquatic environment (Shaw and Connell, 1986; Bruggerman et al., 1981; Chiou et al., 1977; Moriarty, 1975). Because there are tremendous political and policy ramifications that stem from resolution of this issue and because the role of lipids in bioaccumulation takes on several added dimensions if accumulation from food is significant, the major points of argument are presented here.

Identification of bioconcentration as the most relevant route of contaminant exposure stems largely from the observation that uptake of contaminants from water is very rapid and can quickly generate significant body burdens (Fisher et al., 1993; Reynoldson, 1987). Because bioconcentration is a partitioning phenomenon between an aqueous phase and a lipid phase, movement of the contaminant from water into the organism is driven by the lipid content of the organism and is usually predictable from log K_{ow} as described above. Short-term laboratory assays are effective in measuring bioconcentration. However, in aquatic systems, most of the contaminant load is retained in sediment (DiPinto et al., 1993). Thus, it is highly likely that slower transfer processes such as accumulation from sediment or from food are more relevant to long-term ecosystem dynamics. Dietary accumulation requires contaminant desorption from a lipophilic site in food into water in the gut, then subsequently to a lipophilic phase in the intestinal membrane. This two-phase process is less kinetically favorable and less predictable than a simple water-to-lipid partitioning. In addition, long-term assays are needed to quantify bioaccumulation and trophic transfer. However, it now appears that bioaccumulation is a dominant process at lower trophic levels, where benthic invertebrates play an ecologically pivotal role in removing contaminants from storage in sediments and funneling the contaminants into aquatic food chains (Landrum and Robbins, 1990; Swartz and Lee, 1980). Subsequent trophic transfer of contaminants following lower-tier food chain introduction may account for as much as 90–99% of the contaminant load in top predators (Thomann, 1989; Rubenstein et al., 1984; Thomann and Connolly, 1984). It is now acknowledged, in a large number of studies that have used organisms as diverse as invertebrates and fish, that diet is the primary source of contaminant for highly lipophilic (log K_{ow} >5), nonmetabolized chemicals (Landrum et al., 1991; Rasmussen et al., 1990).

Establishing the reality of trophic transfer of contaminants in food chains has important implications for hazard assessment. For example, if the primary source of contaminant accumulation is not water, as suggested by the proponents of bioconcentration, then merely removing contaminants from the water column cannot necessarily be equated with a reduction in hazard. Exposure of living organisms to contaminants will continue as long as contaminants are present in any biologically active medium. In addition, contaminants can move from aquatic to terrestrial environments via consumption of contaminated prey. Food chain length is also important when trophic transfer is accompanied by biomagnification. Lengthening the food chain by introducing species can exacerbate this problem (Rasmussen et al., 1990). Thus, exposure of humans and fish-eating wildlife is, thus, an ongoing health concern (Rasmussen et al., 1990).

9.6.2. Role of Lipids in Food Chain Accumulation

9.6.2.1. Fugacity Model

In theory, the type and amount of lipid in an animal should determine the extent to which a lipid-soluble contaminant will accumulate at each trophic level in a food chain in relation to the chemical activity of the contaminant in the source compared with the organism (the sink) in question. However, to accurately assess the contaminant movement via ingestion, it is necessary to account for lipid levels in both the food (the source of contamination) and the predator (the sink). Additionally, several factors, which do not directly relate to lipid levels, are potentially very important in determining food chain accumulation. These include feeding rates, digestibility of prey, gut retention time, length and morphology of the digestive tract, the rate of fecal egestion, and the growth rate and age of the organism (Sijm et al., 1992; Clark and Mackay, 1991; Thomann, 1989; Serafin, 1984). These factors may obscure the role of lipids in trophic transfer.

The complexity of assessing food chain accumulation calls for an organizing principle that can cull factors of little relevance and highlight processes or factors that are germane. The fugacity concept is particularly useful for discerning the role of lipids in trophic transfer.

The term *fugacity* was originally used to describe the thermodynamic tendency for a gas to escape from one phase and to enter another. The term comes from the Latin "fugal" to flee. Fugacity (f) is defined in units of pressure (pascals or Pa) and is related to chemical concentration, C (mol $\cdot$ m^{-3}), through the fugacity capacity, Z (mol $\cdot$ m^{-3} $\cdot$ Pa^{-1}) where

$$Zf = C \text{ or } f = C \cdot Z^{-1}$$

Thus, fugacity increases linearly with contaminant concentration. The fugacity capacity is specific to a particular compartment and assesses the ability of that compartment to hold the chemical (i.e., prevent it from escaping). The fugacity capacity of a compartment can be considered analogous to the heat capacity of a material. The fugacity capacity is effectively half a partition coefficient and is often determined empirically (Mackay, 1991). A chemical will always move from high to low fugacity unless active transport is occurring (Gobas and Mackay, 1987). At equilibrium, the fugacity in all compartments will be equal (Landrum et al., 1992), and flux, N (mol $\cdot$ h^{-1}), of a chemical between phases will be zero.

The simplest fugacity-based contaminant accumulation models are bioconcentration models that assume lipids are the driving force in accumulation. The fugacity approach to determining bioconcentration is analogous to determining a lipid-based partition coefficient. The lipid-normalized BCF is defined as the ratio of the concentration of contaminant in organism lipid to the contaminant concentration in water at steady state:

$$BCF = (C_{\text{organism lipid}}) \cdot (C_{\text{water}})^{-1}$$

Similarly,

$$(C_{\text{organism lipid}}) \cdot (C_{\text{water}})^{-1} = (Zf_{\text{organism lipid}}) \cdot (Zf_{\text{water}})^{-1}$$

Because at equilibrium, f is equal in all phases,

$$BCF = (Z_{\text{organism lipid}}) \cdot (Z_{\text{water}})^{-1}$$

In fugacity terms, the lipid-normalized BCF can be calculated simply as the ratio of the two fugacity capacities for the respective lipid and water phases.

Although the transfer processes involved in trophic transfer are more complex than in bioconcentration, simple fugacity-based (thermodynamic) models can still be used to describe accumulation. Additionally, parameters can be included in the model to account for uptake from water as well as accumulation from food. For instance, Thomann (1989) constructed a simplified food chain consisting of phytoplankton, zooplankton, small fish, and predatory fish. Uptake of contaminants by algae consisted of simple lipid-normalized bioconcentration. However, accumulation in planktivores and predators included terms describing not only uptake from water but assimilation from food as well. The latter were all lipid-normalized to correct for variability in accumulation due to lipid content. This model works well but is very dependent on the concentrations in the lower food web that are predicted from equilibrium partitioning theory (Thomann, 1989).

9.6.2.2. Mechanism for Trophic Transfer

The fugacity-based approach to understanding trophic transfer has yielded important information about the mechanisms of trophic transfer and their relationship to lipid levels. One implication of the fugacity model, for instance, is that the fugacity of the food must be higher than the fugacity of the organism for trophic transfer to occur. This can occur if the predator is kinetically limited in achieving steady state compared with the prey. Then the prey will be at a higher thermodynamic chemical potential than the predator. Such a situation can easily occur with large growing predators such as large fish that may never attain the thermodynamic chemical potential of the system, and thus transfer of the contaminant will continue down a chemical activity gradient. It may also be the case that the prey is exposed to sources not available to the predator (e.g., the prey come from the sediment and the predator only experiences the overlying water). This could readily occur because fine sediments are focused into depositional areas that are often out of thermodynamic equilibrium with the overlying water.

More often, however, the predator and prey may be at the same thermodynamic potential relative to the major source (e.g., water). How then can bioaccumulation occur from the prey to the predator? By empirical observation, most organisms in a food web appear to have similar fugacities (Gobas et al., 1993a). Thus, the required lipid–water–lipid partition (chemical activity gradient) that is needed to move a contaminant from ingested food into an aqueous phase and then back to a lipid phase in the organism would seem to be absent. Thus, if trophic transfer were to occur, contaminants would have to move without or against a fugacity gradient. In the absence of active uptake, there is no mechanism to account for this.

A focus on lipid and its influence on contaminant transfer has been invoked to explain this apparent contradiction. When contaminated food is ingested by an organism, the contaminants move into the gastrointestinal tract (GIT) in association with lipids. Once present in the gut, digestion begins to take place. Lipids are dissociated from the bulk of ingested material, acted on by specific enzymes, and transported across the gut wall via specific lipoprotein carriers. Two hypotheses can be invoked for contaminant transfer: (1) the contaminants may stay associated with the lipid material and move with the lipids actively, and (2) the contaminants are left behind while the lipids are digested and absorbed. This has two important consequences: (1) there is a smaller food volume remaining in the gut, resulting in a momentary increase in the contaminant concentration; and (2) the food remaining in the GIT has lost part of the lipid component that sequesters contaminants. Thus, the ability of the remaining food to dissolve the contaminant is reduced and the fugacity capacity of the food goes down, causing the fugacity of the food to become higher than that of the organism. The increase in concentration and fugacity of the contaminant in the food relative to the organism causes the contaminants to passively diffuse food across the gut wall into the organism. Thus, net movement of the contaminant between predator and prey, which appear to have equal fugacity, becomes possible (LeBlanc, 1995; Gobas et al., 1993a; Thomann, 1989). In either mechanism, lipids become the driving force for the bioaccumulation of contaminants from food.

Because lipids are the driving force behind trophic transfer (Gobas et al., 1989), it is important to account for the influence of both lipid in the food and lipid in the organism, which may have contrasting effects. Organism lipid levels have been previously discussed. How will contaminant transfer from a high-lipid food differ from a low-lipid food, for instance? According to the fugacity concept, if the primary phase responsible for dissolving the contaminant is lipid, then the fugacity capacity of high-lipid food should be high, resulting in low fugacity at a given food concentration. In other words, less contaminant should move into the consumer's tissues from a high-lipid food than a low-lipid food. In a single study of this issue, uptake of most hydrophobic contaminants from a low-fat diet was much higher than from a high-fat diet (Gobas et al., 1993b). Although the fugacity-based explanation for this observation may be correct, in the case in which the fugacities in the two food sources are equal, it may be true that the digestibility of the high-lipid food is decreased and fugacity undergoes a greater elevation for the low-lipid food during the transit through the GIT. Discerning the role of food lipid in trophic transfer has been identified as a critical research need (Clark and Mackay, 1991).

9.6.3. Factors Affecting Trophic Transfer

9.6.3.1. Assimilation Efficiency

Trophic transfer of contaminants is frequently assessed and quantified by using an assimilation efficiency (AE) as the critical measurement. At its most basic level, %AE is simply the ratio of the contaminant retained in the consumer's tissues compared with the contaminant ingested (Harkey et al., 1994b). A question in anticipating how AE values will vary in nature is, how will AE be affected when tracking the fate of contaminants with different hydrophobicities? Measurement of the AE for organisms can be extremely difficult unless the food source can be isolated and the exact contaminant concentration determined. For many selective feeding organisms, such as benthic amphipods (e.g., *Diporeia* spp.), isolating the food supply can be problematic and exact measurement of the AE difficult to determine (Harkey et al., 1994b). Thus, interpretation of measured AEs must be performed with care.

Clearly, as the log K_{ow} of a contaminant increases, there is a greater propensity to partition into lipid-rich tissues. However, when the contaminant must be transferred from a lipid phase (food) instead of an aqueous phase, then the relationship may be less straightforward, as described previously. Indeed, several studies indicated that there is an inverse relationship between AE and contaminant hydrophobicity (Bierman, 1990; Muir and Yarechewski, 1988). Thus, as the hydrophobicity of the contaminant increases, the lower the tendency to move from food lipid into the consumer's tissues.

9.6.3.2. Miscellaneous Factors Affecting Assimilation

The relationship between contaminant characteristics and AE can be modified by a variety of factors. The feeding rate and the amount of food ingested, for instance, can exert considerable influence on AE. When identical amounts of contaminants were fed to guppies, *Poecilia reticulata,* in two different volumes of food, the AE appeared to be lower in the case in which more food was used (Clark and Mackay, 1991). In reality, the AE declined because fecal egestion increased when a higher food volume was used. Thus, the thermodynamic tendency for the contaminant to move from food into the organism was not altered between exposures, but the processing of the food was quicker in the case of the high food volume, resulting in diminished contact time in the gut and an apparent reduction in absorption efficiency. Gut retention time is a key factor in determining absorption. Bruner et al. (1994b) found that AEs of several contaminants from sediment into zebra mussels were much lower than AE of the same contaminants from algae in part because the residence time of the sediment in the GIT was much lower.

Gut morphology can also be important in determining AE. For birds, the proximal part of the GIT appears to absorb more contaminants (Serafin, 1984). This may indicate a reduction in absorption efficiency in the distal portions of the GIT. Because the fugacity of the contaminant may be highest when the greatest digestion and removal of lipid has occurred in distal portion of the GIT, AE may decrease as a function of intestinal length.

In short, AE is known to be critical in assessing trophic transfer. Organism lipid levels are clearly important in determining AE. However, a variety of other factors that are not related to the adiposity of the organism is also influential. These interacting factors may obscure the relationship between AE and lipids.

9.7. Biomagnification and Organism Lipids

9.7.1. Is Biomagnification Real?

The issue of biomagnification has been a thorny one for years and has always revolved around lipid levels of each food chain element. An early report of biomagnification was the description of DDT levels in a Lake Michigan food chain (Harrison et al., 1970). In a simple food chain consisting of sediment, amphipods, fish, and herring gulls, they detected DDT concentrations of 14, 410, 3,000–6,000, and 99,000 ppb, respectively. The increase in DDT concentration with each trophic level was attributed to the fact that biomass conversion at each trophic link was <50% whereas DDT transfer was close to 80%. The concentration of DDT, thus, increased with each successive food chain link. Additional concentrating mechanisms were later identified. These included resistance to metabolism, high lipid solubility, and increasing lipid levels with each trophic link (Bierman, 1990).

Early reports of biomagnification were contested as a greater mechanistic understanding of the processes involved in accumulation evolved. In particular, the finding that the contaminant uptake directly from water was by far the fastest route of accumulation, and that contaminant levels in top trophic levels could often be produced by bioconcentration alone (LeBlanc, 1995) caused doubt. In addition, failure to find biomagnification occurring in all ecosystems or in all components of a food chain was problematic. Highly hydrophobic contaminants such as PAHs, for instance, do not biomagnify (Burns and Teal, 1979) primarily as a result of biotransformation. Burrows and Whitton (1983) found that contaminant loads were higher in mayflies than in the predators that ate them. In many cases, contaminant loads varied randomly between trophic levels without discernible patterns (Biddinger and Gloss, 1984; Kay, 1984; Macek et al., 1979). In the absence of a mechanism to account for biomagnification, there was only inconsistent evidence to support its validity.

In a recent review of hundreds of studies on biomagnification, Suedel et al. (1994) concluded that, although the occurrence of biomagnification is much more limited than previously suggested, it is nonetheless a reality in some systems. Biomagnification appears to be limited to a small but very important group of highly lipophilic (log K_{ow} >5), nonmetabolized contaminants that include DDT, DDE, PCBs, toxaphene, and organic forms of mercury and arsenic. Biomagnification is precluded by, among other things, susceptibility to metabolism and high elimination rates and can be enhanced when contaminant elimination rates are slow compared with the energetics of the organism.

9.7.2. A Lipid-Based Model for Biomagnification

Mechanistically, biomagnification appears to consist of a continuation of the same process that leads to trophic transfer of contaminants. That is, as contaminated food is digested in the gut, lipids are absorbed and the food volume decreases. The fugacity of the contaminant in the food, thus, increases above the fugacity of the chemical in the organism and provides the force necessary to drive the contaminant across the gut by diffusion. The process can be increased by digestion of nonlipids (e.g., proteins and carbohydrates) because that process increases the surface area from which sorbed contaminants can be released (DiPinto et al., 1993). In addition, it is possible that the compositional changes of partially digested lipids in the gut will lead to a reduction in fugacity capacity (and increase in fugacity) without a reduction in food volume (LeBlanc 1995; Gobas et al., 1993a).

Although the currently accepted models for both trophic transfer and biomagnification in aquatic systems invoke passive diffusion as the force that transports contaminants across the GIT into the organism, models derived from terrestrial mammals have highlighted the possibility that contaminants can also be co-assimilated with dietary lipids. In co-assimilation, contaminants move across the GIT in association with lipids that are being absorbed; the lipids become the vehicle for contaminant transport. In the second model, the contaminant does not move by diffusion, because the chemical concentration and fugacity in the organism are similar or higher than levels in the food, but rather via an active process that uses lipoprotein carriers. If the passive diffusion model is true, in the strictest sense, then the mode of transport is diffusion created by a fugacity gradient and the site of magnification is the GIT (Gobas et al., 1993b). If, however, the second model is true, then the mode of accumulation is lipid co-assimilation and the site of magnification is in the organism's (or predator's) tissues.

Gobas et al. (1993b) evaluated these two models experimentally in goldfish (*Carassius auratus*) by using chlorinated benzenes and PCBs. In this experiment, groups of goldfish were fed identical concentrations of contaminant in food, with lipid levels ranging from 0 to 13.5%. If the contaminants were transferred from food to fish strictly via lipid co-assimilation, then contaminant uptake should be very low in a diet deficient in lipid because lipids are the primary transport vehicle for the contaminants. By contrast, contaminant uptake in high-lipid diets should be significantly greater than from low-lipid diets if lipid co-assimilation predominates. If passive diffusion is responsible for contaminant transfer, then a dramatic change in dietary lipid content should occasion only a small alteration in the amount of contaminant taken up, assuming that fecal egestion rates do not change between treatments.

The results of the experiment suggested that although diffusion resulting from increased fugacity caused by digestion was the main force driving accumulation, lipid co-assimilation also played a role depending on the hydrophobicity of the contaminant. For the lower log K_{ow} contaminants, there was no difference in uptake efficiencies from low- versus high-lipid diets. That is, diffusion appeared to be the driving force behind accumulation, and magnification was taking place in the GIT. However, for hexa-, octa-, and decachlorobiphenyl, the uptake efficiency was significantly higher from low-fat food than from high-fat food due to greater digestibility (Gobas et al., 1993b). This indicated that the amount of transfer was tied to dietary lipid levels as suggested by the lipid co-assimilation model, albeit in the direction opposite that which was initially predicted. This was attributed to the higher digestibility of the low-lipid food. Still, the fact that contaminant uptake was associated with dietary lipid levels was viewed as evidence for the lipid co-assimilation model.

In fact, experimental results exist to support both models. For instance, a tenfold increase in dietary triglycerides had no effect on benzo[*a*]pyrene (BaP), 3-methylcholanthrene, or aminostilbene accumulation from food in rats (Laher et al., 1984; Kamp and Neumann, 1975). This lends credence to the passive diffusion model. Evidence in support of the lipid co-assimilation model has also been adduced, particularly for extremely hydrophobic contaminants such as BaP (Vetter et al., 1985; Rees et al., 1971).

Given the contradictory experimental results, Gobas et al. (1993b) proposed a model that includes aspects of both the diffusion and lipid co-assimilation mechanisms. In this view, contaminants remain associated with dietary lipids as they move from food into contact with intestinal mucosa. This association is particularly significant in the case of extremely hydrophobic contaminants for which disassociation into the aqueous milieu of the gut is thermodynamically and kinetically unfavorable. Thus, attaching to lipid material during the transport avoids dissolution in an aqueous phase. Lipids are then digested and absorbed across the gut. The concentration and fugacity of the contaminants remaining in the gut then increases, providing the diffusive force needed to transfer the contaminants, as single molecules, into the organism. Thereafter, the contaminants may become reassociated with the products of lipid digestion, particularly triglycerides or lipoproteins, in which form they may circulate in the blood prior to deposition and storage in adipose tissues. Although both diffusion and lipid co-assimilation are components of the revised model, diffusion is identified as the rate-limiting step in the transfer process. As a consequence, the actual magnification of the contaminant residue level is seen as taking place in the GIT, not in the consumer's tissues (Gobas et al., 1993b).

9.7.3. Current Issues in Biomagnification and Relationship to Lipids

Despite the availability of a detailed mechanism to explain biomagnification and an abundance of residue data demonstrating biomagnification for a limited group of contaminants with fairly specific physicochemical requirements, the reality of biomagnification continues to be questioned. LeBlanc (1995) has suggested that much of what is misinterpreted as biomagnification is, in fact, simply bioconcentration. In support of this hypothesis, LeBlanc notes, in his limited survey, that lipid levels tend to increase with each trophic levels on a wet weight basis, lipids in phytoplankton average 0.5% ($n = 1$); in invertebrates, lipids average 1.8% ($n = 8$); and in fish, lipids average 5.4% ($n = 10$) (LeBlanc, 1995). Thus, contaminant loads would be expected to be higher in each successive trophic level due simply to passive uptake from water. Additionally, LeBlanc (1995) argues that the composition of lipid changes between trophic levels. If these compositional changes cause the fugacity capacity of lipids at higher trophic levels to increase, then even an increase in lipid levels would not be required to cause an increase in bioconcentration (or bioaccumulation). Finally, LeBlanc (1995) argues that differences in lipid elimination mechanisms between trophic levels could account for elevated contaminant loads in top trophic levels. That is, animal size tends to increase with trophic level. In addition, elimination of contaminants from lipid storage (and subsequent excretion from the body) requires that the contaminant be removed from lipid storage and transferred to an excretory organ, most likely the gill (respiratory organ) or liver (or functional equivalent). Liver cells have specific membrane sites across which contaminants must pass, and these elimination sites constitute a small fraction of the total organ surface area. As animal size increases, the relative abundance of storage-to-elimination sites rises. Also, as organism size increases, the relative size of the respiratory organ decreases. As a result, excretion via passive diffusion either via a liver function or through the respiratory system is much lower in large animals than in smaller animals at lower trophic levels. Similar findings of limited biomagnification were presented for freshwater benthic organisms (Bierman, 1990).

Although the arguments against biomagnification cited above are compelling, there are some situations in which accumulation of a contaminant in an organism above residues found in its food cannot be explained in any way other than biomagnification. For instance, Rasmussen et al. (1990) compared PCB concentrations in a single predatory fish species, lake trout (*Salvelinus namaycush*) from more than 80 different sites throughout Ontario. Despite the fact that the lake trout in all sites experienced similar environmental exposure levels to PCBs, the actual concentrations of PCBs in lake trout varied from 15 to 10,000 ppb. The between-lake variation in PCB content in lake trout was attributed to two factors: length of the food chain and increase in lipid levels as a function of food chain length. Each trophic link in the food chain resulted in a magnification of PCB concentrations by a factor of approximately 3–5. Thus, when the length of the food chain was relatively short, the lake trout were much less contaminated than when the food chain was longer. Lipid levels also increased with food chain length. However, lipid levels increased by a factor of 1.5 with each trophic link and cannot, therefore, account for the magnification of PCBs at each step. In other words, biomagnification of PCB residues from food must be occurring (Rowan and Rasmussen, 1992). A trophic position model for lake trout has been developed for both PCB and mercury biomagnification (Vanderzanden and Rasmussen, 1996). Similarly, biomagnification was found to account for high K_{ow} PCB congeners in white bass (*Morone chysops*) in the Lake Erie food web based on a fugacity analysis (Russell et al., 1995) and in fresh water, Lake Nieuwe Meer, in The Netherlands (van der Oost et al., 1988). Further, the presumed increase in total lipid content with each trophic step does not always occur while biomagnification is observed (Broman et al., 1992), thus refuting the arguments that it is increases in lipid content driving the presumed biomagnification put forth by LeBlanc (1995). In situ biomagnification has been difficult to demonstrate, in part because of the difficulty in placing organisms at appropriate levels within a food chain. However, with the use of stable isotope analysis, the biomagnification of polychlorinated dibenzodioxins and polychlorinated dibenzofurans was clearly demonstrated for a Baltic Sea food chain (Broman et al., 1992).

In short, the relative contributions of contaminated food versus contaminated media may vary between ecosystems or even between organisms within an ecosystem. However, residue data suggest that biomagnification of specific contaminants occurs in nature. This, combined with knowledge of a mechanism for understanding of the process of biomagnification and data that show magnification of residues from food when other avenues of contaminant exposure are precluded, provides strong support for the validity of biomagnification.

9.8. Lipids and Transgenerational Transfer of Contaminants

Loss of stored contaminant residues in female animals through reproduction is now understood as a key elimination mechanism, particularly in long-lived species (Sijm et al., 1992). For instance, contaminant residues in adult birds may be passed to embryos via the egg (Burger and Gochfeld, 1993; Jarman et al., 1993; Heinz et al., 1989). In some cases, the concentrations of contaminants in eggs exceed the concentrations found in parental tissues (Heinz, 1993; Tillitt et al., 1992). Similar results have been obtained for reptiles (e.g., snapping turtles, in which lipid storage of contaminants in adult adipose tissue often protects the adult turtle from overt symptoms of poisoning). However, liberal transfer of maternal residues to eggs has been documented several times (Loganathan et al., 1995; Bishop et al., 1994; Struger et al., 1993). Physiological effects in contaminated embryos of both birds and reptiles such as depressions in the titers of key hormones (e.g., estradiol or enzymes) (Chen et al., 1994; Trust et al., 1994), gross structural abnormalities (Hansen, 1994), embryo death, and complete reproductive failure (van den Berg et al., 1994; Bishop et al., 1991) have been attributed to transgenerational transfer of contaminant loads.

Although reproductive processes are infrequently studied with contaminant fate as a focus, it is clear from what is known of reproductive physiology that lipid metabolism is likely to play a key role in transgenerational contaminant transfer. In birds, reptiles, amphibians, and fish, developing oocytes take up large amounts of maternally derived vitellogenin, which consists of about 20% lipid, to serve as an energy source during subsequent development. In addition, lipids are also taken up directly from maternal stores or synthesized de novo by the embryo from maternal extrahepatic lipid (Mommen and Walsh, 1988). The development of the embryo can be reasonably viewed as taking place in a lipid-rich environment. If the maternal lipids are contaminated, then transfer of the contaminants to the embryo could easily take place following the normal pathways of lipid deposition in the embryo during reproduction. Further, because the lipid content of the embryo is high, relative to maternal tissues, the increase in lipid content will increase the fugacity capacity of the embryo, decrease its fugacity (for a given concentration, increasing Z decreases f), and increase the tendency of the contaminant to move into the embryo. That is, a mechanism for explaining observed "reproductive magnification" of residues in embryonic tissues exists and relates directly to lipid content.

The above scenario is largely unstudied but presents several testable hypotheses that could serve as the focus for additional research. For instance, maternal contaminant loads (lipid normalized) should decrease after a brood has been produced if maternal lipid reserves are the source of embryonic contamination. There is at least one report confirming this assertion in fish (Sijm et al., 1992), but data in other species are needed. In addition, the hypothesis suggests that residues in embryos should increase in proportion to the lipid content of the embryo. Monitoring those changes throughout larval development should be informative. In any event, the primacy of lipid levels in determining all aspects of contaminant uptake highlights the need to examine its role in reproduction because the biological impacts on reproduction appear to be detrimental.

9.9. Conclusions

Lipids are the dominant force in determining organic contaminant accumulation in aquatic organisms. Lipid normalization eliminates most variability between species in bioconcentration studies and is, thus, useful in making predictions about bioconcentration from physical parameters such as log K_{ow}. Normalizing contaminant loads to lipid levels is also helpful in cases in which the contaminant must leave a lipid compartment and negotiate an intermediate aqueous phase before final deposition in a second lipid compartment (e.g., bioaccumulation and trophic transfer). Evidence varies as to whether total lipids or the abundance of specific lipid subclasses are the most relevant referent, but it is clear that lipid normalization will significantly reduce variation for comparing contaminant loads between species. This is an area that should be studied further.

Studies of the relationships between lipids and contaminant accumulation have led to new insights on obviously important but seldom studied phenomena. The role of lipids, for instance, in producing membrane narcosis is pivotal, and its study may elucidate important details of the mechanism for narcosis and how toxins can affect lipid levels, resulting in alterations in energy (lipid) stores and perhaps membrane function. In addition, lipid metabolism appears to account for biomagnification in which contaminants appear to accumulate in predators against a concentration gradient. Finally, study of lipid mobilization and deposition in reproduction may help to define the ability of contaminants to move into filial generations during reproduction; it may also provide insight into the sorts of toxin-induced biological effects that can be expected in contaminated offspring. In short, our ability to predict the movement of organic contaminants in aquatic systems and to project the effects of that contamination depends on an understanding of the importance of lipids. Because these are the key elements of risk assessment, our ability to conduct accurate assessments may reasonably be seen to depend on our knowledge of lipids and their dynamics.

Acknowledgments. We thank Duane Gossiaux for his help in generating the graphics for Figure 9.1. This chapter is GLERL contribution 1024.

References

Arts, M.T.; Headley, J.V.; Peru, K.M. Persistence of herbicide residues in *Gammarus lacustris* (Crustacea: Amphipoda) in prairie wetlands. Environ. Toxicol. Chem. 15:481–488; 1996.

Arts, M.T.; Ferguson, M.E.; Glozier, N.E.; Robarts, R.D.; Donald, D.B. Spatial and temporal variability in lipid dynamics of common amphipods: assessing the potential for uptake of lipophilic contaminants. Ecotoxicology 4:91–113; 1995.

Axelman, J.; Broman, D.; Naf, C.; Pettersen, H. Compound dependence of the relationship log K_{ow} and log BCF_L. Environ. Sci. Pollut. Res. 2:33–36; 1995.

Banerjee, S.; Baughman, G.L. Bioconcentration factors and lipid solubility. Environ. Sci. Technol. 25:536–539; 1991.

Barron, M.G. Bioconcentration. Environ. Sci. Technol. 24:1612–1618; 1990.

Bickel, M.H. The role of adipose tissue in the distribution and storage of drugs. Prog. Drug Res. 28:273–303; 1994.

Biddinger, G.R.; Gloss, S.P. The importance of trophic transfer in the bioaccumulation of chemical contaminants in aquatic systems. Residue Rev. 91:103–145; 1984.

Bierman, V.J. Equilibrium partitioning and biomagnification of organic chemicals in benthic animals. Environ. Sci. Technol. 24:1407–1412; 1990.

Bishop, C.A.; Brown, G.P.; Brooks, R.J.; Lean, D.R.S.; Carey, J.H. Organochlorine contaminant concentrations and their relationship to the body size and clutch characteristics of the female common snapping turtle in Lake Ontario, Canada. Arch. Environ. Contam. Toxicol. 27:82–87; 1994.

Bishop, C.A.; Brooks, R.J.; Carey, J.H.; Ng, P.; Norstrom, R.J.; Lean, D.R.S. The case for cause–effect linkage between environmental contamination and development in eggs of the common snapping turtle from Ontario, Canada. J. Toxicol. Environ. Health 33:521–547; 1991.

Bligh, E.G.; Dyer, W.J. A rapid method of total lipid extraction and purification. Can. J. Biochem. Physiol. 39:911–917; 1959.

Brannon J.M.; Price, C.B.; Reiley, F.J., Jr., Pennington, J.C.; McFarland, V.A. Effects of sediment organic carbon on distribution of radiolabeled fluoranthene and PCBs among sediment, interstitial water and biota. Bull. Environ. Contam. Toxicol. 51:873–880; 1993.

Broman, D.; Näf, C.; Rolff, C.; Zebühr, R.; Fry, B.; Hobbie, J. Using ratios of stable nitrogen isotopes to estimate bioaccumulation and flux of polyclorinated dibenzo-p-dioxins (PCDDs) and dibenzofurans (PCDFs) in two food chains from the northern Baltic. Environ. Toxicol. Chem. 11:331–345; 1992.

Bruger, J.; Gochfeld, M. Lead and cadmium accumulation in eggs and fledgling seabirds in the New York Bight. Environ. Toxicol. Chem. 12:261–267; 1993.

Bruggerman, W.A.; Martron, L.B.J.M.; Koolman, D.; Hutzinger, O. Accumulation and elimination kinetics of di-, tri-, and tetrachlorobiphenyls by goldfish after dietary and aqueous exposure. Chemosphere 10:811–815; 1981.

Bruner, K.A.; Fisher, S.W.; Landrum, P.F. The role of the zebra mussel, *Dreissena polymorpha,* in contaminant cycling. I. The effect of body size and lipid content on the bioconcentration of PCBs and PAHs. J. Great Lakes Res. 20:725–734; 1994a.

Bruner, K.A.; Fisher, S.W.; Landrum, P.F. The role of the zebra mussel, *Dreissena polymorpha,* in contaminant cycling. II. Contaminant accumulation from ingested algal and suspended sediment particles and contaminant trophic transfer from zebra mussel feces to the benthic invertebrate, *Gammarus fasciatus.* J. Great Lakes Res. 20:735–750; 1994b.

Burns, K.A.; Teal, J.M. The West Falmouth oil spill: hydrocarbons in the salt marsh ecosystem. Est. Coastal Mar. Sci. 8:349–360; 1979.

Burrows, I.G.; Whitton, B.A. Heavy metals in water, sediments and invertebrates from a metal contaminated river free of organic pollution. Hydrobiologia 106:263–273; 1983.

Chen, S.W.; Dzuik, P.J.; Francis, B.M. Effect of four environmental toxicants on plasma Ca and estradiol 17B and hepatic P450 in laying hens. Environ. Toxicol. Chem. 13:789–795; 1994.

Chiou, C.T.; Freed, V.H.; Schmedding, D.W.; Kohnert, R.L. Partition coefficient and bioaccumulation of selected organic chemicals. Environ. Sci. Technol. 11:475–478; 1977.

Clark, K.E.; Mackay, D. Dietary uptake and biomagnification of four chlorinated hydrocarbons by guppies. Environ. Toxicol. Chem. 10:1205–1217; 1991.

Connell, D.W. Bioaccumulation behavior of persistent organic chemicals with aquatic organisms. Rev. Environ. Contam. Toxicol. 101:117–154; 1988.

Cravedi, J.P.; Tulliez, J. Metabolism of n-alkanes and their incorporation into lipids in rainbow trout. Environ. Res. 39:180–187; 1986.

de Boer, J. Chlorobiphenyls in bound and non-bound lipids of fishes: comparison of different extraction methods. Chemosphere 17:1803–1810; 1988.

DiPinto, L.M.; Coull, B.C.; Chandler, G.T. Lethal and sublethal effects of sediment-associated PCB Arochlor 1254 on a meiobenthic copepod. Environ. Toxicol. Chem. 12:1909–1918; 1993.

DiToro, D.M.; Zarba, C.S.; Hansen, D.J.; Berry, W.J.; Swartz, R.C.; Cowan, C.E.; Pavlou, S.P.; Allen, H.E.; Thomas, N.A.; Paquin, P.R. Technical basis for establishing sediment quality criteria for nonionic organic chemicals by using equilibrium partitioning. Environ. Toxicol. Chem. 12:1541–1583; 1991.

Ellis, G.S.; Huckins, J.N.; Rostad, C.E.; Schmitt, C.J.; Petty, J.D.; MacCarthy, P. Evaluation of lipid-containing semipermeable membrane devices for monitoring organochlorine contaminants in the Upper Mississippi River. Environ. Toxicol. Chem. 14:1875–1884; 1995.

Ewald, G. Role of lipids in the fate of organochlorine compounds in aquatic ecosystems. Doctoral dissertation, Department of Ecology, Lund University, Lund, Sweden; 1996.

Ewald, G.; Larsson, P. Partitioning of ^{14}C-labelled 2,2′, 4,4′-tetrachlorobiphenyl between water and fish lipids. Environ. Toxicol. Chem. 13:1577–1580; 1994.

Fisher, S.W.; Gossiaux, D.C.; Bruner, K.A.; Landrum, P.F. Investigations of the toxicokinetics of hydrophobic contaminants in the zebra mussel (*Dreissena polymorpha*). In: Nalepa, T.F.; Schloesser, D.W., eds. Zebra Mussels: Biology, Impacts and Control. Boca Raton, FL: CRC Press; 1993:p. 453–464.

Folch, J.; Lees, M.; Cloane Stanley, G.H. A simple method for the isolation and purification of total lipids from animal tissues. J. Biol. Chem. 226:497–509; 1957.

Gardner, W.S.; Landrum, P.F.; Chandler, J.F. Lipid-partitioning and disposition of benzo(a)pyrene and hexachlrorbiphenyl in Lake Michigan, *Pontoporeia hoyi*. Environ. Toxicol. Chem. 10:35–46; 1990.

Gardner, W.S.; Frez, W.A.; Cichocki, E.A.; Parrish, C.C. Micromethod for lipids in aquatic invertebrates. Limnol. Oceanogr. 30:1100–1105; 1985.

Geyer, H.J.; Scheunert, I.; Bruggeman, R.; Matthies, M.; Steinberg, C.E.W.; Zitko, V.; Kettrup, A.; Garrison, W. The relevance of aquatic organisms lipid content to the toxicity of lipophilic chemicals: toxicity of lindane to different fish species. Ecotoxicol. Environ. Safety 28:53–70; 1994.

Geyer, H.J.; Scheunert, I.; Rapp, K.; Gebefugi, I.; Steinberg, C.; Kettrup, A. The relevance of fat content in toxicity of lipophilic chemicals to terrestrial animals with special reference to dieldrin and 2,3,7,8-tetrachlorodibenzo-p-dioxin. Ecotoxicol. Environ. Safety 26:45–60; 1993.

Gobas, F.A.P.C.; Mackay, D. Dynamics of hydrophobic organic chemical bioconcentration in fish. Environ. Toxicol. Chem. 6:495–504; 1987.

Gobas, F.A.P.C.; Zhang, X.; Wells, R. Gastrointestinal magnification: the mechanism of biomagnification and food chain accumulation of organic chemicals. Environ. Sci. Technol. 27:2855–2864; 1993a.

Gobas, F.A.P.C.; McCorquodale, J.R.; Haffner, G.D. Intestinal absorption and biomagnification of organochlorines. Environ. Toxicol. Chem. 12:567–577; 1993b.

Gobas, F.A.P.C.; McNeil, E.J.; Lovett-Doust, L.; Haffner, G.D. Bioconcentration of chlorinated aromatic hydrocarbons in aquatic macrophytes. Environ. Sci. Technol. 25:924–929; 1991.

Gobas, F.A.P.C.; Clark, K.E.; Shiu, W.Y.; Mackay, D. Bioconcentration of polybrominated benzenes and biphenyls and related superhydrophobic chemicals in fish: role of bioavailability and fecal elimination. Environ. Toxicol. Chem. 8:231–247; 1989.

Håkansson, H.; Sudlin, P.; Andersson, T.; Brunström, B.; Dencker, L.; Engwall, M.; Ewald, G.; Gilek, M.; Holm, G.; Honkassalo, S.; Idestam-Almquist, J.; Jonsson, P.; Kautsky, N.; Lundburg, G.; Lund-Kvernheim, A.; Martinsen, K.; Norrgren, L.; Pesonen, M.; Rundgren, M.; Sålberg, M.; Tarkpea, M.; Wesén, C. In vivo and in vitro toxicity of fractionated fish lipids, with particular regard to their content of chlorinated organic compounds. Pharmacol. Toxicol. 69:344–345; 1991.

Hansen, L.G. Halogenated aromatic compounds. In: Hansen, L.G.; Shane, B.S., eds. Basic Environmental Toxicology. Ann Arbor, MI: CRC Press; 1994:p. 199–230.

Harkey, G.A.; Landrum, P.F.; Kaline, S.J. Comparison of whole sediment, elutriate, and porewater for use in assessing sediment-associated organic contaminants in bioaccumulation assays. Environ Toxicol. Chem. 13:1315–1329; 1994a.

Harkey, G.A.; Lydy, M.J.; Kukkonen, J.; Landrum, P.F. Feeding selectivity and assimilation of PAH and PCB in *Diporeia* spp. Environ. Toxicol. Chem. 13:1445–1455; 1994b.

Harrison, H.L.; Loucks, O.L.; Mitchell, J.W.; Parkhurst, D.F.; Tracy, C.R.; Watts D.G.; Yannacone, V.J., Jr. System studies of DDT transport. Science 170:503–508; 1970.

Hawker, D.W.; Connell, D.W. Bioconcentration of lipophilic compounds by some aquatic organisms. Ecotoxicol. Environ. Safety 11:184–197; 1986.

Heinz, G.H. Selenium accumulation and loss in mallard eggs. Environ. Toxicol. Chem. 12:775–778; 1993.

Heinz, G.H.; Hoffamn, D.J.; Gold, L.G. Impaired reproduction of mallards fed and organic form of selenium. J. Wildl. Manage. 53:418–428; 1989.

Huckins, J.N.; Tubergen, M.W.; Manuweera, G.K. Semipermeable membrane devices containing model lipid: a new approach to monitoring and estimating their bioconcentration potential. Chemosphere 20:533–552; 1990a.

Huckins, J.N.; Tubergen, M.W.; Lebo, J.A.; Gale, R.W.; Schwartz, T.R. Polymeric film dialysis in organic solvent media for cleanup of organic contaminants. J. Assoc. Off. Anal. Chem. 73:290–293; 1990b.

Hunn, J.B.; Allen, J.L. Movement of drugs across the gills of fishes. Annu. Rev. Pharmacol. 14:47–55; 1974.

Jarman, W.M.; Burns, S.A.; Chang, R.R.; Stephens, R.D.; Norstrom, R.J.; Simon, M.; Linthicum, J. Determination of PCDDs, PCDFs and PCBs in California peregrine falcons (Falco peregrinus) and their eggs. Environ. Toxicol. Chem. 12:105–114; 1993.

Kaiser, K.L.E.; Vladmanis, I. Apparent octanol/water partition coefficients of pentachlorophenol as a function of pH. Can. J. Chem. 60:2104–2106; 1982.

Kamp, J.D.; Neumann, H.G. Absorption of carcinogens into the thoracic duct lymph of the rat: aminostilbene derivatives and 3-methylchloanthrene. Xenobiotica 5:717–727; 1975.

Kay, S.H. Cadmium in food webs. Residue Rev. 96:13–43; 1984.

Kenega, E.E. Correlation of concentration factors of chemicals in aquatic and terrestrial organism with their physical and chemical properties. Environ. Sci. Technol. 14:553–556; 1980.

Kucklick, J.R.; Harvey, H.R.; Ostrom, P.H.; Ostrom, N.E.; Baker, J.E. Organochlorine dynamics in the pelagic food web of Lake Baikal. Environ. Toxicol. Chem. 15:1388–1400; 1996.

Kucklick, J.R.; Bidelman, T.F.; McConnell, L.L.; Walla, M.D.; Ivanov, G.P. Organo-chlorines in the water and biota of Lake Baikal, Siberia. Environ. Sci. Technol. 28:31–37; 1994.

Kukkonen, J.; Landrum, P.F. Distribution of organic carbon and organic xenobiotics among different particle-size fractions in sediments. Chemosphere 32:1063–1076; 1996.

Kukkonen, J.; Landrum, P.F. Effects of sediment-bound polydimethylsiloxane on the bio-availability and distribution of benzo(a)pyrene in lake sediment to Lumbriculus variegatus. Environ. Toxicol. Chem. 14:523–531; 1995.

Laher, J.M.; Rigler, M.W.; Vetter, R.D.; Barrowman, J.A.; Patton, J.S. Similar bio-availability and lymphatic transport of benzo(a)pyrene when administered to rats in different amounts of dietary fat. J. Lipid Res. 25:1337–1342; 1984.

Lake, J.L.; Rubenstein, N.I.; Lee, H., II; Lake, C.A.; Heltshe, J.; Pavignano, S. Equilibrium partitioning and bioaccumulation of sediment-associated contaminants by infaunal organisms. Environ. Toxicol. Chem. 9:1095–1106; 1990.

Lal, B.; Singh, T.P. Impact of pesticides on lipid metabolism in the freshwater catfish, *Clarias batrachus,* during the vitellogenic phase of its annual reproductive cycle. Eco-toxicol. Environ. Safety 13:13–23; 1987.

Landrum, P.F. Toxicokinetics of organic xenobiotics in the amphipod, *Pontoporeia hoyi:* role of physiological and environmental variables. Aquat. Toxicol. 12:245–271; 1988.

Landrum, P.F.; Dupuis, W.S. Toxicity and toxicokinetics of pentachlorophenol and carbaryl to *Pontoporeia hoyi* and *Mysis* relicta. In: Landis, W.G.; Van der Schalie, W. H., eds. Aquatic Toxicology and Risk Assessment, 13th vol. ASTM STP 1096. Philadelphia: American Society for Testing and Materials; 1990:p. 278–289.

Landrum, P.F.; Faust, W.R. The role of sediment composition on the bioavailability of laboratory-dosed sediment-associated contaminants to the amphipod, *Diporeia* spp. Chem. Speciat. Bioavail. 6:85–92; 1994.

Landrum, P.F.; Faust, W.R. Effect of variation in sediment composition on the uptake rate coefficient for selected PCB and PAH congeners by the amphipod, *Diporeia* spp. In: Mayes, M.A.; Barron, M.G., eds. Aquatic Toxicology and Risk Assessment, vol. 14. ASTM STP 1124. Philadelphia: American Society for Testing and Materials; 1991:p. 263–279.

Landrum, P.F.; Robbins, J.A. Bioavailability of sediment associated contaminants: a review and simulation model. In: Baudo, R.; Giesy, J.P.; Muntau, H., eds. Sediments: Chemistry and Toxicity of In-Place Pollutants. Chelsea, MI: Lewis Publishers; 1990:p. 237–263.

Landrum, P.F.; Dupuis, W.S.; Kukkonen, J. Toxicity and toxicokinetics of sediment-associated pyrene in *Diporeia* spp.: examination of equilibrium partitioning theory and residue effects for assessing hazard. Environ. Toxicol. Chem. 13:1769–1780; 1994.

Landrum, P.F.; Lee, H.; Lydy, M.J. Toxicokinetics in aquatic systems: model comparisons and use in hazard assessment. Environ. Toxicol. Chem. 11:1709–1725; 1992.

Landrum, P.F.; Eadie, B.J.; Faust, W.R. Toxicokinetics and toxicity of a mixture of sediment-associated polycyclic aromatic hydrocarbons to the amphipod *Diporeia* spp. Environ. Toxicol. Chem. 10:35–46; 1991.

Lassiter, R.R.; Hallam, T.G. Survival of the fattest: Implications for acute effects of lipophilic chemicals on aquatic populations. Environ. Toxicol. Chem. 9:585–595; 1990.

LeBlanc, G.A. Trophic level differences in bioconcentration of chemicals: Implications in assessing environmental biomagnification. Environ. Sci. Technol. 29:154–160; 1995.

Lee, H., II. Models, muddles and mud: predicting bioaccumulation of sediment associated pollutants. In: Burton, G.A., ed. Sediment Toxicity Assessment. Ann Arbor, MI: Lewis Publishers; 1992:p. 73–94.

Leo, A.; Hansch, C.; Elkins, D. Partition coefficients and their uses. Chem. Rev. 71:525–616; 1971.

Lien, G.J.; McKim, J.M. Predicting branchial and cutaneous uptake of 2,5,2',5'-[14]tetrachlorobiphenyl in fathead minnows (*Pimephales promelas*) and Japanese medaka (*Oryzias latipes*): rate limiting factors. Aquat. Toxicol. 27:15–32; 1993.

Lien, G.J.; Nichols, J.W.; McKim, J.M.; Gallinat, C.A. Modeling the accumulation of three waterborne chlorinated ethanes in fathead minnows (*Pimephales promelas*): a physiologically based approach. Environ. Toxicol. Chem. 13:1195–1205; 1994.

Loganathan, B.; Kannan, K.; Watanabe, I.; Kawano, M.; Irvine, K.; Kumar, S.; Sikka, H. Isomer-specific determination and toxic evaluation of polychlorinated biphenyls and dioxins. Environ. Sci. Technol. 29:1832–1838; 1995.

Ma, L.; Taraschi, T.F.; Janes, N. Nuclear magnetic resonance partitioning studies of solute action in lipid membranes. Bull. Mag. Reson. 14:293–98; 1992.

McCarty, L.S.; Mackay, D. Enhancing ecotoxicological modeling and assessment. Environ. Sci. Technol. 27:1719–1728; 1993.

Macek, K.J.; Petrocelli, S.R.; Sleight, B.H., III. Consideration in assessing the potential for and significance of, biomagnification of chemical residues in aquatic foodchains. In: McFarland, V.A. Activity-based evaluation of potential bioaccumulation from sediments. In: Montgomery, R.L.; Leach, J.W., eds. Dredging and Dredged Material Disposal Proceedings of the Conference Dredging '84. New York: American Society of Civil Engineering; 1984:p. 461–466.

McFarland, V.A.; Clark, J.U. Environmental occurrence, abundance, and potential toxicity of polychlorinated biphenyl congeners: considerations for a congener-specific analysis. Environ. Health Perspect. 81:225–239; 1989.

Mackay, D. Multimedia Environmental Models: The Fugacity Approach. Chelsea, MI: Lewis Publishers; 1991.

Mackay, D. Correlation of bioconcentration factors. Environ. Sci. Technol. 16:274–278; 1982.

Marking L.L.; Kimerle, R. A., eds. Aquatic Toxicology. ASTM STP 667. Philadelphia: American Society for Testing and Materials; 1979:p. 251–268.

Mommen, T.P.; Walsh, P. J. Vitellogenesis and oocyte assembly. Vol. 11. In: Hoar, W.S.; Randall, D. J., eds. Fish Physiology. New York: Academic Press; 1988:p. 347–406.

Moriarty, F. Exposure and residues. In: Moriarty, F., ed., Organochlorine Insecticides: Persistent Organic Pollutants. New York: Academic Press; 1975:p. 29–72.

Muir, D.C.G.; Yarechewski, G.R.B. Dietary accumulation of four chlorinated dioxin congeners by rainbow trout and fathead minnows. Environ. Toxicol. Chem. 7:227–235; 1988.

Mullins, L.J. Some physical mechanisms in narcosis. Chem. Rev. 54:289–323; 1954.

Neely, W.B.; Branson, D.R.; Blau, G.E. Partition coefficient measure bioconcentration potential of organic chemicals in fish. Environ. Sci. Technol. 8:1113–1115; 1974.

Nichols, J.W.; McKim, J.M.; Lien, G.J.; Hoffman, A.D.; Bretelsen, S.L. Physiologically-based toxicokinetic modeling of three waterborne chloroethanes in rainbow trout (*Oncorhynchus mykiss*). Toxicol. Appl. Pharmacol. 110:374–389; 1991.

Nichols, J.W.; McKim, J.M.; Anderson, M.E.; Gargas, H.J.; Clewell, H.J., III; Erickson, R.J. A physiologically-based toxicokinetic model for the uptake and disposition of waterborne organic chemicals in fish. Toxicol. Appl. Pharmacol. 106:433–447; 1990.

Opperhuizen, A.; Damen, H.W.J.; Asyee, G.M.; Van der Steen, J.M.D.; Hutzinger, O. Uptake and elimination by fish of polydimethylsiloxanes (silicones) after dietary and aqueous exposure. Toxicol. Environ. Chem. 13:265–285; 1987.

Opperhuizen, A.; Velde, E.W.; Gobas, F.A.P.C.; Liem, D.A.K.; Van der Steen, J.M.D.; Hutzinger, O. Relationship between bioconcentration in fish and steric factors of hydrophobic chemicals. Chemosphere 14:1871–1896; 1985.

Pawlisz, A.V.; Peters, R.H. A radioactive tracer technique for the study of lethal body burdens of narcotic organic chemicals in *Daphnia magna*. Environ. Sci. Technol. 27:2795–2800; 1993a.

Pawlisz, A.V.; Peters, R.H. A test of the equipotency of internal burdens of nine narcotic chemicals using *Daphnia magna*. Environ. Sci. Technol. 27:2801–2806; 1993b.

Prest, H.F.; Jarman, W.M.; Burns, S.A.; Weismüller, T.; Martin, M.; Huckins, J.N. Passive water sampling via semipermeable membrane devices (SPMDS) in concert with bivalves in the Sacramento/San Joaquin river delta. Chemosphere 25:1811–1823; 1992.

Randall, R.C.; Lee, H., II; Ozretich, R.J.; Lake, J.L.; Purell, R.J. Evaluation of selected lipid methods for normalizing pollutant bioaccumulation. Environ. Toxicol. Chem. 10:1431–1436; 1991.

Rasmussen, J.B.; Rowan, D.J.; Lean, D.R.S.; Carey, J. H. Food chain structure in Ontario lakes determines PCB levels in lake trout (*Salvelinus namaycush*) and other pelagic fish. Can. J. Fish Aquat. Sci. 47:2030–2038; 1990.

Rees, D.E.; Mandelstam, P.; Lowry, J.Q.; Lipscomb, L.N. A study of the mechanism of intestinal absorption of benzo(a)pyrene. Biochim. Biophys. Acta 225:96–107; 1971.

Reynoldson, T.B. Interactions between sediment contaminants and benthic organisms. Hydrobiology 149:53–66; 1987.

Rowan, D.J.; Rasmussen, J.B. Why don't Great Lakes fish reflect environmental concentrations of organic contaminants?—An analysis of between-lake variability in the ecological partitioning of PCBs and DDT. J. Great Lakes Res. 18:724–741; 1992.

Rubenstein, N.I.; Gilliam, W.T.; Gregory, N.R. Dietary accumulation of PCBs from a contaminated sediment source by a demersal fish (*Leiostomus xanthurus*). Aquat. Toxicol. 5:331–342; 1984.

Russell, R.W.; Lazar, R.; Haffner, G.D. Biomagnification of organochlorines in Lake Erie white bass. Environ. Toxicol. Chem. 14:719–724; 1995.

Saito, S.; Tateno, C.; Tanoue, A.; Matsuda, T. Electron microscope autoradiographic examination of uptake behavior of lipophilic chemicals into fish gill. Ecotoxicol. Environ. Safety 19:184–191; 1990.

Schneider, R. Polychlorinated biphenyls (PCBs) in cod tissues from the western Baltic: significance of equilibrium partitioning and lipid composition in the bioaccumulation of lipophilic pollutants in gill-breathing animals. Meeresforsch. 29:69–79; 1981.

Schultz, I.R.; Hayton, W.L. Body size and the toxicokinetics of trifluralin in rainbow trout. Toxicol. Appl. Pharmacol. 129:138–145; 1994.

Serafin, J.A. Avian species differences in intestinal absorption of xenobiotics (PCBs, Dieldrin, Hg^{2+}). Comp. Biochem. Physiol. 78:491–496; 1984.

Shaw, G.R.; Connell, D.W. Factors controlling bioaccumulation in food chains. In: Waid, J.S., ed. PCBs and the Environment. vol. I. Boca Raton, FL: CRC Press; 1986:p. 135–141.

Shaw, G.R.; Connell, D.W. Physicochemical properties controlling polychlorinated biphenyl (PCB) concentrations in aquatic organisms. Environ. Sci. Technol. 18:18–23; 1984.

Sijm, D.T.H.M.; Seinen, W.; Opperhuizen, A. Life-cycle biomagnification study in fish. Environ. Sci. Technol. 26:2162–2174; 1992.

Södergren, A. Solvent-filled dialysis membranes simulate uptake of pollutants by aquatic organisms. Environ. Sci. Technol. 21:855–863; 1987.

Södergren, A.; Okla, L. Simulation of interfacial mechanisms with dialysis membranes to study uptake and elimination of persistent pollutants in aquatic organisms. Verh. Int. Verein. Limnol. 23:1633–1638; 1988.

Stange, K.; Swackhamer, D.L. Factors affecting phytoplankton species-specific differences in accumulation of 40 polychlorinated biphenyls (PCBs). Environ. Toxicol. Chem. 13:1849–1860; 1994.

Stehly, G.R.; Hayton, W.L. Effect of pH on the accumulation kinetics of pentachlorophenol in goldfish. Arch. Environ. Contam. Toxicol. 19:464–470; 1990.

Struger, J.; Elliot, J.E.; Bishop, C.A.; Obbard, M.E.; Norstrom, R.J.; Weseloh, D.V.; Simon, M.; Ng, P. Environmental contaminants of the common snapping turtle from the Great Lakes-St. Lawrence River basin of Ontario, Canada (1981–1984). J. Great Lakes Res. 19:681–694; 1993.

Suedel, B.C.; Boraczek, J.A.; Peddicord, R.K.; Clifford, P.A.; Dillon, T.M. Trophic transfer and biomagnification potential of contaminants in aquatic ecosystems. Rev. Environ. Contam. Toxicol. 136:21–84; 1994.

Swackhamer, D.L.; Hites, R.A. Occurrence and bioaccumulation of organochlorine compounds in fishes from Siskiwit Lake, Isle Royal, Lake Superior. Environ. Sci. Technol. 22:543–548; 1988.

Swartz, R.C.; Lee, H., II. Biological processes affecting distribution of pollutants in marine sediments. Part I. Accumulation, trophic transfer, biodegradation and migration. In: Baker, R.A., ed. Contaminants and Sediments. Ann Arbor, MI: Ann Arbor Science Publishers; 1980:p. 534–563.

Thomann, R.V. Bioaccumulation model of organic chemical distribution in aquatic food chains. Environ. Sci. Technol. 23:699–707; 1989.

Thomann, R.V.; Connolly, J.P. Model of PCB in the Lake Michigan lake trout food chain. Environ. Sci. Technol. 18:65–71; 1984.

Tillitt, D.E.; Ankley, G.T.; Giesy, J.P.; Ludwig, J.P.; Kurita-Matsuba, H.; Weseloh, D.V.; Ross, P.S.; Bishop, C.A.; Sileo, L.; Stromborg, K.L.; Larson, J.; Kubiak, T.J. Polychlorinated biphenyl residues and egg mortality in double-crested cormorants from the Great Lakes. Environ. Toxicol. Chem. 11:1281–1288; 1992.

Trust, K.A.; Fairbrother, A.; Hooper, M.J. Effects of 2,3,7,8-tetrachlorodibenz(a)anthracene on immune function and mixed function oxidase in the European starling. Environ. Toxicol. Chem. 13:821–830; 1994.

Tulasi, S.J.; Reddy, P.U.M.; Ramana Rao, J. V. Accumulation of lead and effects on total lipids and lipid-derivatives in the freshwater fish *Anabas testudineus* (Bloch). Ecotoxicol. Environ. Safety 23:33–38; 1992.

van den Berg, M.E.J.; Craane, B.L.H.J.; Sinnige, T.; van Mourik, S.; Dirksen, S.; Boudewijn, T.; van der Gaag, M.; Lutke-Schipholt, I.J.; Spenkelink, B.; Brouwer, A. Biochemical and toxic effects of polychlorinated biphenyls (PCBs), dibenzo-p-dioxins (PCDDs), dibenzofurans (PCDFs) in the cormorant (*Phalacrocorax carbo*) after in ovo exposure. Environ. Toxicol. Chem. 13:803–816; 1994.

van den Heuvel, M.R.; McCarty, L.S.; Lanno, R.P.; Hickie, B.E.; Dixon, D.G. Effect of total body lipid on the toxicity and toxicokinetics of pentachlorophenol in rainbow trout (*Oncorhynchus mykiss*). Aquat. Toxicol. 20:235–252; 1991.

van der Oost, R.; Heida, H.; Opperhuizen, A. Polychlorinated biphenyl congeners in sediments, plankton, mollusks, crustaceans, and eel in a freshwater lake: Implications of using reference chemicals and indicator organisms in bioaccumulation studies. Arch. Environ. Contam. Toxicol. 17:721–729; 1988.

Vanderzaden, M.J.; Rasmussen, J.B. A trophic position model of pelagic food webs—

impact on contaminant bioaccumulation in lake trout. Ecol. Monogr. 66:451–477; 1996.

van Wezel, A.P.; Opperhuizen, A. Narcosis due to environmental pollutants in aquatic organisms: residue-based toxicity, mechanisms and membrane burdens. Crit. Rev. Toxicol. 25:255–279; 1995.

van Wezel, A.P.; de Vries, D. A.M.; Kostense, S.; Sijm, D.T. H.M.; Opperhuizen, A. Intraspecies variation in lethal body burdens of narcotic compounds. Aquatic Toxicol. 33:325–342; 1995.

Veith, G.D.; Macek, K.J.; Petrocelli, S.R.; Carroll, J. An evaluation of using partition coefficients and water solubility to estimates bioconcentration factors for organic chemicals in fish. In: Eaton, J.G.; Parrish, P.R.; Hendricks, A. C., eds. Aquatic Toxicolgy. ASTM STP 707. American Society for Testing and Materials; 1980:p. 116–129.

Veith, G.D.; DeFoe, D.L.; Bergstedt, B.V. Measuring and estimating the bioconcentration factor of chemicals in fish. J. Fish. Res. Bd. Can. 36:1040–1048; 1979.

Vetter, R.D.; Carey, M.C.; Patton, J.S. Co-assimilation of dietary fat and benzo(a)pyrene in the small intestine; an absorption model using killifish. J. Lipid Res. 26:428–434; 1985.

Walter, A.; Gutknecht, J. Permeability of small nonelectrolytes through lipid bilayer membranes. J. Membrane Biol. 90:207–217; 1986.

Westhall, J.C. Influence of pH and ionic strength on the aqueous-nonaqueous distribution of chlorinated phenols. Environ. Sci. Technol. 19:193–198; 1985.

Zitko, V. Metabolism and distribution by aquatic animals. In: Hutzinger, O., ed. The Handbook of Environmental Chemistry. vol. 2, part A. Berlin: Springer; 1980:p. 221–229.

10
Lipids in Water-Surface Microlayers and Foams

Guillermo E. Napolitano and Daniel S. Cicerone

10.1. Introduction

The water-surface microlayers of lakes and streams are unique environments with a different chemical composition and physical properties from the underlying water. The surface layer acts as an interface to the exchange of gases between air and water and is a vehicle for the transport of inorganic and organic materials between the atmosphere and the water column. The definition, boundaries, and the measurement of the thickness of the water microlayer have been elusive, but the microlayer is generally acknowledged to be limited to the uppermost 30–100 μm of the water (Hardy et al., 1988, and references therein). These dimensions, however, reflect more the characteristics and capabilities of the sampling devices than rigorous measurements of the microstructure of the surface waters.

Dissolved organic matter originates from several autochthonous and allochthonous sources, including the hydrolysis of animal and plant materials, metabolic products of algae and vascular plants. It has been estimated that 10–30% of the aquatic primary production is "lost" as exudates, contributing to the pool of dissolved organic matter (Fogg, 1977; Sharp, 1977). Autochthonous sources of dissolved and particulate organic material determine, to a large extent, the characteristics of the marine surface microlayer (Meyers and Kawka, 1982). In the freshwater environment, however, the properties of the watershed and hydrological regimes probably have a stronger influence on the microlayer composition than does the contribution from aquatic primary production.

The surface microlayer of streams and lakes accumulates biogenic surfactants, primarily consisting of humic and fulvic materials, leached from soils by precipitation events (Wissmar and Simenstad, 1984; Pojasek and Zajicek, 1978). The presence of these natural surfactants is important because they favor the uptake of biogenic lipids and hydrophobic pollutants by the water microlayer. In fact, an extensive database now exists that shows an enrichment of natural and anthropogenic hydrophobic compounds in the surface microlayer with respect to the underlying water (Napolitano and Richmond, 1995; Hardy et al., 1988; Sodergren, 1979; Duce et al., 1972). Many studies have also demonstrated that the concentrations of metals of environmental concern (e.g., Zn, Cd, Pb, Cu) (Elzerman and Armstrong, 1979; Eisenreich et al., 1978; Szekielda et al., 1971) and the densities of certain microorganisms (Johnson et al., 1989; Means and Wijayarante, 1982) can be up to 4 orders of magnitude higher in the surface microlayers than in the subsurface waters.

Bubble production is a common natural phenomena at the surface water of marine (Hunter and Liss, 1981) and freshwater environments and a critical process in surface-layer chemistry. Bubbles may be caused by the activity of organisms, breaking waves, and/or turbulent flows. Experiments have shown that as bubbles rise through the water column, selected organic molecules and particles are adsorbed to their surfaces (Blanchard, 1963). At the water surface, the bubbles may burst, ejecting aerosol droplets into the air. This process has been called the "bubble microtome" because the droplets eject not only the materials contained in the bubbles' surface but also those contained in the water-surface microlayer (MacIntyre, 1968). Under certain conditions, emerging bubbles may not burst instantly and may accumulate on the water surface, producing foam. Thus, foam formations represent an excellent opportunity for studying the composition of the water-surface microlayer and the advection of materials from deeper in the water column to the surface.

The aqueous phase of foam contains high concentrations of surface active materials and is therefore expected to concentrate natural lipids and hydrophobic contaminants. Although we briefly discuss the fate and the distribution of some biogenic and petroleum hydrocarbons, owing to space limitations we do not cover the literature on hydrophobic pollutants of the water microlayers (Kucklick and Bidleman, 1994; Butler and Sibbald, 1987; Duce et al., 1974, 1972). This chapter is a critical evaluation of the research on the distribution and dynamics of biogenic lipids in the surface microlayer and foams of freshwater systems. Attention is focused on sampling techniques and on the incorporation of biogenic lipids into these particular environments.

10.2. Basic Physicochemistry of Surface Microlayers

The air–water interface, like all discontinuities in aquatic environments, is a relative high free-energy system (thermodynamically not stable) in which singularities take place. In its simplest form, the air–water interface of a natural system can be visualized as a four-layer model consisting (from top to bottom) of turbulent air, stagnant air, stagnant water, and mixed water (Fig. 10.1). The dimensions of the layers forming the air–water interface presented in Figure 10.1 reflect typical values found in both laboratory (Schwarzenbach, 1983; Genereux, 1991; Duran and Hemond, 1984; Wilcock, 1984; Mackay and Yeun, 1983; Rathbun and Tai, 1983, 1982; Münnich et al., 1978; Liss, 1973) and field studies (Kucklick and Bidleman, 1994; Daumas et al., 1976; Liss, 1975; Jarvis, 1967; Garret, 1965). The monomolecular slice of surface active materials and the layer of stagnant water constitute the actual water-surface microlayer and are the subject of this chapter. The surface microlayer is a region of 0.01–100 μm of thickness, in which the

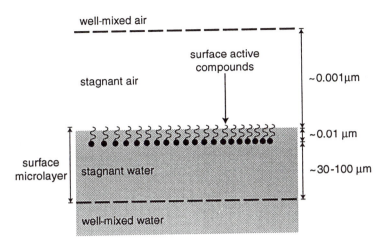

FIGURE 10.1. Schematic view of the water-surface microlayer and the air–water interface. The dotted line represents the ideal contact between the two phases.

importance of turbulent processes is considered to be small (Schwarzenbach et al., 1993). This phenomenon is explained by the highly viscous behavior of fluids in the surface microlayer and the concomitant reduction of wind effects at the boundary layer, just above the surface water. Therefore, the transport of material across this region is dominated by molecular diffusion. The fluid dynamic that governs the transport of chemicals between the different layers of the air–water interface led to the development of two conceptual frameworks, the *stagnant boundary layer model* (Whitman, 1923) and the *surface renewal model* (Danck-werts, 1951; Higbie, 1935), which provide satisfactory descriptions of the micro-layers of lentic and lotic environments, respectively.

A complete assessment of these models is beyond the scope of this chapter (please see the review by Schwarzenbach et al., 1993). Briefly, the stagnant boundary layer model assumes that water and air turbulence are not strong enough to stir the thin air and water layers adjacent to the interface. However, the surface renewal model takes into account the continual turnover of air and water "parcels" with their associated material load at the air–water interface (Danckwerts, 1951). The rate of mass transfer of substances across the surface microlayer for the stagnant and the renewal models are shown in Equations 1 and 2, respectively.

Stagnant boundary layer model:
$$F = \frac{D_w}{1,000 \times z_w}(c_{wa} - c_w) \qquad (1)$$

Surface renewal model:
$$F = \sqrt{\frac{D_w}{10^6 \times r_w}}(c_w - c_{wa}) \qquad (2)$$

where F (mol $\cdot$ cm^{-2} $\cdot$ s^{-1}) is the exchange flux of material per unit time and area, D (cm^2 $\cdot$ s^{-1}) is the diffusion coefficient of the solute, z (cm) the thickness of the surface microlayer, c (mol · dm^{-3}) is the concentration of the solute, and r the surface renewal rate due to microscopic eddies above and below the interface. The subscripts w and wa indicate water and the water–air interfaces, respectively.

Although Equations 1 and 2 have the same mathematical form, they have different adjustable parameters (z and r) and different dependence on the empirical diffusion coefficient, proportional to D in the former, proportional to $\sqrt{D}$ in the latter. It has been found that F depends on D^α, where α varies between 0.5–1. The stagnant boundary layer model assumes that the concentration gradient across the layers of thickness z (the driving force of the flux) must be constant. This assumption implies that the concentration profile has time to reach a steady-state condition and is not significantly modified by chemical reactions. The surface renewal model assumes that the concentration gradient across the water–air interface is time dependent. It couples the rate of delivery of new bulk media "packets" to the interface, a consequence of the level of turbulence in the air and water reservoirs, with the local molecular diffusive exchange out of, or into, the interfacial region. This dependence is incorporated in the model by the surface renewal rate parameter (r), which represents the mean frequency of creating contact surfaces.

The surface renewal rate r_w is an idealized model parameter determined experimentally. This parameter is a "fitting" coefficient used to adjust for wind and water turbulence local effects. O'Connor and Dobbins (1958) proposed an empirical relation to determine r_w, which is shown in Equation 3.

$$r_w \simeq \frac{u_w}{d_w} \tag{3}$$

where u_w and d_w are the water velocity and water depth of a turbulent aquifer, respectively.

A great variety of natural lipidic surfactants is produced by many different organisms for a variety of functions (Table 10.1). These compounds profoundly affect the physicochemical properties of the water-surface microlayer (Adamson, 1990). The top layer of surfactants (Fig. 10.1) influences the solubility and the flux of hydrophobic compounds across the region. They also modify the diffusion characteristics due to hydrophobic and electrostatic interactions with other compounds present in the microlayer. When a surfactant is dissolved in water, the hydrophobic tail of the molecule produces a reordering of water molecules in its surroundings. This results in the development of repulsive forces and an increase in the free energy of the system. Minimization of the repulsion forces is achieved by the orientation of the surfactant molecules at the interface: the hydrophilic head of the surfactant readily dissolves in water, whereas the hydrophobic tail goes to the air layer. The net result is a reduction in the free energy per unit area of the interface and an increase of the interfacial viscosity.

A different way in which a surfactant can reduce the free energy of the system is by forming micelles, bilayers, and vesicles (Fig. 10.2). Surfactant micelles form assemblages in which the hydrophobic portion of the surfactant orients inward and the hydrophilic portion remains in contact with the water. The forces that hold these structures together include hydrophobic, van der Waals, electrostatic, and

TABLE 10.1. Summary of the occurrence and CMC of lipidic biogenic surfactants in various microorganisms.[a]

Surfactant		Organism	CMC (mg · L^{-1})	Reference
	Phospholipids	*Corynebacterium insidiosum*	—	Akit et al., 1981
		Pseudomona aeruginosa	46	Churchill et al., 1995
	Rhamnolipids		18	Thangamani and Shreve, 1994
		Bacillus sp. *AB-2*	—	Banat, 1993
Polar lipids	Glycolipids	*Rhodococcus aurantiacus*	—	Ramsay et al., 1988
		Rhodococcus sp. *Strain H13A*	1,500	Finnerty and Singer, 1984
	Pentasaccharide lipid	*Nocardia corynebacteroides*	25	Powalla et al., 1989
	Sophoros lipid	*Candida lipolytica Y-917*	—	Lesik et al., 1989
	Rubiwettins	*Serratia rubidaea*	10	Matsuyama et al., 1990
		Bacillus licheniformis JF2	0.02	McInerney et al., 1990
			10	Lin et al., 1994
	Lipopeptides	*Bacillus licheniformis 86*	10	Horowitz et al., 1990
		Arthrobacter MIS38	—	Morikawa et al., 1993
	Neutral lipids	*Clostridium pasteurianum*	—	Cooper et al., 1980
		Nocardia erythropolis	—	McDonald et al., 1981
	Fatty acids	*Corynebacterium lepus*	150	Cooper et al., 1979
	Fatty acids + neutral lipids	*Nocardia erythropolis*	—	McDonald et al., 1981
	Protein–carbohydrate complex	*Pseudomonas fluorescens 378*	<10	Persson et al., 1988
	Phosphatidylethanol amines	*Rhodococcus erythropolis*	30	Kretschmer et al., 1982

[a] —, Not reported; CMC, critical micelar concentration.

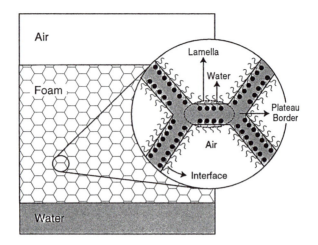

FIGURE 10.2. Slice of a typical foam formation at the air–water interface and insert showing the details of the structure of the air–water interface within the foam. (Adapted from Schramm and Wassmuth, 1994.)

hydrogen bonding interactions. Because no chemical bonds are formed, these structures are fluid-like and are easily transformed from one state to another, as certain conditions, such as electrolyte concentration and temperature, change. Lipids can form micelles (spherical or cylindrical) or bilayers, depending on the size of the hydrophilic head group and the chain length of the hydrophobic tail (Georgiu et al., 1992). The concentration of surfactant at which micelles begin to form is known as the critical micelle concentration (CMC). The CMC, although not extensively used in freshwater studies, shows the ability of surfactants compounds to form micelles, bilayers, and vesicles. The presence of these structures has profound effects on the solubilization of organic compounds in the surface microlayer (Banat, 1995), the fractionation of metals and inorganic nutrients (Chiu and Huang, 1991) and on the biodegradation of natural and anthropogenic compounds (Aronstein and Paterek, 1995; Roch and Alexander, 1995). Furthermore, the CMC is a good indicator of the ability of a surfactant to produce and stabilize foam. For comparative purposes, Tables 10.1 and 10.2 present the CMC of a number of biogenic and industrial surfactants, respectively. These data show that the capability of lipidic surfactants (isolated from microorganisms) and synthetic formulations to concentrate lipophilic compounds and stabilize foam varies widely, as indicated by their different CMC values.

10.3. Basic Structure of Foams

Foams are colloidal systems in which a gas is dispersed in a continuous liquid phase. Foam can be formed in a liquid if gas bubbles are injected at a rate that is higher than the rate at which the liquid between bubbles can drain. Like other

TABLE 10.2. Summary of the characteristics of common anthropogenic surfactants found in freshwaters.[a]

Surfactants	Formula	CMC (mg · L^{-1})	Reference
Neodol 25–35 (AES)	$C_nH_{2n+1}(OCH_2\text{-}CH_2)_xOSO_3Na$ $n = 12\text{–}15\ x \approx 3$	0.380	Roch and Alexander, 1995
SDS (LS)	$CH_3\text{-}(CH_2)_n\text{-}CH_2\text{-}O\text{-}SO_3Na$ $n = 10$	2,332.8	Williams et al., 1955
LXS-814 (LAS)	$CH_3\text{-}(CH_2)_n\text{-}CH(\phi\text{-}SO_3Na)\text{-}CH_3$ $n = 8\text{–}14$	0.390	Roch and Alexander, 1995
p-1Methyldecylbenzene (LA)	$CH_3\text{-}(CH_2)_n\text{-}CH_2\text{-}\phi\text{-}CH_3$ $n = 10$	184.6	Gershman, 1957
Potassium oleate (soap)	$CH_3\text{-}(CH_2)_7CH{=}CH\text{-}(CH_2)_7COOK$	480	Flockhart, 1953
Triton X-100 (APE)	$CH_3\text{-}C(CH_3)_2\text{-}CH_2\text{-}C(CH_3)_2\text{-}\phi\text{-}(O\text{-}CH_2\text{-}CH_2)_xOH$ $n = 2\ x \approx 10$ average	0.195	Roch and Alexander, 1995
Sucrose monolaurate (nonionic surfactants)		97	Wachs and Hayano, 1962

[a]AES, alcohol ethoxy sulfate; LS, linear alquil sulfate; LAS, linear alquilbenzene sulfonate; LA, linear alquilbenzene; APE, alkylphenol ethoxilate.

colloidal systems, foams are thermodynamically unstable because bubbles tend to coalesce rapidly. However, the presence of surfactants at the air–water interface reduces the surface tension and promotes stability between the neighboring bubbles, contributing to the persistence of the foam structure.

In carefully controlled environments, it is possible to produce surfactant-stabilized static bubbles and foams with lifetimes of months or even years (Schramm and Wassmuth, 1994). Natural foam formations, however, have a life span of hours to a few days. Fig. 10.2 represents a slice of a typical foam formation at the air–water interface and a magnified region showing its various internal structures. Of interest for the purpose of this chapter is the thin, liquid film dividing the gas phase, denominated *lamella,* the thin aqueous film that provides the structure of the foam, and the *plateau border,* which is the connection of three lamellae at an angle of 120°. These structures contains the materials that form the water-surface microlayer of the body of water, where the bubbles emerge at the surface and the foam is formed. The plateau border plays an important role in the mechanism of film drainage and may collect different types of materials, such as inorganic particles, oil droplets, and microorganisms (Wasan et al., 1994; Adamson, 1990). Assuming that a typical foam consists of approximately 90% air (Schramm and Wassmuth, 1994) and that the mean thickness of the surface microlayer is 50 μm (Hardy et al., 1988), 1 L of foam water (destabilized or collapsed foam) would represent 2 m^2 of surface microlayer.

10.4. Sampling Techniques

Sampling foam for lipid analysis is a rather simple manual operation that requires nothing more than a clean jar and a piece of solvent-rinsed aluminum foil (Napolitano and Richmond, 1995). Nevertheless, the sampling procedure should be performed carefully to avoid drainage of water or contamination of the foam with subsurface water, practices that could bias the results of foam constituents.

In contrast to the simple procedures involved in the collection of foam samples, the past few decades have witnessed an extensive development of tools for sampling water-surface microlayers. Despite these efforts, no standard methods have yet been established for collecting the water-surface microlayer. Consequently, considerable variations in the reported values of the thickness of the microlayers, the enrichment factors of specific compounds, and even the definition of the microlayer itself have arisen, partly as a reflection of the disparity between the different sampling techniques. Table 10.3 presents a compilation of methods and devices frequently used for sampling the microlayer. The design and materials used for the construction of these sampling instruments affect not only the definition of the physical boundaries of the microlayer but may probably also affect the results of chemical analysis, due to differential affinity and selective extraction of certain compounds (Marty and Choiniere, 1979).

Garret (1965) developed the first tool for collecting slick-forming materials from the sea surface. This simple device consisted of a metal screen (0.14-mm-diameter wire, with ~60% open space) mounted in a 0.75 × 0.60-m aluminum frame. During the sampling operation, the screen was lowered vertically to the proximity of the water, then reoriented horizontally to enable contact with the surface film, thus entrapping it between the wires of the mesh. The screen was then immediately returned to its vertical position, and the liquid was drained into an appropriate collection bottle. Several variations of the original metal screen were used by other workers, including a stainless-steel model used by Sieburth (1965) and Williams et al. (1986). Duce et al. (1972) and Pojasek and Zajicek (1978) used a polyethylene screen, particularly suitable for the analyses of metals in sea-surface microlayers. This later screen material, however, reportedly changes the surface potential of the surface water and therefore may not be an adequate sampling device (MacIntyre, 1974).

Screen samplers are simple and inexpensive, but their use is limited due to the time and effort involved in obtaining a reasonable volume of water. Moreover, there are concerns about whether a thin, unmixed, and uncontaminated layer is effectively sampled with these devices (Liss, 1975). To solve these problems, two methods based on completely different operation principles were introduced: the rotating drum (Hardy et al., 1988; Harvey, 1966) and the freezing probe (Hamilton and Clifton, 1979).

Harvey (1966) pioneered the construction of a series of continuous samplers of different materials (steel, ceramic, polyvinyl chloride [PVC]) based on the principle of a rotating hydrophilic drum. The first rotating microlayer collector (Harvey, 1966) consisted of a ceramic-coated stainless-steel drum (0.38-m diameter; 0.60 m long). The drum collected a microlayer (60–100 μm thick) that adhered to its surface, which was then removed by a neoprene wiper and finally collected into a plastic cup. This rotating drum collected about 300 ml of water per minute. The main problem with this sampler is that it only works effectively under calm conditions. Most earlier versions of the rotating samplers were coated with relatively hydrophilic materials, such as ceramic (Harvey, 1966), PVC (Brockmann et al., 1976), and glass (Carlson et al., 1988), which obviously were not ideal for collecting lipids. Furthermore, the use of neoprene and plastic materials for the collection of lipids and lipophilic substances is always questionable due to a great potential for introducing contaminants. Using the same basic principle of the rotating drum, Hardy et al. (1988) developed a new large-volume hydrophobic sampler coated with Teflon (polytetrafluoroethylene [PTFE]). The PTFE rotating drum efficiently sampled the biogenic lipids and anthropogenic contaminants from the top 34 μm of the surface microlayer (Hardy et al., 1988).

Hamilton and Clifton (1979) developed the unorthodox freezing probe technique to sample the surface microlayer of estuarine and coastal waters. This device consists of a polymethylmethacrylate (acrylic plastic) disk (0.25-m diameter), encased in a thin (0.1 mm) PVC membrane. The upper surface of the disk is prechilled in liquid N_2 and attached to a plastic rod with freezing water. The freezing probe is brought into contact with the water surface, which immediately freezes to a depth of about 1,000 μm. The frozen surface microlayer can be detached from the PVC membrane, stored, and later sectioned with a sledge microtome for further analysis of sections of the desired thickness. Due to the speed of sampling (the surface microlayer is frozen in <1 s), this technique is particularly useful under rough water conditions.

Another unconventional sampling device designed to measure chemicals at the air–water interface consists of a prism of optically polished germanium (Baier, 1970). Surface films are captured by a simple dip mechanism of the prism into the water. The polar head groups of the organic material are adsorbed onto the hydrophilic surface of the prism, obtaining a thin (0.01 μm) film of the microlayer.

The simplest device to sample the water microlayer is probably the glass plate. A small (~0.20 × 0.20 m) glass plate can be applied and then withdrawn vertically from the water surface by manual motion at a rate of 0.20 m · s^{-1} and wiped off with a windshield wiper blade (Harvey and Burzell, 1972). This device samples the top 60–100 μm of the water surface (Table 10.3). According to Hardy et al. (1988), the glass plate has collection characteristics that are similar to those of the PTFE-coated rotating devices. However, the former requires a considerably larger effort to obtain an acceptable sample size. In a variation of the glass plate technique, Larsson et al. (1974) used a perforated PTFE plate to collect lipids from the water-surface microlayer along the Swedish west coast. Although a quantitative assessment of the efficiency of this type of plate is not reported, the authors postulate that the recovery of lipids from the water microlayer can be augmented by the use of a densely perforated PTFE surface.

In another attempt to modify the simple glass plate method, Gever et al. (1996) used a silanized borosilicate glass plate to sample the surface microlayer of a rice field. The plates (silanized with 5% solution of dichlorodimethylsylane) are used in a way similar to conventional glass plates in that the hydrophobic materials in the microlayer are extracted by running an organic solvent from a Pasteur pipette across the plate.

This brief assessment of the different instruments and methods used in sampling the surface microlayer demonstrates the wide spectrum of available materials and designs. Because each technique has its own advantages and drawbacks, the optimal method cannot be recommended without taking into consideration the type of environment, the prevailing flow conditions, and the size of sample needed. In general, the most important characteristics to consider in evaluating a particular sampling equipment are the inertness and hydrophobicity of the materials, speed of operation, low potential for introducing contaminants from the craft and the support equipment, and consistency in the thickness of the film sampled. Due to inertness and versatility considerations, PTFE screens, surfaces, and accessories (wipers and collecting cups) are highly recommended for the sampling of biogenic lipids and hydrophobic contaminants from surface waters. A PTFE screening material (~360–500-μm mesh) and PTFE sheets are recommended by the American Society for Testing Materials (1995) to sample oil spills and slicks from surface waters. Unfortunately, we are not aware of any environmental or ecological study that has used or tested this supposedly standard technique. Furthermore, authoritative and widely used handbooks of water analysis do not provide a standard method for sampling the surface microlayer (Greenberg et al., 1992).

10.5. Physicochemical Processes at the Surface Microlayers

The fate and distribution of materials in the surface microlayers and foam depend on the material's physical and chemical properties (e.g., aqueous solubility, diffusivity in air and water, viscosity, specific gravity, vapor pressure, and reactivity) and a series of environmental factors that influence their transport and transformation (Fig. 10.3). The aqueous solubility of a substance depends on the dielectric constant (ϵ) of the solvent (i.e., on the polar characteristic of the water molecules). Studies of the dielectric properties of water adsorbed on gels and porous and nonporous substrates demonstrated that the dielectrict constant changes in the range of 78.5–4.5 (at 25°C and 1 atm) for film thicknesses

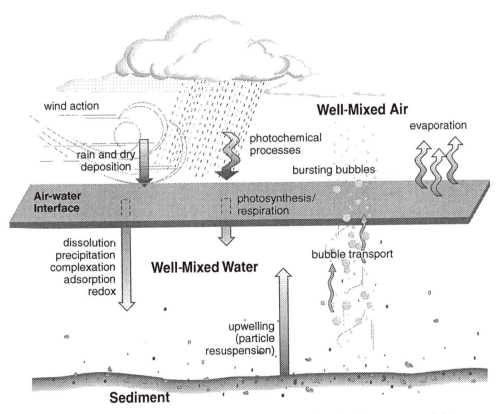

FIGURE 10.3. Physicochemical and biological processes related to the transport and alteration of materials occurring at the air–water interface of aquatic ecosystems.

<1.4 μm (Stobbe and Peschel, 1997; Metzik et al., 1973; McCafferty et al., 1970). The reduction of the dielectric constant of the water in thin films affects its characteristics as a solvent and should result in an increment of the solubility of lipids and other organic hydrophobic substances. As ϵ decreases, the intermolecular forces in the liquid water decrease and so does the energy needed for the organic molecules to "insert" themselves into water. To the best of our knowledge, the changes in the dielectric constant of the water-surface microlayer, the associated changes of the solubility of hydrophobic organic matter, and the resulting implications for the assessment of the biogeochemistry of natural waters have been overlooked.

Low aqueous solubility (hydrophobic character), a low relative density (0.8–0.9), low vapor pressure, and the presence of natural and anthropogenic surfactants stabilize organic material in the surface microlayer. Studies of the behavior of natural films in the microlayers have shown that, at low lipid densities, the lipid layers resemble a two-dimensional gas, in which loosely arranged molecules move freely on the water surface (Gaines, 1966). At higher lipid concentrations, the film can be in the liquid or solid state, depending on molecular interactions leading to cohesive forces between molecules. Straight-chain saturated fatty acids in microlayers manifest strong hydrophobic interactions because of the overlap of the lineal hydrocarbon chains, which produces a tightly packed film. Unsaturated fatty acids, however, produce less cohesive films even at relatively high concentrations, because of the bends in the hydrocarbon chain produced by the double bonds (Norkrans, 1980).

The processes that influence the distribution of material in the surface microlayer described by Garret (1972, 1967) and Jarvis (1967) are summarized in Figure 10.3. Bubble production is a common natural phenomena at the surface water of marine and freshwater environments and a critical process in surface-layer chemistry. Bubbles may be caused by the activity of organisms, breaking waves, and/or turbulent flows. They serve as a two-way transport system, removing salt nuclei, surface active material, and particles into the deeper water as well as collecting material from the near-surface layers (air phase) into the surface film (Blanchard, 1968, 1964, 1963; Baylor et al., 1962). At the water surface, the bubbles may burst, ejecting aerosol droplets into the air (Lai and Shemdin, 1974; Blanchard and Syzdek, 1972; Paterson and Spillane, 1969). In the process of production of aerosols, some fractionation of salts and the preferential removal of some fractions of the organic matter by production of aerosols and by evaporation are well documented (Volz, 1972; Blanchard, 1964; Wilson, 1959). Transport by bubbles is more extensively described in the chemical engineering literature (Karger and DeVivo, 1968; Dorman and Lemlich, 1965).

Wind and wave action contribute to the equilibration of atmospheric gases into the surface layer, enhancing the rates of evaporation and dissolution altered by surface active compounds in the surface microlayer (Wu, 1974; Mallinger and Mickelson, 1973; Healy and La Mer, 1964; Garrett and Bultman, 1963). Wind and waves also contribute to the formation of bubbles and the cycling of particles through the surface microlayer. Accumulation of bacteria on the surfaces of bubbles, with the subsequent transfer to the surface film, has been demonstrated

(Blanchard and Syzdek, 1972). The material collected in the surface film is subject to photochemical processes driven by ultraviolet light, which promotes the decomposition of nitrate, iodine compounds, peptides, esters, and organic acids (Zepp et al., 1975; Antia and Landymore, 1974; Zafiriou, 1974; Bennett-Corriea et al., 1970; Miyake and Tsunogai, 1963).

Atmospheric wet and dry deposition provide dust particles to the surface micro-layer (Hardy, 1982), which interact with the surface material (heavy metals, organic compounds, and bacteria). Sinking of these particles and injection of bubbles to the subsurface waters are the major mechanisms of material transport out of the surface microlayer.

The relative importance of each of these factors varies widely and depends on the particular environment factors and the prevailing meteorological and hydro-logical conditions. In addition, biologically driven processes, such as photo-synthesis, respiration, bioaccumulation, and biodegradation, can alter the con-centration and speciation of compounds present at the interface (Norkrans, 1980).

10.6. Lipids in the Water-Surface Microlayers and Foams

10.6.1. Total Lipids and Major Lipid Classes

Lipids, polysaccharides, and polypeptides are the major organic constituents of the surface microlayers and foams (Velimirov, 1982; Barger and Garret, 1976). Total dissolved organic carbon in the surface microlayer is estimated to range from 0.5 to 2 g · L^{-1} (Stumm and Morgan, 1981), whereas the concentrations of fatty acids (often the main dissolved lipid) may vary from about 10 to 90 µg · L^{-1} in the microlayer of lakes (Meyers and Owen, 1980) and from about 1 to 20 mg · L^{-1} in the microlayer of streams (Napolitano and Richmond, 1995). Many studies have demonstrated that the concentration of lipids in the surface micro-layers and foams is consistently higher than in the subsurface waters (Napolitano and Richmond, 1995; Johnson et al., 1989; Kattner and Brockmann, 1978; Larsson et al., 1974; Garrett, 1967; Harvey, 1966). The enrichment of lipids in the surface microlayer is a consequence of their hydrophobicity, a low relative den-sity (0.8–0.9), low vapor pressure, the presence of natural and anthropogenic surfactants, and the special characteristics of the water microlayer as a solvent for hydrophobic substances. Lipids in the surface microlayers are normally not only more concentrated than in the subsurface water, but the chain length and satura-tion pattern of their fatty acid constituents are different (Meyers and Owen, 1980). This observation implies that fatty acids in the microlayer and those from subsur-face waters have a different origin, or alternatively, the observed differences arise from physicochemical and biological processes that alter the lipid composition during transport and accumulation into the microlayer.

Lipid contents and compositions of the surface microlayers and subsurface waters of Lake Michigan were measured in an attempt to evaluate the rela-tive importance of fluvial and autochthonous sources of organic matter (Meyers and Owen, 1980). In this study, lipids were also used to investigate the partition of the total organic matter pool between the dissolved and the particulate fractions. This work, and earlier studies, have shown that the concentration of lipids in the particulate organic matter of lakes may sometimes be larger and at other times smaller than the corresponding dissolved fraction (Meyers and Owen, 1980; Larsson et al., 1974). These differences highlight the importance of local physicochemical and biological processes in the fractionation of lipids and other organic constituents between the dissolved and the particulate fractions.

In general, 40–65% of the lipids of unpolluted surface microlayers are fatty acids and triacylglycerols, whereas 15–30% are phospholipids and hydrocarbons (Norkrans, 1980). Structural analysis of these lipids, in particular the examination of the fatty acid composition, can provide valuable information for elucidating the major sources of organic matter in the aquatic environment (e.g., algae, zooplankton, terrestrial plants; see below and Napolitano, this volume). Due to their chemical stability, aliphatic hydrocarbons originating from algae and terrestrial plant waxes may be an important lipid constituent in pristine environments (Napolitano and Richmond, 1995; Napolitano et al., 1992; Marty and Choiniere, 1979). Detailed analyses of individual components, however, often reveal a petroleum origin of the hydrocarbons.

Analyses of ^{13}C isotope ratios in hydrophobic and hydrophilic organic compounds in the foam and subsurface water from an eelgrass (*Zostera marina*) and macroalgal habitat in the Duckabush River estuary (near Puget Sound, Washington) indicated that a substantial portion of the dissolved organic matter in foams originated from sea grasses and macroalgal exudates (Wissmar and Simenstard, 1984). Hydrophobic organic compounds comprised between 6–48% of the total dissolved organic matter of the Duckabush River. The highest concentrations of these compounds were found in the neritic subsurface water (48%), macroalgal foam (47%), and foam and subsurface waters of the eelgrass habitat (38%). A crude separation (Leenheer and Huffman, 1976) of the hydrophobic organic matter of the estuary indicated that they mainly consisted of hydrocarbons and high-molecular-weight organic acids, possibly fulvic material (Wissmar and Simenstard, 1984).

Most biogeochemical studies of the lipid, especially those involving fatty acids and biogenic hydrocarbons, indicate that algal exudates and terrestrial plant detritus are the major sources of dissolved and particulate lipids in natural waters. Analysis of the hydrophobic compounds in the surface microlayer (top 440 μm) and foam in the St. Lawrence River estuary (Marty and Choiniere, 1979) showed fatty acids and alkanes as the major lipid components, at concentrations ranging from 8 to 25 μg · L^{-1} and 0.5 to 1.8 μg · L^{-1}, respectively. Hydrocarbons consisted of a series of n-alkenes, with dominance of odd-carbon numbered chains over even-carbon numbered chains, where the major components were n-C25 and n-C27. Petroleum hydrocarbons show no preference between odd-carbon numbered chains and even-carbon numbered chains (Napolitano et al., 1992; and references therein). Therefore, the hydrocarbon constituents of foam and surface microlayer materials from the St. Lawrence River pointed to a specific biogenic origin (Marty and Choiniere, 1979) and were typical of the cuticular lipids of terrestrial vegetation (Johnson and Calder, 1973).

In a similar study of a wind-generated foam in a remote high-latitude Andean Lake (Laguna Lejia, Antofagasta, Chile), it was reported that lipids consisted of hydrocarbons and a series of not-well-defined acyl esters, which included phospholipids (Simoneit et al., 1980). The quantitation of the total hydrocarbon content in the foam of this apparently pristine environment is of particular interest because it may represent baseline levels or biogenic constituents. Although the total lipid content of the foam was $\sim$2 mg $\cdot$ g^{-1} of dry foam, hydrocarbons accounted for $\sim$1.4% of the total lipids. The hydrocarbon constituents of the foam were mainly n-alkanes, ranging from n-C15 to n-C33 with maxima at C18 and C29 and a strong odd-to-even carbon preference index (CPI, % hydrocarbons containing an odd number of carbon atoms/% hydrocarbons containing an even number of carbon atoms) of 3.5. This hydrocarbon profile is consistent with a biogenic origin and is thought to derive, in part, from vascular plant waxes (Simoneit et al., 1980). Other hydrocarbons present in the foam were 7- and 8-methylheptadecane, which has been shown to be a major component of cyanobacteria species (Han et al., 1968). The analysis of the water-surface microlayer of Lake Sisjon, an unpolluted reservoir in Sweden, showed that triacylglycerols and free fatty acids, but not hydrocarbons, were the major lipid components of the hydrophobic organic matter (Larsson et al., 1974).

Studies of the fractionation of organic matter and other chemicals between the surface and the subsurface waters of large water falls are also of interest due to the magnitude of gas injection into the water flow that occurs at these sites. For example, the analysis of trace metals and lipids in the water and foam formations below Niagara Falls was of particular environmental concern, due to the large inflow of municipal and industrial waste into the system (Warry and Chan, 1981). Concentrations of metals and various lipid classes were measured in the foam and the subsurface water just below Niagara Falls (Johnson et al., 1989). The concentrations of total particulate (50 μg $\cdot$ L^{-1}) and total dissolved lipids (200 μg $\cdot$ L^{-1}) in Niagara Falls were similar to those reported in marine coastal environments (Parrish, 1987). Foams presented a substantial enrichment of lipids with respect to the bulk water, showing concentrations >12,000 μg $\cdot$ L^{-1}. The foam lipids consisted of (in decreasing order of concentrations) "acetone mobile polar lipids" (which include plant pigments and other polar compounds), sterols, triacylglycerols, hydrocarbons, free fatty acids, phospholipids, and minor amounts of wax esters, methyl esters, ketones, and fatty alcohols. Although hydrocarbons were not analyzed in detail by gas chromatography, their relatively high concentration in the foam ($\sim$11% of the total lipids) suggested a significant contribution from petroleum sources. Regardless of the concentration of a particular lipid class in the bulk water, its enrichment factor in the foam (concentration of a lipid in the foam divided by its concentration in the subsurface water) depended on the polarity of the compounds; less polar lipids (hydrocarbons, wax esters, and triacylglycerols) were enriched in foam by a factor of <25, whereas more polar lipid classes (free fatty acids, sterols, and phospholipids) were enriched by a factor

>50 (Johnson et al., 1989). Using ^{14}C- and ^{32}P-labeled lipids, Sodergren (1979) also found that the polar lipid classes of the water microlayer of an experimental microcosm had an enrichment factor greater than that of neutral lipids.

Lipid content and lipid class compositions were recently analyzed in the subsurface water and foam samples from a series of streams in eastern Tennessee (Napolitano and Richmond, 1995). Concentrations of total lipids were always greater in the foam than in the subsurface water from the same site. Concentrations of lipids in foam ranged from 10 to 110 mg · L^{-1}, showing enrichments factors of up to 100. Eight lipid classes were detected in water and foam samples: free fatty acids, phospholipids, hydrocarbons, and acetone mobile polar lipids accounted for >50% of the total lipids. Other detectable components were sterols, diacylglycerols, fatty alcohols, triacylglycerols, and sterol esters. Relatively large concentrations of phospholipids (labile cellular lipids that decompose rapidly after the organism dies) measured in some stream foams, indicating that viable bacteria, protozoa, and/or algae were probably major contributors to the pool of particulate organic matter. A more detailed analysis of taxonomically specific signature lipids (i.e., fatty acids and sterols) by gas chromatography could be used to identify which of these groups or organisms predominate (Tunlid and White, 1992). The main by-products of the enzymatic hydrolysis of phospholipids are free fatty acids. Large quantities of free fatty acids coinciding with low levels of phospholipids were observed in water from some streams (Napolitano and Richmond, 1995), which provides an indication that biological decay of organic matter was a dominant process in the foam at these sites.

The metabolic processes that can be inferred from the analysis of biochemical constituents in the foam are not necessarily linked to the conditions in the entire body of water; instead, they reflect highly localized and specific biological processes that occur within the foams and in the water-surface microlayer. A more general evaluation of stream condition arises from measurements of hydrocarbons. Biogenic hydrocarbons are commonly extracted from uncontaminated waters or organisms. However, the natural hydrocarbons typically account for only a small proportion (typically <5%) of the total lipids (Nevenzel, 1989). The presence of hydrocarbons at 10–30% of lipids in foams (Johnson et al., 1989) strongly suggests petroleum contamination. Hydrocarbons present in foam from small streams in eastern Tennessee (Napolitano and Richmond, 1995) consisted of an intricate mixture of many compounds, dominated by a series of aliphatic, straight-chain alkanes, containing 13–32 carbon atoms, with a maximum at C-25. Also present were minor amounts of the isoprenoid hydrocarbons pristane (2,6,10,14-tretramethyl-pentahexane), phytane (2,6,10,14-tetramethyl-hexadecane), and a CPI of 2.49. This hydrocarbon profile suggests combined input of two different sources of organic materials; the presence of a complete series of normal alkanes and phytane and the large number of hydrocarbon components strongly indicates petroleum contamination. However, the dominance of compounds containing an odd number of carbon atoms over compounds containing an even number of carbon atoms suggested the importance of leaf hydrocarbons originating from terrestrial vegetation (Johnson and Calder, 1973).

Foam formed in the tidal channels and at the edge of the water during rising tides in the Bay of Fundy showed a high lipid content (made up mainly of triacylglycerols, phospholipids, sterols, hydrocarbons, and pigments), typically at a concentration of 65 mg · L^{-1} (Napolitano et al., 1992). The hydrocarbons present in this foam consisted of a series of n-alkanes, from C-20 to C-39, with no odd/even-chain length preference (CPI = 1.008), reflecting a very low level of petroleum contamination.

10.6.2. Fatty Acids

Analysis of fatty acids in surface microlayers and foams is of special interest because they provide information of sources and sinks of organic matter in aquatic systems and about the type of organisms involved in their production (see Napolitano, this volume). Furthermore, the abundance of saturated and polyunsaturated fatty acids in a sample is an indicator of the relative contribution of detritus and viable biomass, respectively (White, 1995).

Despite some anomalies in the chemical composition of surface waters, due to highly localized processes, some generalizations about the distribution of fatty acid types are valid. For example, the concentration of short-chain fatty acids in subsurface waters is usually higher than in the microlayer (Garret, 1967). Short-chain fatty acids are displaced from the surface film by longer-chain species due to the competitive/adsorptive process. Long-chain fatty acids are less water-soluble and more surface-active. Also, the concentration of polyunsaturated fatty acids is typically lower at the surface microlayer due to photoxidation at the air–water interface (Armstrong et al., 1966).

There are very few detailed analyses of fatty acids in foams or surface microlayers in fresh water. Table 10.4 compares available data on the fatty acid composition of surface microlayers and foams from freshwater and marine environments. The fatty acid composition of foam and surface waters from the different environments obviously reflects unique chemical and biological processes occurring at each site, and therefore generalizations are difficult to make. It is interesting to observe, however, that in studies of both marine and freshwater samples, saturated and monounsaturated fatty acids accounted for 80–90% of the total fatty acids, and long-chain polyunsaturated fatty acids (PUFA; characteristic components of organismal aquatic lipids) were not detected. Although the surface microlayers from fresh water and marine waters lack or have only trace concentrations of PUFA (Simoneit et al., 1980), estuarine and marine foam has significant concentrations of ω6 and ω3 PUFA (Napolitano et al., 1992; Marty and Choiniere, 1979).

In estuarine waters, the concentrations of saturated and monounsaturated fatty acids in the lipids of the particulate fraction of the surface microlayers more closely resemble those of the foam than those of the dissolved lipids in the microlayer itself (Marty and Choiniere, 1979). These results suggest that particulate organic matter in the surface microlayer is effectively captured during the process of foam formation. However, lipids in foams collected during a 1-year

TABLE 10.4. Comparison of the fatty acid constituents (percentage of total fatty acids) of lipids in the water-surface microlayer and foam from different environments.[a]

Fatty acid	Surface microlayer			Foams		
	Fresh water	Estuary	Marine	Fresh water	Estuary	Marine
8:0	—	5.7	—	—	—	—
10:0	—	2.8	7.1	—	—	—
12:0	2.7	4.6	26.4	1.5	2.2	—
14:0	10.9	6.3	28.6	20.7	8.4	4.6
16:0	37.3	26.3	13.6	39.0	20.5	17.0
16:1ω7	7.9	11.6	2.1	11.7	7.4	9.2
18:0	18.6	11.7	2.9	10.1	13.9	7.0
18:1ω9	18.1*	10.2*	2.4*	11.7**	14.5*	25.7
18:1ω7	—	—	—	—	—	2.1
18:2ω6	—	4.3	1.3	—	4.5	9.9
18:3ω3	—	—	0.5	—	5.1	1.0
18:4ω3	—	—	—	—	—	0.4
20:5ω3	—	—	—	—	—	7.3
22:6ω3	—	—	—	—	—	2.1
Others	4.5	16.2	14.7	5.3	23.5	13.5
Σ SAFA	69.5	57.4	78.6	71.3	45.0	28.6
Σ MUFA	26.0	21.8	4.5	23.4	21.9	37.0
Σ ω3	—	—	0.5	—	5.1	10.8
Σ ω6	—	4.3	1.3	—	4.5	9.9
Reference[b]	1	2	3	4	3	5

[a]—, Not reported; *, all isomers combined; **, all C-18 unsaturated combined.
[b]References: 1, Meyers and Owen, 1980; 2, Gever et al., 1996; 3, Marty and Choiniere, 1979; 4, recalculated from Simoneit et al., 1980; 5, Napolitano et al., 1995.

period (Marty and Choiniere, 1979) consistently contained higher concentrations of the PUFA 18:2ω6 (linoleic) and 18:3ω3 (linolenic), when compared with both dissolved and particulate lipids of the surface microlayers (Table 10.4). Linoleic and linolenic acids are regarded as biochemical markers for fungal and green algal biomass (see Napolitano, this volume, and references therein). Considering that PUFA are susceptible to rapid degradation, their enrichment in the lipids of foam suggests that they are produced by active microbial populations that colonize foams. High concentrations of PUFA in natural foams are not surprising, because stable foam formations can serve as microenvironments that favor the growth and dispersal of bacterial, algal, and fungal populations (Descals et al., 1995a,b; Webster et al., 1994; Meneses, 1993).

Fatty acids in foams from Laguna Lejia (Simoneit et al., 1980) mainly contained saturated (range C-10–C-28 and maximum at 16:0) and monounsaturated (maximum at 16:1ω7) species and only trace concentrations of polyunsaturated fatty acids. Although details about the presence of low concentrations of bacterial fatty acids were not provided, it can be assumed that this freshwater foam was not particularly receptive for microbial proliferation. This fatty acids profile appears to have been derived from epicuticular plant lipids and hydrolyzed microbial biomass (Jamieson and Reid, 1972).

In contrast to foams from lakes and estuaries, foams from coastal marine environments reveal a very different fatty acid composition (Table 10.4). Foam formed in the tidal channels and at the edge of the water during the rising tides in the Bay of Fundy, Canada, showed saturates of mainly 16:0 (17.0%) and 18:0 (7.0%), monounsaturates of especially 18:1ω9 (25.7%) and 16:1ω7 (9.3%), and also PUFA, accounting for 25% of the total fatty acids (Napolitano et al., 1992). PUFA consisted of almost equal proportions of fatty acids of the ω6 (mainly 18:2ω6) and the ω3 series (including 20:5ω3 and 22:6ω3). The 18:2ω6 probably indicates a direct terrestrial plant input, whereas the presence of 20:5ω3 obviously derives directly or indirectly from microalgae, principally diatoms (Napolitano et al., 1992).

Several unusual features were reported in the lipids of the surface microlayers of California rice fields by Gever et al. (1996). These authors reported relatively high concentrations of short-chain fatty acids (8:0 and 10:0) and the presence of very large concentrations of fatty acid methyl esters (FAME; approximately 95% of the fatty acids ≥20 carbon atoms long were found as methyl esters). FAMEs are usually a very minor components of the lipids of natural waters, and their presence, with a few exceptions (Conte et al., 1994), is normally regarded as an artifact of the lipid extraction or a result of environmental contamination. Considering that rice lipids (Gunstone et al., 1994) and monocotlyledonean cuticular waxes in general (Kolattukudy et al., 1976) do not contain appreciable quantities of FAME and owing to the absence of other obvious natural sources of these compound, the origin of FAME remains unknown. One can speculate that FAMEs are the product of contamination of the surface microlayer via deposition of particles from an unidentified distant source or, more likely, that they may be an ingredient in the formulation of pesticides applied to the rice field.

10.7. Research Needs

Testing and validation of a standard method for sampling of the water microlayer for lipid analysis is probably the more critical research need. Devices based on the principle of the rotating drum, equipped with PTFE-coated surfaces and accessories of the same material, are the more appropriate and accepted tools for the collection of the surface microlayer. Data are still needed, however, to fully evaluate their performance, especially those aspects concerning the selective absorption of different lipidic and lipophilic substances.

Information presented in the preceding sections clearly showed the marked qualitative and quantitative differences between the chemical composition of the surface microlayer and the underlaying water. As inferred from studies performed in thin water films (Metsik et al., 1973; McCafferty et al., 1970), the dielectric constant of the top few micrometers of the water is much lower than that of the bulk of the water. Therefore, it is expected from the surface microlayer to present a much higher affinity for hydrophobic substances. To the best of our knowledge, the measurement of this unusual property of natural waters and the resulting geochemical implications have not been evaluated.

Lipid markers are widely used in geochemical and ecological studies of freshwater and marine environment (see Napolitano, this volume). Although foam can hold a diverse microbial community of bacteria, algae, and fungi, ecological studies of this microenvironment are rare. Furthermore, the trophic structure of foam could be far more complex due to the possible inclusion of neustonic organisms commonly found in surface water (i.e., rotifers, nematodes, microarthropods). The biomass, trophic interactions, and the physiological status of these neustonic communities could be assessed by the analysis of lipid content and lipid composition.

Although foams are a natural phenomenon, the sudden development foam patches in lakes and streams may also be related to industrial operations and, therefore, can raise serious regulatory and compliance concerns. Some government and industrial regulations require the reporting of the characteristic and the origin of foam formations in lakes and streams near industrial sites. However, the differences between natural and artificial foams are subtle, and they are based on subjective criteria such as appearance, color, and texture (Valentine, Environmental Compliance, ORNL, U.S. Department of Energy, personal communication). The presence of foam as result of a water pollution event is obviously associated with the release of detergents or other potent surface active products. Therefore, the definitive characterization of a particular foam as of natural or man-made origin will require the isolation and identification of the contaminant surfactant. This procedure may sometimes become time-consuming and expensive. The analysis of lipids in foam and surface microlayers has the potential to be used as an alterative method to distinguish between natural and artificial foam. The proposed method does not require the identification of the surfactants. Instead, it consists of extracting the surface active compounds from the water by means of current standard methods, their subsequent hydrolysis, and a chromatographic analysis of their hydrophobic chains. Hypothetically, industrial surfactants would show a relatively simple chromatographic profile, mainly consisting of the C-12 and C-14 acyl or alkyl moieties normally used in the formulation of detergents and emulsifiers. Contrarily, biogenic surfactants would present a much wider spectrum of acyl and alkyl moieties and the dominance of the C-16 and C-18 fatty acids typically present in natural lipids.

10.8. Final Remarks

In the preceding sections, we have shown that lipids, derived from the decay and exudates of aquatic and terrestrial biota and from autochthonous neustonic populations, accumulate in the water-surface microlayer. We have defined the surface microlayer as the top 30–100 μm of the water column and showed how the transport of material across this environment can be qualitatively described by the stagnant boundary layer model (Whitman, 1923) and the surface renewal model (Danckwerts, 1951; Higbie, 1935). Hydrophobic pollutants and metals are also concentrated in the surface microlayer by the production of bubbles and other transport agents, contributing to the load of organic and inorganic matter deposited at the air–water interface (Johnson et al., 1989; Hardy, 1982; Elzerman and Armstrong, 1979; Eisenreich et al., 1978; Szekielda et al., 1971). The presence of biogenic surfactants in the microlayer, consisting of fulvic and humic acids draining from land and also from the aquatic microbial origin, enhances the solubility of hydrophobic compounds. The enrichment factor of lipophilic constituents in the surface microlayer varies enormously because it depends on the interplay of many factors such as the proximity of the sources (decaying organic material, industrial imput), transport agents (advection currents, wind, drainage), and the suitability of the particular microlayer or foam to sustain microorganisms. An additional factor contributing to the variability in the measure of the concentrations of substances in the water microlayer is the disparity of tools and materials used for sampling. The different sampling devices not only collect a different thickness of the microlayer, but some of the collecting surfaces (e.g., glass, metal, ceramic, PTFE) may have a differential affinity for the organic matter in natural waters. Due to inertness considerations, PTFE screens, collecting surfaces, and accessories (wipers and bottles) are recommended for the sampling of lipids and hydrophobic contaminants from surface waters.

Although foam formations are often associated with pollution events, they are a common occurrence in uncontaminated waters. Foams are constructed of the materials formed and accumulated at the surface microlayer, and therefore, they offer an excellent opportunity to study the chemical composition of the microlayer. Surface microlayers and foams of natural waters present a large list of organic constituents, which include humic and fulvic acids, carbohydrates, proteinaceous materials, and lipids. All lipid classes found in aquatic and terrestrial organisms are also found in the surface microlayers and foams (Napolitano et al., 1995; Johnson et al., 1989). The lipid class composition of the water microlayer, however, differs significantly from that of aquatic organisms. Lipids of the microlayers are, in general, enriched in those fractions resistant to chemical degradation, such as hydrocarbons and saturated free fatty acids. Hydrocarbons normally constitute a very small fraction of the lipids of most organisms (Nevenzel, 1989), but they can be the major constituent of the water microlayer. In most of these cases, however, a structural analysis reveals that the hydrocarbons are largely anthropogenic in origin.

Although foams normally concentrate the most refractory fraction of the biogenic lipids (Simoneit et al., 1980), some foam formations may also contain relatively large proportions of labile phospholipids or PUFA of both $\omega 3$ and $\omega 6$ series. The presence of these chemically unstable lipids reflects the capacity of foam to host microbial life.

Considering the spatial continuity of the freshwater and marine environments, it is easy to assume that the processes occurring at the water-surface microlayers, which cover ~70% of the area of the planet, have profound implications for a complete assessment of the cycles of the aquatic organic matter. Unfortunately, owing to the lack of standard methods for the consistent and reproducible sampling of the surface microlayer, the examination of its properties and chemical composition is not normally included in environmental studies. We look forward to a renewed interest in using the surface microlayer as a sensor and amplifier of processes that affect the whole aquatic system.

Acknowledgments. Writing of this chapter was supported in part by the appointment of G.E.N. to the Oak Ridge National Laboratory (ORNL) Research Associates Program administered jointly by ORNL and by the Oak Ridge Institute for Science and Education. ORNL is managed by Lockheed Martin Energy Research Corp. for the U.S. Department of Energy under Contract DE-ACO5-96OR22464. D.S.C. is supported by a fellowship from Consejo Nacional de Investigaciones Científicas y Técnicas (CONICET), Argentina. Arthur J. Stewart and Tommy Phelps (ORNL) provided constructive comments on a previous version of this work. This chapter is Environmental Sciences Division publication 4683.

References

Adamson, A.W. Physical Chemistry of Surfaces. New York: John Wiley & Sons; 1990.

Akit, J.; Cooper, D.J.; Manninen, K.I.; Zajic, J.E. Investigation of potential biosurfactant production among phytopathogenic *Corynebacteria* and related soil microbes. Curr. Microbiol. 6:145–150; 1981.

American Society for Testing and Materials (ASTM). Annual Book of ASTM Standards, vol. 11.01. Philadelphia, PA: ASTM; 1995.

Antia, N.J.; Landymore, A.F. Physiological and ecological significance of the chemical instability of uric acid and related purines in sea water and marine algal culture medium. J. Fish. Res. Bd. Can. 31:1327–1335; 1974.

Armstrong, F.A.J.; Williams, P.M.; Strickland J.D.H. Photo-oxidation of organic matter in sea water by ultraviolet radiation, analytical and other applications. Nature 211:481–483; 1966.

Aronstein, B.N.; Paterek, J.R. Effect of nonionic surfactant on the degradation of glass-sorbed PCB congeners by integrated chemical-biological treatment. Environ. Toxicol. Chem. 14:749–754; 1995.

Baier, R.E. Surface quality assessment of natural bodies of water. Proceedings of the 13th Conference on Great Lakes Research: International Association of Great Lakes Research 114–127; 1970.

Banat, I.M. Biosurfactants production and possible uses in microbial enhanced oil recovery and oil pollution remediation: a review. Bioresource Technol. 51:1–12; 1995.

Banat, I.M. The isolation of a thermophilic biosurfactant producing *Baciullus* sp. Biotech. Lett. 15:591–594; 1993.

Barger, W.R.; Garret, W.D. Surface active organic material in air over the Mediterranean and over the eastern equatorial Pacific. J. Geophys. Res. 81:3151–3157; 1976.

Baylor, E.R.; Sutcliffe, W.H.; Hirschfeld, D.S. Adsorption of phosphates onto bubbles. Deep Sea Res. 9:120–124; 1962.

Bennett-Corriea, W.; Sokol, H.A.; Garrison, W.M. Reductive Deamination in the Radiolysis of Oligopeptides in Aqueous Solution and in the Solid State. AEC rep. UCRL-19504. Berkeley: University of Calif. Radiation Laboratory; 1970.

Blanchard, D.C. Surface-active organic material on airborne salt particles. Proceedings of the International Conference on Cloud Physics, Toronto 1:25–29; 1968.

Blanchard, D.C. Sea-to-air transport of surface active material. Science 146:396–397; 1964.

Blanchard, D.C. The electrification of the atmosphere by particles from bubbles in the sea. Progr. Oceanogr. 1:71; 1963.

Blanchard, D.C.; Syzdek, L.D. Concentration of bacteria in jet drops from bursting bubbles. J. Geophys. Res. 77:5087–5099; 1972.

Brockmann, U.H.; Kattner, G.; Hentzschel, G.; Wandschneider, K.; Junge, H.D.; Huehnerfuss, H. Natuerliche Oberflaechenfilme im Seegebiet vor Sylt. Mar. Biol. 36:135–146; 1976.

Butler, A.C.; Sibbald, R.R. Sampling and GC-FID, GC/MS analysis of petroleum hydrocarbons in the ocean surface microlayer off Richards Bay, South Africa. Estuarine Coastal Shelf Sci. 25:27–42; 1987.

Carlson, D.J.; Cantey, J.L.; Cullen, J.J. Description of and results from a new surface microlayer sampling device. Deep-Sea Res. 35:1205–1213; 1988.

Chiu, H.L.; Huang, S.D. Adsorptive bubble separation of heptachlor and hydroxychlordene. Sep. Sci. Tech. 26:73–84; 1991.

Churchill, S.A.; Griffin, R.A.; Jones, L.P.; Churchill, P.F. Biodegradation rate enhancement of hydrocarbons by an oleophilic fertilizer and a rhamnolipid biosurfactant. J. Environ. Qual. 24:19–28; 1995.

Conte, M.H.; Volkman, J.K.; Eglinton, G. Lipid biomarkers of the Haptophyta. In: Green, J.C.; Leadbeater, S.C., eds. The Haptophyta Algae. Systematics Association Special Publication 51. Oxford: Clarendon Press; 1994:p. 351–377.

Cooper, D.G.; Zajic, J.E.; Gerson, D.F.; Manninen, D.I. Isolation and identification of biosurfactants produced during anaerobic growth of *Clostridium pasteurianum*. J. Ferment. Tech. 58:83–86; 1980.

Cooper, D.G.; Zajic, J.E.; Gerson, D.F. Production of surface active lipids by *Corynebacterium lepus*. Appl. Environ. Microbiol. 37:4–10; 1979.

Danckwerts, P.V. Significance of liquid-film coefficients in gas absorption. Ind. Eng. Chem. 43:1460–1467; 1951.

Daumas, R.A.; Laborde, P.L.; Marty, J.C.; Saliot, A. Influence of sampling method on the chemical composition of water surface film. Limnol. Oceanogr. 21:319–326; 1976.

Descals, E.; Peláez, F.; López Lorca, L.V. Fungal spora of stream foam from central Spain. I. Conidia identifiable to species. Nova Hedwigia 60:533–550; 1995a.

Descals, E.; Peláez, F.; López Lorca, L.V. Fungal spora of stream foam from central Spain. II. Chorology, spore frequency and unknown forms. Nova Hedwigia 60:551–569; 1995b.

Dorman, D.C.; Lemlich, R. Separation of liquid mixtures by non-foaming bubble fractionation. Nature 207:145–146; 1965.

Duce, R.A.; Quin, J.G.; Olney, C.E.; Piotrowicz, S.R.; Ray, B.J.; Wade, T.L. Enrichment of heavy metals and organic compounds in the surface microlayer of Narragansett Bay, Rhode Island. Science 176:161–163; 1974.

Duce, R.A.; Stumm, W.; Prospero, J.M. Working symposium on sea-air chemistry: summary and recommendations. J. Geophys. Res. 77:5059–5061; 1972.

Duran, A.P.; Hemond, H.F. Dichlorodifluoromethane (freon 12) as a tracer for nitrous oxide release from a nitrogen-enriched river. In: Brutsaert, W.; Jirka, G.H., eds. Gas Transfer at Water Surfaces. Boston: Reidel; 1984:p. 421–429.

Eisenreich, S.J.; Elzerman, A.W.; Armstrong, D.E. Enrichment of micronutrients, heavy metals, and chlorinated hydrocarbons in wind-generated lake foam. Environ. Sci. Technol. 12:413–417; 1978.

Elzerman, A.W.; Armstrong, D.E. Enrichment of Zn, Cd, Pb and Cu in the microlayer of Lakes Michigan, Ontario and Mendota. Limnol. Oceanogr. 24:133–144; 1979.

Finnerty, W.R.; Singer, M.E. A microbial biosurfactant-physiology, biochemistry, and applications. Dev. Ind. Microbiol. 25:31–46; 1984.

Flockhart, B.D.; Graham, H.J. Dilute solutions of sodium oleate. J. Colloid Sci. 8:105–115; 1953.

Fogg, G.E. Excretion of organic matter by phytoplankton. Limnol. Oceanogr. 22:576–577; 1977.

Gaines, G.L. Insoluble Monolayers at Liquid–Gas Interfaces. New York: John Wiley & Sons; 1966.

Garrett, W.D. Impact of natural and man-made surface films on the properties of the air-sea interface. In: Dyrssen, D.; Jagner, D., eds. The Changing Chemistry of the Oceans. Nobel Symposium 20. Stockholm: Almqvist and Wiksell; 1972:p. 75–91.

Garrett, W.D. The organic chemical composition of the ocean surface. Deep-Sea Res. 14:221–227; 1967.

Garrett, W.D. Collection of slick-forming materials from the sea surface. Limnol. Oceanogr. 10:602–605; 1965.

Garrett, W.D.; Bultman, J.D. The Damping of Water Waves by Insoluble Organic Mono-layers. Nav. Res. Lab. Rep. 6003. Washington, DC: Nav. Res. Lab.; 1963.

Genereux, D.P. Field studies of stream flow generation using natural and injected traces on Bickford and Walker Branch watersheds. PhD thesis, Massachusetts Institute of Technology, Cambridge; 1991.

Georgiu, G.; Lin, S-C.; Sharma, M.M. Surface-active compounds from microorganisms. Bio/Technology 10:60–65; 1992.

Gershman, J.W. Physico-chemical properties of solutions of para long chain alkylbenzene-sulfonates. J. Phys. Chem. 61:581–584; 1957.

Gever, J.R.; Mabury, S.A.; Crosby, D.G. Rice field surface microlayers: collection, composition and pesticide enrichment. Environ. Toxicol. Chem. 15:1676–1682; 1996.

Greenberg, A.E.; Clesceri, L.S.; Eaton, A.D., eds. Standard Methods for the Examination of Water and Wastewater. 18th ed. Washington, DC: American Public Health Association; American Water Works Association; Water Environment Federation; 1992.

Gunstone, F.D.; Hardwood, J.L.; Padley, F.B. The Lipid Handbook. New York: Chapman & Hall; 1994.

Hamilton, E.I.; Clifton, R.J. Techniques for sampling the air–sea interface for estuarine and coastal waters. Limnol. Oceanogr. 24:188–193; 1979.

Han, J.; McCarthy, E.D.; Calvin, M.; Benn, M.D. Hydrocarbon constituents of the blue-green algae *Nostoc muscorum, Anacystis nidulans, Phormidium luridum* and *Chlorogloea frischii*. J. Chem. Soc. C:2785–2791; 1968.

Hardy, J.T. The sea surface film microlayer: biology, chemistry and anthropogenic enrichment. Progr. Oceanogr. 11:307–328; 1982.

Hardy, J.T.; Coley, J.A.; Antrim, L.D.; Kiesser, S.L. A hydrophobic large-volume sampler for collecting aquatic surface microlayers: characterization and comparison with the glass plate method. Can. J. Fish. Aquat. Sci. 45:822–826; 1988.

Harvey, G.W. Microlayer collection from the sea surface: a new method and initial results. Limnol. Oceanogr. 11:608–613; 1966.

Harvey, G.W.; Burzell, L.A. A simple microlayer method for small samples. Limnol. Oceanogr. 17:156–157; 1972.

Healy, T.W.; La Mer, V.K. The effect of mechanically produced waves on the properties of monomolecular layers. J. Phys. Chem. 68:3535–3539; 1964.

Higbie, R. The rate of adsorption of a pure gas into a still liquid during short periods of exposure. Trans. Am. Inst. Chem. Eng. 35:365–389; 1935.

Horowitz, S.; Gilbert, J.N.; Griffin, W.M. Isolation and characterization of a surfactant produced by *Bacillus licheniformis* 86. J. Ind. Microbiol. 6:243–248; 1990.

Hunter, K.A.; Liss, P.S. Organic sea surface films. In: Duursma, E.K.; Dawson, R., eds. Marine Organic Chemistry, Evolution, Composition, Interactions and Chemistry of Organic Matter in Seawater. Elsevier Oceanography Series 31. New York: Elsevier Scientific Publishing Company; 1981:p. 259–298.

Jamieson, G.R.; Reid E.H. The component fatty acids of some marine algal lipids. Phytochemistry 11:1423–1432; 1972.

Jarvis, N.L. Adsorption of surface-active material at the sea-air interface. Limnol. Oceanogr. 12:213–221; 1967.

Johnson, B.D.; Zhou, X.; Parrish, C.C.; Wangersky, P.J.; Kerman, B.R. Fractionation of particulate matter, the trace metals Cu, Cd, and Zn, and lipids in foam and water below Niagara Falls. J. Great Lakes Res. 15:189–196; 1989.

Johnson, R.W.; Calder, J.A. Early diagenesis of fatty acids and hydrocarbons in salt marsh environments. Geochim. Cosmochim. Acta 37:1943–1955; 1973.

Karger, B.L.; DeVivo, D.G. General survey of adsorptive bubble separation processes. Sep. Sci. 3:393–424; 1968.

Kattner, G.G.; Brockmann, U.H. Fatty-acid composition of dissolved and particulate matter in surface films. Mar. Chem. 6:233–241; 1978.

Kjelleberg, S.; Stenström, T.A.; Odham, G. Comparative study of different hydrophobic devices for sampling lipid surface films and adherent microorganisms. Mar. Biol. 53:21–25; 1979.

Kolattukuddy, P.E.; Croteau, R.; Walton, T.J. Biochemistry of plant waxes. In: Kolattukuddy, P.E., ed. Chemistry and Biochemistry of Natural Waxes. Amsterdam: Elsevier; 1976:p. 315–329.

Kretschmer, A.; Bock, H.; Wagner, F. Chemical and physical characterization of inter-facial-active lipids from *Rhodococcus erythropolis* grown on n-alkanes. Appl. Environ. Microbiol. 44:864–870; 1982.

Kucklick, J.R.; Bidleman, T.F. Organic contaminants in Winyah Bay, South Carolina. I: pesticides and polycyclic aromatic hydrocarbons in subsurface and microlayer waters. Mar. Environ. Res. 37:63–78; 1994.

Lai, R.J.; Shemdin, O.H. Laboratory study of the generation of spray over water. J. Geophys. Res. 79:3055–3063; 1974.

Larsson, K.; Odham, G.; Södergren, A. On lipid surface films on the sea. I. A simple method for sampling and studies of composition. Mar. Chem. 2:49–57; 1974.

Leenheer, J.A.; Huffman, E.W.D., Jr. Classification of organic solutes in water by using macroreticular resins. J. Res. U.S. Geol. Survey 4:737–751; 1976.

Lesik, O.Y.; Karpenko, E.V.; Elysseev, S.A.; Turovsky, A.A. The surface-active and emulsifying properties of *Candida lipolytica* Y-917 grown on n-hexadecane. Microbiol. J. 51:56–59; 1989.

Lin, S.C.; Carswell, K.S.; Sharma, M.M.; Georgiou, G. Continuous production of the lipopeptide biosurfactant of *Bacillus licheniformis* JF-2. Appl. Microbiol. Biotech. 41:281–285; 1994.

Liss, P.S. Chemistry of the sea surface microlayer. In: Riley, J.; Skirrow, G.; Chester, R., eds. Chemical Oceanography. London: Academic Press; 1975:p. 193–243.

Liss, P.S. Processes of gas exchange across an air-water interface. Deep Sea Res. 20:221–238; 1973.

MacIntyre, F. Chemical fractionation and sea-surface microlayer processes. In: Goldberg, E.D., ed. The Sea, vol. 5. New York: Wiley-Interscience; 1974:p. 245–299.

MacIntyre, F. Bubbles: a boundary-layer "microtome" for micron-thick samples of liquid surfaces. J. Phys. Chem. 72:589–592; 1968.

Mackay, D.; Yeun, A.T.K. Mass transfer coefficients correlations for volatilization of organic solutes from water. Environ. Sci. Technol. 17:211–233; 1983.

Mallinger, W.D.; Mickelson, T.P. Experiments with monomolecular films on the surface of the open sea. J. Phys. Oceanogr. 3:328–336; 1973.

Marty, J.C.; Choiniere, A. Acides gras et hydrocarbures de l'écume marine et de la micro-couche de surface. Naturaliste Can. 106:141–147; 1979.

Matsuyama, T.; Kaneda, K.; Ishizuka, I.; Toida, T.; Yano, I. Surface-active novel glycolipid and linked 3-hydroxy fatty acids produced by *Serratia rubidaea*. J. Bacteriol. 172:3015–3022; 1990.

McCafferty, E.; Pravdic, V.; Zettlemoyer, A.C. Dielectric behavior of adsorbed water films on the alfa-iron III oxide surface. Trans. Faraday Soc. 66:1720–1731; 1970.

McDonald, C.R.; Cooper, D.G.; Zajic, J.E. Surface-active lipids from *Nocardia erythropolis* grown on hydrocarbons. Appl. Environ. Microbiol. 41:117–123; 1981.

McInerney, M.J.; Javaheri, M.; Nagle, D.P. Properties of the biosurfactant produced by *Bacillus licheniformis* strain JF-2. Ind. Microbiol. 5:95–102; 1990.

Means, J.C.; Wijayaratne, R. Role of natural colloids in the transport of hydrophobic pollutants. Science 215:968–970; 1982.

Meneses, I. Foam as a dispersal agent in the rocky intertidal of central Chile. Eur. J. Phycol. 28:107–110; 1993.

Metsik, J.S.; Perevertaev, V.D.; Liopo, V.A.; Timoshchenko, G.T.; Kiselev, A.B. New data on the structure and properties of thin films on mica crystals. J. Colloid Interface Sci. 43:662–669; 1973.

Meyers, P.A.; Kawka, D.E. Fractionation of hydrophobic organic materials in the surface microlayers. J. Great Lakes Res. 8:288–298; 1982.

Meyers, P.A.; Owen, R.M. Sources of fatty acids in Lake Michigan surface microlayers and subsurface waters. Geophys. Res. Lett. 7:885–888; 1980.

Miyake, Y.; Tsounogai, S. Evaporation of iodine from the ocean. J. Geophys. Res. 68:3989–3993; 1963.

Morikawa, M.; Daido, H.; Takao, T.; Murata, S.; Shimonishi, T.; Imanaka, T. A new lipopeptide biosurfactant produced by *Arthrobacter* sp. strain MIS38. J. Bacteriol. 175:6459–6466; 1993.

Münnich, K.O.; Clarke, W.B.; Fischer, K.H.; Flothmann, D.; Kromer, B.; Roether, W.; Siegenthaler, U.; Top, Z.; Weiss, W. Gas exchange and evaporation studies in a circular wind tunnel, continuous radon-222 measurements at sea, and tritium/helium-3 measurements in a lake. In: Favre, H.; Hasselmann, K., eds. Turbulent Fluxes through the Sea Surface, Wave Dynamics and Predictions. New York: Plenum; 1978:p. 151–165.

Napolitano, G.E.; Richmond, J.E. Enrichment of biogenic lipids, hydrocarbons and PCBs in stream-surface foams. J. Environ. Toxicol. Chem. 14:197–201; 1995.

Napolitano, G.E.; Ackman, R.G.; Parrish, C.C. Lipids and lipophilic pollutants in three species of migratory shorebirds and their food in Shepody Bay (Bay of Fundy, New Brunswick). Lipids 27:785–790; 1992.

Nevenzel, J. Biogenic hydrocarbons in marine organisms. In: Ackman, R.G., ed. Marine Biogenic Lipids, Fats and Oils. Boca Raton, FL: CRC Press; 1989:p. 3–71.

Norkrans, B. Surface microlayers in aquatic environments. In: Alexander M., ed. Advances in Microbial Ecology, vol. 4. New York: Plenum Press; 1980:p. 495–514.

O'Connor, D.J.; Dobbins, W.E. Mechanisms of reaeration in natural streams. Trans. Am. Soc. Civ. Eng. 123:641–684; 1958.

Parrish, C.C. Time series of particulate and dissolved lipid classes during spring phytoplankton blooms in Bedford Basin, a marine inlet. Mar. Ecol. Prog. Ser. 35:129–139; 1987.

Paterson, M.P.; Spillane, K.T. Surface films and the production of sea salt aerosol. Q. J. R. Meteorol. Soc. 95:526–534; 1969.

Persson, A.; Österberg, E.; Dostalek, M. Biosurfactant production by *Pseudomonas fluorescens* 378: growth and product characteristics. Appl. Microbiol. Biotechnol. 29:1–4; 1988.

Pojasek, R.B.; Zajicek, O.T. Surface microlayers and foams-source and metal transport in aquatic systems. Water Res. 12:7–10; 1978.

Powalla, M.; Lang, S.; Wray, V. Penta- and disaccharide lipid formation by *Nocardia corynebacteroides* grown on n-alkanes. Appl. Microbiol. Biotechnol. 31:473–479; 1989.

Ramsay, B.; McCarthy, J.; Guerra-Santos, L.; Käppeli, O.; Feichter, A.; Margaritis, A. Biosurfactant production and diauxic growth of *Rhodococcus aurantiacus* when using n-alkanes as the carbon source. Can. J. Microbiol. 34:1209–1212; 1988.

Rathbun, R.E.; Tai, D.J.J. Gas-film coefficients for streams. J. Environ. Eng. Div. ASCE 109:1111–1127; 1983.

Rathbun, R.E.; Tai, D.J.J. Volatilization of organic compounds from streams. Environ. Eng. Div. ASCE 108:973–989; 1982.

Roch, F.; Alexander, M. Biodegradation of hydrophobic compounds in the presence of surfactants. Environ. Toxicol. Chem. 14:1151–1158; 1995.

Schramm, L.L.; Wassmuth, F. Foams: basic principles. In: Schramm, L.L., ed. Foams: Fundamentals and Applications in the Petroleum Industry. Washington, DC: American Chemical Society; 1994:p. 3–45.

Schwarzenbach, R.P. Assessing the behaviour and fate of hydrophobic organic compounds in the aquatic environment—general concepts and case studies emphasizing volatile halogenated hydrocarbons. Habilitation thesis, Swiss Federal Institute of Technology, Zurich; 1983.

Schwarzenbach, R.P.; Gschwend, P.M.; Imboden, D.M. Environmental Organic Chemistry. New York: John Wiley & Sons, Inc.; 1993.

Sharp, J.H. Excretion of organic matter by marine phytoplankton: do healthy cells do it? Limnol. Oceanogr. 22:381–399; 1977.

Sieburth, J. Bacteriological samplers for air–water and water–sediment interfaces. Trans. Joint Conf. MTS and ASLO, Washington, DC; 1965:p. 1064–1068.

Simoneit, B.R.T.; Halpern, H.I.; Didyk, B.M. Lipid productivity of a high Andean lake. In: Trudinger, P.A.; Walter, M.R.; Ralph, B.J., eds. Biogeochemistry of Ancient and Modern Environments. Berlin: Springer-Verlag; 1980:p. 201–210.

Södergren, A. Origin and composition of surface slicks in lakes of different trophic status. Limnol. Oceanogr. 32:1307–1316; 1987.

Södergren, A. Origin of ^{14}C and ^{32}P labelled lipids moving to and from freshwater surface microlayers. Oikos 33:278–289; 1979.

Stobbe, H.; Peschel, G. Experimental determination of static permittivity of extreme thin liquid layers of water dependent on their thickness. Colloid Polym. Sci. 275:162–169; 1997.

Stumm, W.; Morgan, J.J. Aquatic Chemistry. New York: John Wiley & Sons; 1981.

Szekielda, K.H., Kupferman, S.L.; Klemas, V.; Polis, D.F. Element enrichment in organic films and foams associated with aquatic frontal systems. J. Geophys. Res. 77:5278–5282; 1971.

Thangamani, S.; Shreve, G.S . Effect of anionic biosurfactant on hexadecane partitioning in multiphase systems. Env. Sci. Tech. 28:1993–2000; 1994.

Tunlid, A.; White, D.C. Biochemical analysis of biomass, community structure, nutritional status, and metabolic activity of microbial communities in soil. In: Stotzky, G.; Bollag J-M., eds. Soil Biochemistry. New York: Marcel Dekker; 1992:p. 229–262.

Velimirov, B. Sugar and lipid components in sea foam near kelp beds. Mar. Ecol. 3:97–107; 1982.

Volz, F.E. Infrared absorption by atmospheric aerosol substances. J. Geophys. Res. 77: 1017–1031; 1972.

Wachs, W.; Hayano, S. Über die kritische Micellkonzentration (CMC) von fettsäuremonoestern der Saccharose und ihre Beziehung zum HLB-WERT. Kolloid-Z 181:139–144; 1962.

Warry, N.D.; Chan, C.H. Organic contaminants in the suspended sediments of the Niagara River. J. Great Lakes Res. 7:394–403; 1981.

Wasan, D.T.; Koczo, K.; Nikolov, A.D. Mechanisms of aqueous foam stability and antifoaming action with and without oil. In: Schramm, L.L., ed. Foams: Fundamentals and Applications in the Petroleum Industry. Washington, DC: American Chemical Society; 1994:p. 48–114.

Webster, J.; Marvanová, L.; Eicker, A. Spores from foam from South African rivers. Nova Hedwigia 59:379–398; 1994.

White, D.C. Chemical ecology: possible linkage between macro- and microbial ecology. Oikos 74:177–184; 1995.

Whitman, W.G. The two film theory of gas absorption. Chem. Metal. Eng. 29:146–148; 1923.

Wilcock, R.J. Reaction studies on some New Zealand rivers using methyl chloride as a gas tracer. In: Brutsaert, W.; Jirka, G.H., eds. Gas Transfer at Water Surfaces. Boston: Reidel; 1984:p. 413–420.

Williams, P.M.; Carlucci, A.F.; Henrichs, S.M.; Van Vleet, E.S.; Horrigan, S.G.; Reid, F.M.H.; Robertson, K.J. Chemical and microbial studies of sea-surface films in the southern Gulf of California and off the west coast of Baja California. Mar. Chem. 19:17–98; 1986.

Williams, R.J.; Phillips, J.N.; Mysels, K. Critical micelle concentration of sodium dodecylsulfate at 25°C. J. Trans. Faraday Soc. 51:728–737; 1955.

Wilson, A.T. Surface of the ocean as a source of air-borne nitrogenous material and other plant nutrients. Nature 184:99–101; 1959.

Wissmar, R.C.; Simenstad, A. Surface foam chemistry and productivity in the Duckabush River estuary, Puget Sound, Washington. In: Kennedy, V.S., ed. The Estuary as a Filter. San Diego, CA: Academic Press; 1984:p. 331–348.

Wu, J. Evaporation due to spray. J. Geophys. Res. 79:4107–4109; 1974.

Zafiriou, O.C. Photochemistry of halogens in the marine atmosphere. J. Geophys. Res. 79:2730–2732; 1974.

Zepp, R.G.; Wolfe, N.L.; Gordon, J.A.; Baughman, G.L. Dynamics of 2,4-D esters in surface waters. Hydrolysis, photolysis, and vaporization. Environ. Sci. Technol. 9:1144–1150; 1975.

11

Comparison of Lipids in Marine and Freshwater Organisms

Robert G. Ackman

11.1. Introduction

Lipids, carbohydrates, and proteins are the basic components of aquatic organisms, and all have distinct roles. For example, the carbohydrates can be structural components in phytoplankton and macrophytes and energy reserves in bivalve mollusks. Carbohydrates are barely mentioned in fish biochemistry except for glucose, which has a function in muscle energy metabolism (Kiessling et al., 1995). The proteins do not seem important in primitive organisms, but some invertebrates use free amino acids for ionic balance and, in moving up the evolutionary scale, the role of proteins in muscle of mobile life forms becomes very important. The role of lipids in photosynthetic carbon fixation is not obvious although the photosynthetic apparatus depends on certain fatty acids and lipid classes (Gurr and Harwood, 1991). In most organisms, fatty acids are commonly three-quarters of the mass of phospholipids, which are critical in membranes. Aquatic bacteria survive in a highly stressful environment because they are encased in lipids. Keweloh and Heipieper (1996) point out that stable saturated fatty acids are present in these lipids, but the high-melting *trans* monoethylenic fatty acids may be formed in parallel to, or from, the common *cis* isomers, to adapt this type of organism to a hostile environment. More intriguing is the recent but repeated discovery of eicosapentaenoic acid ($20:5\omega3$, popularly designated EPA) in marine bacteria (Nichols et al., 1996; Yazawa, 1996; Henderson et al., 1995a; Akimoto et al., 1990). Hitherto, this fatty acid, sensitive to oxidation, would have been associated with invertebrates that accumulate it from phytoplankton (Ackman and Kean-Howie, 1995). In moving up the evolutionary scale, lipids also play a major role in the neurotransmission system critical to mobile animals and in sensory perception organs such as the retina or *tapetum lucidum* of the fish eye, where docosahexaenoic acid ($22:6\omega3$, popularly designated DHA) is important (Nicol et al., 1972). However, marine invertebrates do not seem to have this specific requirement for high levels of DHA, and in the case of *Artemia* sp. eye phospholipids, there is no $22:6\omega3$ (Navarro et al., 1992).

11.2. Discussion

11.2.1. Lipid Classes

Lipids are universal, but careful recovery for study needs to be promoted for both freshwater (Herbes and Allen, 1983) and marine organisms (Roose and Smedes, 1996; Smedes and Thomasen, 1996). Space does not permit a critical review of this subject, nor of minor but biochemically important lipid classes such as tocopherols. The subject has recently been reviewed elsewhere (Ackman, 1993), and despite objections to the use of chloroform, it is clear that chloroform/methanol extracts such as the Bligh and Dyer (1959) system are preferred solvent systems for total lipid recovery.

The major lipid classes of most aquatic animals are shown in Figure 11.1. The phospholipids are both structural and functional elements of cell membranes, and

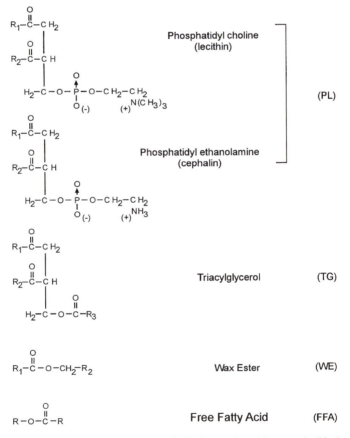

FIGURE 11.1. Some of the more important lipid classes found in aquatic life forms. Free fatty acids are, in fact, hardly observed in living animals but may be liberated from any of the more complex lipids by postmortem enzymatic hydrolysis.

the roles of triacylglycerols and wax esters are basically those of energy storage (Albers et al., 1996). Free fatty acids are very minor components. They can be transported around bodies and passed through membranes into cells for catabolism but are usually generated "on site" by appropriate enzymes acting on lipids. This is clear in a comprehensive chromatographic illustration of most of the above-mentioned lipid classes (Fig. 11.2). In this Iatroscan thin-layer chromatography with flame ionization detection (TLC-FID) analysis of fresh muscle of silver hake (*Merluccius bilinearis*) after 24-h exposure to the natural autolytic enzymes in the body during storage on ice, the important role of the triacylglycerols is obvious, even though the total lipid was only 2.4%. The very small free fatty acid peak is only 0.8% of that lipid. In whole bodies of the hypolimnetic copepod *Diaptomus sicilis* of Lake Michigan, the lipids were mainly triacylglycerols, but free fatty acids were 1.1–5.5% of the total lipid, which was 30–40% of dry weight (Vanderploeg et al., 1992). During part of a starvation study, the proportion of free fatty acids did rise temporarily in digestive gland lipids of the tiger prawn *Penaeus esculentus* (Chandumpai et al., 1991). Other-

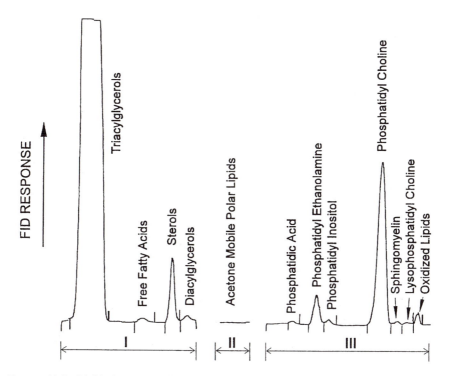

FIGURE 11.2. Lipid classes revealed by Iatroscan thin-layer chromatography—flame ionization detector (TLC-FID) profiles from three development steps for the lipid classes extracted from silver hake muscle tissue. (From Zhou and Ackman, 1996, by permission of the American Oil Chemists' Society.)

wise, free fatty acids in important proportions of extracted lipids should be regarded with doubt and probably the result of enzymatic hydrolysis of original lipids in situ. The proportions of the dominant phosphatidylcholine and of the lesser phosphatidylethanolamine, found to be approximately 2:1 in most tissues of marine organisms (Vaskovsky, 1989), are also illustrated in the chromatographic profile.

The rearrangement of the proportions of these lipid classes may be observed under stress conditions. For example, lipid classes were studied over a whole year with emphasis on the sexual maturation period for the freshwater crustacean *Macrobrachium borellii* (Gonzalez-Baro and Pollero, 1988). The different lipids of the muscle remained almost unaltered, but the shift of lipid, especially triacylglycerols, into the female gonadal tissue from the hepatopancreas was obvious in the Southern Hemisphere summer period of October–March. In many invertebrates, triacylglycerols are a major part of the oocyte. The effects of sexual maturation in the marine shrimp *Penaeus kerathurus* have also been followed by Mourente and Rodríguez (1991) in considerable detail. On a dry-weight basis, ovary total lipid doubled, and triacylglycerol content increased fourfold. Polar lipids approximately doubled, but proportions of fatty acids among important subclasses of polar lipids did not alter except in phosphatidylethanolamine, suggesting a simple transfer of total lipid to the oocytes. The continuing fascination with the Antarctic *Euphausia superba* showed some very interesting results (Saether et al., 1985), as in different samples both phospholipids and triacylglycerols varied with total lipid content, a situation unlike fish, in which the muscle phospholipids are considered stable and only the depot fat triacylglycerols vary. Females contained twice the total lipids of males; histological examinations of the distribution of lipid in tissues were included in this report. Their observations over a range of 1–6% lipid (wet weight), including females with roe, appear to confirm that in such crustacea total body lipid is involved in a depot lipid function. Similar conclusions were reached by Teshima et al. (1988) for induced maturation by destalking in the prawn *Penaeus japonicus*. This group also examined the lipid classes of the hemolymph, clearly involved in the vitellogenesis process and rich in free sterol during this stage. The tiger prawn *Penaeus esculentus* was the subject of experimental starvation (Chandumpai et al., 1991) to explore the changes in digestive gland lipids. Very little triacylglycerol (8% of body lipid) was present in that tissue and less in the muscle; the latter tissue preserved its relative proportions of lipid classes during molting as well as during starvation. This inability to store much lipid could lead to difficulty in molting or reproduction if food was not readily available.

By contrast, the freshwater amphipods *Diporeia* spp. (formerly including *Pontoporeia hoyi*) deposit an extensive proportion of triacylglycerols, as much as 84% of total lipid. Females have more lipid than males and transfer this lipid to eggs for brooding (Cavaletto et al., 1996). An extensive survey of energy reserve lipids in a saline lake over 2 years is especially interesting as it shows that lag periods of lipid deposition were of the order of 1–2 months (Arts et al., 1993). This may explain the inexact correlations between productivity and zooplankton lipid classes in smaller freshwater lakes (Wainman et al., 1993).

11.2.2. Sterols and Cholesterol

The sterol lipid class is about 95% cholesterol (sterols in Fig. 11.2) in higher organisms, but cholesterol may be only about one-half of the total sterols in molluskan filter feeders (Napolitano et al., 1993). It is known that the American lobster (*Homarus americanus*) and related crustacea have an absolute requirement for cholesterol and do not thrive on phytosterols (Kean et al., 1985; D'Abramo et al., 1984; Castell et al., 1975). In invertebrates, sterols circulate in the blood or hemolymph (Giese, 1966) in analogy to their role in fish (Waellert and Babin, 1994) and in mammalian food digestion and transport of lipids. Sterols also possibly have a functional role in maintaining the fluidity of membranes. The cholesterol content of membranes does not appear to be related to salinity (Hazel and Williams, 1990).

Studies on the mollusk species *Diplodom patagonicus* (freshwater) and *D. variabilis* (estuarine, subject to marine influence) of Argentina showed a much higher proportion of cholesterol in the freshwater lake animals than in the estuarine relative (Pollero et al., 1983). In original experiments on three other freshwater bivalves collected in Bulgaria and in a discussion and review, other authors (Popov et al., 1981) found evidence of control over the proportion of cholesterol in such animals. Pollero et al. (1983) also found the sterol composition in freshwater bivalves to be simpler than that in marine analogues. The latter finding may explain the apparent higher level of cholesterol in freshwater bivalves.

11.2.3. Wax Esters and Triacylglycerols

Wax esters in near-surface copepods are accepted as important in energy transfers rather than as structural elements (Graeve and Kattner, 1992). Another role may be for buoyancy control (Phleger and Grigor, 1990). Wax esters are found in very-deep-water fish such as the orange roughy (*Hoplostethus atlanticus*) (Grigor el al., 1990) and also in midwater fish (e.g., the lanternfish *Lampanyctodes hectoris*). These perform diurnal vertical migrations, and this and other similar species contain moderate amounts of wax esters (de Koning and Evans, 1991). In other species of midwater fish, Saito and Murata (1996) have reported either very low (0.5%) or very high (87.9%) amounts of wax esters in different types of myctophid fish. Wax esters are thus more important than is generally realized by many invertebrate specialists. The tropical reef corals use wax ester to store photosynthetic energy (Lee and Patton, 1989). The energy of wax esters is easily assimilated by predatory fish, and after hydrolysis, fatty alcohols are converted to fatty acids (Lie and Lambertsen, 1991; Tocher and Sargent, 1984; Sargent et al., 1979). In the Baltic, where the general salinity is too low for calanoid copepods rich in wax esters, including 22:1 fatty alcohol, the fatty acids of the Baltic herring in consequence have very little 22:1 fatty acid (Linko et al., 1985). This is one of the main fatty acids in marine fish triacylglycerols and results from oxidation of dietary wax ester alcohols. In one copepod sample, this 22:1 alcohol comprised >40% of the fatty alcohols (Ackman et al., 1974). This leads to wax ester chain lengths of up to C_{44}, as the 22:1 alcohol can, in most samples, be complemented by a range of fatty acids up to and including the important fatty acid 22:6ω3 (Kattner et al. 1990). Albers et al. (1996) point out that in omnivorous and carnivorous species, wax esters rich in 14:0 or 16:0 alcohol may be more important.

The presence of wax esters in freshwater zooplankton was not confirmed until 1989. Cavaletto et al. (1989) then showed that in the total lipids of two calanoid copepods, *Limnocalanus macrurus* and *Senecella calanoides,* from Lake Michigan 57–80% of wax esters were stored in a sac, with up to 17.5% triacylglycerols also present in this energy reserve. The coexisting calanoid *Diaptomus sicilis* had a substantial lipid component (36.3% of dry weight), but only 1.1% was made up of wax or sterol esters. The common thin-layer chromatography (TLC) technology for separating total lipids on silica gel-coated plates or by Iatroscan TLC-FID on silica gel Chromarods does not usually separate these two lipid classes, so minor amounts of wax esters may have been overlooked by other researchers. Gas-liquid chromatography (GLC) of neutral lipids does not eliminate such problems, although hydrogenation of the neutral lipids can help resolution and quantitation of several lipid classes by GLC (Yang et al., 1996) and of lipids generally by TLC-FID (Shantha and Ackman, 1990). The farmed (Sri Lanka) marine decapod crustacean *Penaeus monodon* had more steryl esters than free cholesterol (28 and 4.1 $\mu g \cdot g^{-1}$ lipid, respectively) in the hepatopancreas, whereas in gill, muscle, and ovary there was more free sterol than steryl ester (Young et al., 1992). In the latter tissues, the cholesterol in free form would be fitted into phospholipid-rich membranes. In the hepatopancreas storage organ, steryl esters would remain liquid, probably dissolved in triacylglycerols in adipocytes, whereas free cholesterol (m.p. 148°C) has a marked tendency to crystallize under all favorable conditions. The fatty acids of the cholesteryl esters were undistinguished except for 20:4ω6 being equal to or exceeding 20:5ω3. The proportion of 20:4ω6 in fatty acids is, however, higher in lipids of freshwater animals than in those of marine life (see below) and also usually higher in animals of tropical marine waters than in those of higher latitudes.

Polychaetes are a known food resource for in-shore fish and also for migrating shore-feeding birds (Napolitano et al., 1992). Most do not have much storage lipid, and yet lipid droplets are observed in the Pacific species *Sabella pavonina.* Koechlin et al. (1981) demonstrated that these droplets in the nephridial epithelia were roughly half free cholesterol and half cholesterol esters. The popularity of colorimetric assays for free cholesterol could have missed this occurrence of steryl esters, which these authors illustrated clearly by TLC and have confirmed by mass spectrometry.

The impact of subzero seawater on the lipids of one bivalve (*Yoldia hyperborea*) and two polychaetes (*Nephthys ciliata* and *Artacama proboscidea*) has been clarified by Parrish et al. (1996a) by using TLC-FID. By contrast, the lipid variations recorded for *Nereis virens* by Pellerin-Massicotte et al. (1994) were obtained by a colorimetric method developed for human serum lipids. Unfortunately, this reduces the value of the observations of seasonal variations in lipids of this species in the cold climate of the St. Lawrence River estuary of Canada. The actual lipid contents were ~ 1 mg $\cdot$ g^{-1} fresh weight in May and ≥ 2 mg $\cdot$ g^{-1} in September. Whether this represented any deposition of triacylglycerols as energy storage for overwintering or for reproductive tissue could not be determined. There are alternative methods of quantitatively analyzing lipid classes in detail (e.g., by planar TLC; Ackman, 1991) or HPLC (high-performance liquid chromatography; Christie, 1992, 1987), although they may require 10 mg of sample as compared with TLC-FID, which can operate with 20 μg of lipid. Total lipid obtained by a colorimetric method seems an inadequate technology when so many aspects of survival and reproduction are governed by energy reserves of neutral lipid in both freshwater and marine milieus.

A similar problem affects the results of Bruner et al. (1994) for lipids of zebra mussels *Dreissena polymorpha* when they related lipid content to bioconcentration of polychlorinated biphenyls (PCB) and polycyclic aromatic hydrocarbons (PAH). By contrast, a study of lipids in the Pacific lugworm *Abarenicola pacifica*, conducted by TLC-FID, reported data for total lipids, phospholipids, sterols, free fatty acids, triacylglycerols, and wax esters (Taghon et al., 1994). Low levels of lipid, reserves, primarily triacylglycerols with less wax esters, were especially abundant (8% of ash-free dry weight) if eggs were being carried. Comparisons of lipids of the marine benthic amphipods *Pontoporeia femorata* from diverse locations, and of *Monoporeia affinis,* were successfully carried out with this technique (Hill et al., 1992), and lipid classes explained several aspects of biochemistry for this international family. Larger animals always had more lipid, but seasonal effects were different in the lipid storage systems. The two marine species fed and behaved differently, so that *P. femorata* lipids were always dominated by triacylglycerols, whereas in *M. affinis* this lipid class became dominant only in late summer and fall. Females had more lipid than males. The lipids of the closely related freshwater *Diporeia hoyi* have already been discussed (Cavaletto et al., 1996; Vanderploeg et al., 1992).

The view that triacylglycerols are useful as a condition index of fish, bivalve, and crustacean larvae (Fraser, 1989) is based on marine organisms but applies equally to freshwater species. This author's proposition is that triacylglycerol should be expressed as a ratio to sterol content because sterol itself can be related to dry weight. An important part of this proposition extending to reproductive success follows from the observation that the eggs of many aquatic organisms are rich in triacylglycerols. As shown for the bonefish (*Albula* sp.), a great deal of reorganization of such lipids takes place subsequent to emergence from the egg (Padrón et al., 1996). In larvae, phosphatidylethanolamine was replaced by phosphatidylcholine, which was not initially present, and the triacylglycerols were rearranged to provide highly unsaturated fatty acids, especially 22:6ω3, for the new phospholipids. Half of both the triacylglycerol and polar lipids originally present were catabolized in one way or another. Similar lipid class modifications have been shown to occur in herring roe and larvae (Tocher et al., 1985), except that phosphatidylcholine was the dominant polar lipid of all stages of development. Triacylglycerol proportions actually increased with egg development stage and were not affected by hatching. Subsequently, they diminished, possibly as fatty acids were needed for new phospholipids (see below).

In a study of the zooplankton available to vendace (*Coregonus albula* L.), Linko et al. (1992) showed that in Lake Pyhäjärvi the cladoceran *Holopedium gibberum* was dominant in June, whereas *Daphnia* species were abundant in July. Calanoid and cyclopoid copepods were important at all times from June to October. Wax esters were not found in either the plankton or the fish, but arachidonic acid (20:4ω6) made up 4–8% of the fatty acids of composite plankton samples, accounting for the 2–4% of this fatty acid deposited in the triacylglycerols of the vendace flesh. Similar proportions of this fatty acid were found in triacylglycerols of fish from northern Canadian lakes (Ackman et al., 1967). In cold-water marine fish muscle, this fatty acid might be in the range of 0.1–0.5% triacylglycerol fatty acids (Ackman, 1990; see also Table 11.5).

The Cladocera have, in fact, been intensively investigated for freshwater energy transfers. Lipid (oil) droplets have been observed and are reported as triacylglycerols (Goulden and Henry, 1984). Lipid reserves were tested in laboratory feeding experiments. Adult Cladocera with good lipid reserves could better survive low-food situations, but not all species aggressively transferred lipid to eggs in these situations. This process seemed to be limited in one smaller species, possibly thus improving survival of the adults.

The least well-known lipid class of both freshwater and marine invertebrates is acetone-mobile polar lipids (AMPL), first discussed by Parrish (1987). In his classic paper on the Iatroscan TLC-FID method, he included chromatograms for both marine (Bedford Basin, Nova Scotia) and freshwater (Lakes Huron and Michigan) materials. In freshwater sediment trap samples and in those of *Pontoporeia hoyi,* the then-unknown material (AMPL) was found to be an important lipid class. More recently, another AMPL peak has been identified as partly containing galactosyl and other plant lipid diacylglycerols (Parrish et al., 1996b). As AMPL is literally made up of "acetone-mobile polar lipids," it can include chlorophyll and other pigments, so caution in identification and quantitation is indicated. In respect to seston, another important dietary feed for marine bottom dwellers, Parrish et al. (1995) have quantitated hydrocarbons, AMPL, sterols, steryl and wax esters, triacylglycerols, free fatty acids, fatty alcohols, and even phospholipids by TLC-FID analyses. This work illustrates perfectly why the Iatroscan TLC-FID technology is used worldwide by aquatic scientists, including in analyses reported in recent papers by freshwater invertebrate research groups (Cavaletto et al., 1996). Criticism has focused on the total lipid recovery calculated from TLC-FID results sometimes being about 10% less than that from gravimetric methods. However, the lipid class detail available from even very small samples cannot be matched by any other technology. There is also ample opportunity for replicates or parallel analyses of different samples because of the system of handling ten Chromarods as a unit. Other systems such as HPLC can only handle one sample at a time. The ability to analyze lipid or other organic materials particularly for nonvolatile materials, on a microgram level, can be complemented by very careful GLC for the volatile materials (Yang et al., 1996).

11.2.4. Fatty Acids

Invertebrate lipids are broken down by fish in varying degrees (Henderson and Tocher, 1987), and all components are rearranged into appropriate fish lipid classes or catabolized for energy. This process is the same in marine or freshwater milieus. Finnish lakes were examined for invertebrate-to-fish transfers of fatty acids by Muje et al. (1989). They concluded that young fish deposited zooplankton fatty acids in total lipids with little change, whereas older fish were more selective in altering fatty acid deposits for specific bodily needs or functions. As a rule of thumb, young fish have relatively much less (1–2%) total lipid than older fish (5–15%). This indicates that the young fish are building up muscle and require basically only more cellular phospholipids. In older fish, the triacylglycerols are accumulated as energy stores and for gonadal development. Edible fish do not usually contain wax esters, diacyl glyceryl ethers, or hydrocarbons, and sterols (primarily cholesterol) are relatively minor. Attention has recently focused on fatty acids of fish and shellfish for human nutrition. There are firm indications that two fatty acids, 20:5ω3 and 22:6ω3, provided by fish and shellfish, have different roles in the human body (Goodnight, 1996). They can be exploited for better health from modified diets if the lipid class and individual fatty acids, or both, are understood. Studies of aquatic life reporting only total lipids, or fatty acids of total lipids, may be of some value to human nutritionists and dieticians but are of only minor value to biologists. An example is the analysis of white (dorsal) muscle of 56 species of Swedish fish, including those of both Baltic and freshwater origins, by Ahlgren et al. (1994). They concluded that the human nutritional values (i.e., selected fatty acids of the ω6 and ω3 families of polyunsaturated fatty acids) were similar for marine, brackish, and freshwater fish. We now expect that absolutely essential cellular phospholipids rich in DHA will make up ~0.6% of the mass of such muscle tissue (Ackman, 1990), so when less DHA is shown as a percentage of all fatty acids, it can only suggest energy reserves of triacylglycerols that probably have less of this fatty acid (Table 11.1). Because this pattern is not observed in all likely cases, the merit of the data for assessing energy reserve levels in freshwater and marine fish is further diminished. Similarly, the analysis of 35 Icelandic fish species for lipid and fatty acids (Sigurgisladóttir and Pálmadóttir, 1993) is only a convenient record for those promoting long-chain ω3 fatty acids for good health.

The mixtures of fatty acids of the lipids of marine and freshwater fish do not differ radically. Roughly a third are saturated, a third or more are monounsatu-

TABLE 11.1. Principal and biochemically interesting fatty acids (w/w%) of particular lipids of muscle from a retail-farmed Atlantic salmon (*Salmo salar*) and of total lipids of whole meats of laboratory-reared European oysters (*Ostrea edulis*), farmed U.S. catfish (*Ictalurus punctatus*), and oil cooked out of an East African fish.

| | Marine | | | Fresh water | |
| | Salmon[a] | | Oyster[b] | U.S. catfish[c] | Kenyan fish |
Fatty acid	Triglyceride	Phospholipid	Total lipid	muscle lipid	oil[d]
14:0	8.62	3.46	2.3	1.4	5.0
16:0	16.13	23.13	16.6	18.3	23.0
18:0	2.03	1.87	5.3	4.0	6.3
Σ Saturated	26.78	28.46	24.2	23.7	34.3
16:1ω7	8.10	2.39	5.3	4.6	14.0
18:1ω9	11.8	5.23	3.8	50.0	16.8
18:1ω7	2.56	1.8	3.3	—[e]	3.1
Σ 20:1	11.14	2.11	5.6	1.3	0.7
Σ 22:1	10.07	1.05	—	0.2	0.1
Σ Monoenoic	43.67	12.58	18.0	56.1	34.7
18:2ω6	4.67	1.55	2.0	12.0	1.7
18:3ω3	0.91	0.46	1.4	0.9	2.0
18:4ω3	1.44	0.35	1.4	0.1	0.3
20:4ω6	0.27	0.51	6.6	0.5	1.3
20:4ω3	0.93	0.65	—	—	0.5
20:5ω3	5.26	10.65	14.2	0.4	3.7
22:5ω6	0.06	0.29	1.8	—	0.8
22:5ω3	1.78	2.98	1.0	0.6	3.4
22:6ω3	6.60	33.18	16.2	1.2	7.3
Σ Polyenoic	21.92	50.62	44.6	15.7	21.0
Others	7.63	8.33	13.2	4.5	7.0

[a] S.M. Polvi and R.G. Ackman, unpublished results.
[b] From Napolitano et al., 1988.
[c] From Nettleton et al., 1990.
[d] R.G. Ackman, unpublished results.
[e] —, Not reported.

rated, and up to a third can be polyunsaturated. Originally, many of the latter were thought to be biosynthesized by invertebrates of higher trophic levels or by fish, but it is now recognized that they can be conserved from the original phytoplankton via several levels of invertebrates (Ackman and Kean-Howie, 1995). Most farmed fish are fed on diets balanced between what is nutritionally desirable and commercially available (see Olsen, this volume). Tables 11.1 and 11.2 show the differences between fatty acid compositions for phospholipid and triacylglycerols in salmon and compare wild versus farmed fish compositions for this species. Increasingly, published human nutrition data for food fish and shellfish are based on the farmed animals, further distorting the information databases provided by many recent lipid and fatty acid analyses. It is useful to know that lipids of all fish and invertebrates commonly eaten are readily digested by humans and the fatty acids assimilated. Beyond that, the farmed U.S. catfish edible muscle differs in having more 18:1 and 18:2ω6 from the dietary fatty acids and less of the

Table 11.2. Comparison of major or important fatty acids (w/w%) of food fish from different parts of the world with particular reference to fish from the wild and their farmed counterparts.[a]

Fatty acid	Salmon, Atlantic[b] Salmo salar Wild	Salmon, Atlantic[b] Salmo salar Farmed	Milkfish[c] Chanos chanos Brackish	Milkfish[c] Chanos chanos Fresh water	Red sea bream[d] Pagrus major Retail	American eel[e] Anguilla rostrata Wild	American eel[e] Anguilla rostrata Farmed	Ayu[f] Plecoglossus altivelis Wild	Ayu[f] Plecoglossus altivelis Farmed	Tilapia[g] Tilapia sp. Wild
14:0	2.4	4.9	5.9	4.3	2.0	3.3	2.8	5.0	4.4	5.2
16:0	11.2	13.0	14.0	8.9	17.6	15.5	22.3	26.4	30.0	30.2
18:0	3.8	2.8	5.5	16.5	7.8	4.5	5.3	2.9	3.4	5.6
Σ Saturated	17.4	21.9	23.4	29.7	27.4	23.3	30.4	37.0	39.1	40.9
16:1	4.5	6.7	1.6	15.8	8.6	10.5	13.2	18.6	10.0	15.6
18:1	24.0	10.0	10.2	13.0	20.9	27.3	21.5	11.6	20.6	16.5
20:1	4.0	12.0	5.8	1.9	1.6	2.2	2.7	—[h]	4.9	4.7
22:1	5.0	(12.0)	6.1	1.3	0.9	—	—	—	2.8	—
Σ Monoenoic	37.4	40.7	23.6	32.0	32.0	40.0	37.4	31.2	38.3	36.2
18:2ω6	3.1	6.7	1.1	2.5	2.1	8.9	0.7	5.6	7.4	2.6
20:4ω6	4.7	(0.8)	—	—	2.6	4.4	2.2	0.9	tr[i]	1.3
18:3ω3	5.2	1.1	3.8	7.9	—	2.6	0.2	10.4	0.9	0.3
18:4ω3	1.5	1.8	4.2	1.8	—	—	—	—	—	2.3
20:5ω3	5.7	6.9	3.0	0.1	4.3	2.9	6.1	4.1	4.8	1.0
22:5ω3	4.5	—	tr	—	3.5	2.7	2.4	1.5	1.7	3.2
22:6ω3	19.8	14.6	0.4	2.2	17.8	5.4	15.0	1.8	7.0	3.4
Σ Polyenoic	45.2	31.9	12.6	14.4	30.3	30.0	28.6	31.8	22.6	14.7
Percent lipid	6.3	10.9	4.5	4.9	2.8	3.4	14.7	10.7	3.1	4.6

[a]From Ackman, 1995, with permission.
[b]Marine (Exler, 1987). Figures in parentheses adjusted from original report where 22:1 was included in 20:4n-6.
[c]Bautista et al., 1991.
[d]Marine (Ohtsuru et al., 1984).
[e]Otwell and Rickards, 1981.
[f]Muscle triacylglycerol only (Nakagawa et al., 1991).
[g]Fresh water (Steiner-Asiedu et al., 1991).
[h]—, Not reported.
[i]tr, Trace.

TABLE 11.3. Similarities in the total fatty acid compositions of triacylglycerols from different sources.[a]

Fatty acid	Triacylglycerols		Total neutral lipids Bonefish larvae	Triacylglycerols *Penaeus kerathurus*[b]
	Vendace flesh	Composite plankton		
14:0	8.3	7.8	6.4	6.4
16:0	14.6	14.6	19.1	22.0
18:0	2.7	3.5	4.6	1.3
Σ 16:1	7.9	12.9	8.5	8.9
Σ 18:1	13.0	8.1	11.0	8.7
Σ 20:1	0.6	0.4	ND[c]	11.6
Σ 22:1	ND	ND	ND	2.2
18:2ω6	4.5	3.9	2.9	1.6
18:3ω3	6.3	5.0	5.0	1.4
18:4ω3	9.3	6.0	5.3	0.7
20:4ω6	3.5	6.2	3.6[d]	2.8
20:5ω3	9.4	14.8	10.8	10.8
22:5ω3	2.0	0.8	1.6	2.4
22:6ω3	9.7	7.8	17.4	11.0

[a] The freshwater fish flesh (vendace, *Coregonus albula*) in Finland is presented next to a composite plankton sample from the same lake (Linko et al., 1992), both collected on September 5. Neutral lipids of an advanced stage of metamorphosizing larvae of a Gulf of California marine fish (bonefish, *Albula* sp.) (Padrón et al., 1996) and of triacylglycerols from midgut glands of wild-caught female marine shrimp *Penaeus kerathurus* from the Gulf of Cádiz (Mourente and Rodríguez, 1991) are typical of many marine invertebrate fatty acid compositions.
[b] Stage I.
[c] ND, not determined.
[d] No structure specified.

marine 20:1, 22:1, 20:5ω3, and 22:6ω3 found in the salmon. The other species of these two tables are diversified to show that similarities in fatty acid composition found in analyses are the norm rather than the exception. Compared with the numbers of species of the wide variety of marine invertebrates consumed by humans, the freshwater resources are limited, although there is considerable farming of crustaceans in warmer climates.

Fatty acids can occasionally show diet-specific effects (Paradis and Ackman, 1977, 1976), but generally the similarity of fatty acid compositions can be a handicap in assessing food chains. An illustration of this is provided by Table 11.3, a comparison of flesh triacylglycerol fatty acids of a freshwater fish from Finland with total lipids of a composite planktonic prey sample and of two marine lipid samples of diverse origins. The 18:1 and 20:1 of the shrimp gut triacylglycerols are made up of one-half of the common isomer 18:1ω9 and 20:1ω9 and one-half of another series, 18:1ω7 and 20:1ω7, originally derived from 16:1ω7. The longer-chain ω7 monoethylenic fatty acids are of unknown function and totally unrelated to the 22:1 derived from copepod marine lipids (see below). This is an indication of how we can understand part of the fatty acid biochemistry and yet be frustrated by our limited understanding of the biology of the species.

Saturated

Linear Even	Linear Odd	Iso and Anteiso
12:0	(13:0)	Iso 14:0 (Anteiso 14:0)
14:0	15:0	Iso 15:0 Anteiso 15:0
16:0	17:0	Iso 16:0 (Anteiso 16:0)
18:0	19:0	Iso 17:0 Anteiso 17:0
20:0	—	Iso 19:0 (Anteiso 19:0)

Isoprenoid

TMTD   (C_{16})
Pristanic (C_{19})
Phytanic (C_{20})

Monounsaturated

14:1n-5, n-7, n-9
16:1n-9, n-7, n-5
18:1n-11, n-9, n-7, n-5
20:1n-11, n-9, n-7, n-5
22:1n-13+11, n-9, n-7
24:1n-15+13+11, n-9, n-7

Polyunsaturated C_{18} and Polyunsaturated C_{16}

18:4n-3 Δ6,9,12,15	18:4n-6 (Δ3,6,9,12)
18:3n-3 Δ9,12,15	18:3n-6 Δ6,9,12
18:2n-3 (Δ12,15)	18:2n-6 Δ9,12
16:4n-1 Δ6,9,12,15	16:4n-4 (Δ3,6,9,12)
16:3n-1 Δ9,12,15	16:3n-4 Δ6,9,12
16:2n-1 (Δ12,15)	16:2n-4 Δ9,12
18:5n-3 Δ3,6,9,12,15	>18:3n-4 (Δ8,11,14)
>18:4n-1 Δ8,11,14,17	>18:2n-4 Δ11,14
>18:3n-1 Δ11,14,17	16:2n-7 (Δ6,9)

Non-methylene-interrupted dienoic

20:2 Δ5,Δ11   22:2 Δ7,Δ13
20:2 Δ5,Δ13   22:2 Δ7,Δ15

Polyunsaturated C_{20}, C_{21} and Polyunsaturated C_{22}

20:5n-3	20:4n-6	22:6n-3	22:5n-6
20:4n-3	20:3n-6	22:5n-3	22:4n-6
20:3n-3	20:2n-6	(22:4n-3) (22:3n-6)	
(20:3n-9)		21:5n-3	

FIGURE 11.3. Fatty acids likely to be observed in aquatic organisms, including phytoplankton and macrophytes. The n = x notation clearly shows the family relationships between those of different chain lengths.

Figure 11.3 assembles almost all possible fatty acids commonly identified in fish and shellfish lipids in both marine and freshwater environments. Certain generalities in the fatty acid percentages in oils or total lipids can be observed. Tables 11.1 and 11.2 are designed to summarize such similarities in the three types of fatty acids. Modern capillary GLC can easily identify 30–50 fatty acids for marine invertebrates such as sea urchins (Takagi et al., 1986) or fish from either freshwater (Linko et al., 1992) or marine (Sigurgisladóttir and Pálmadóttir, 1993) sources. Except in cases of exotic fatty acids used as tracers in food chains, a minimum of 14 fatty acids usually is sufficient to describe a lipid (Table 11.2). A few extra (e.g., 20:4ω3, 22:5ω6) can be included but tend to be minor components as shown in Table 11.1. However, 16:0 (palmitic acid) invariably is the most

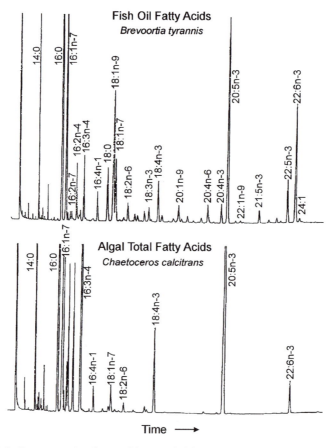

FIGURE 11.4. Representative fatty acids revealed by gas-liquid chromatography in a marine phytoplankton (below) and the mixture typically found in a fish oil (above), in this case from menhaden (*Brevoortia tyrannis*). (From Ackman and Keen-Howie, 1995, by permission of the American Oil Chemists' Society.)

important saturated fatty acid and, with 14:0 (myristic) and 18:0 (stearic) acid, adds up to 20–40% of the total fatty acids. The monounsaturated fatty acids similarly tend to be dominated by 18:1ω9 (oleic). Prior to 1980, most fatty acid analyses were conducted on packed GLC columns and the oleic acid peak always included 18:1ω7 (*cis*-vaccenic). Capillary GLC conveniently reveals many such isomer details. As can be seen from Figure 11.4, some 16:1ω7 is usually elongated to 18:1ω7 but is nevertheless often important by itself (Tables 11.1 and 11.2). It is also rational to expect some 20:1ω9, elongated from 18:1ω9, and yet two shoulders, for 20:1ω11 (by desaturation of 20:0) or 20:1ω7 (by elongation from 18:1ω7), are also commonly observed. The 22:1 family pose a unique problem in that the dominant isomer is 22:1ω11. In the North Atlantic and Pacific, this is derived from the corresponding fatty alcohol in the wax esters of copepods as already discussed. This alcohol is a structural anomaly in respect to biosynthesis of the unusual bond position but is mostly oxidized to the corresponding fatty acid on digestion by fish (Sargent et al., 1979) and, together with 20:1, heavily influences the total monoethylenic fatty acids of species such as herring or capelin feeding on these zooplankton (Ratnayake et al., 1979a,b). In fact, in the production of fish oil and meal by the Atlantic fish meal production industry a ratio of Σ22:1 > Σ20:1 in the oil, totaling 30–40% of fatty acids, is almost a sure sign of herring as the source of a fish oil. Because 22:1 and 20:1 are so abundant, 18:1 and even 16:1 may be present at reduced levels in these oils. The fish feeding directly on phytoplankton that do not have the 20:1 and 22:1 of zooplankton, notably the American menhaden (*Brevoortia tyrannis*) (Fig. 11.4), instead break down the plant polysaccharides and synthesize additional 14:0 and 16:0. The common assembly route to fish triacylglycerols as first proposed by Brockerhoff (Litchfield, 1972) is shown in Figure 11.5. The final insertion of fatty acids in the *sn*-3 position (sometimes marked as the α′-position) depends on whatever fatty acid is circulating and available. Thus, if the menhaden finds 14:0 or 16:0 readily available, or the herring 20:1 or 22:1, or in some cases a polyunsaturated fatty acid is available, then these will be inserted. The DHA is preferred by the phospholipid intermediates for the 2-position in this process. This explains why 22:6ω3 may be primarily in the 2-position and 20:5ω3, originally from phytoplankton, more scattered between the 2-position and the 3-position (Ando et al., 1992; Litchfield, 1972). Certain fish oils show an ability to "winterize" (Ackman, 1988). An oil clear at 30°C may turn cloudy at 20°C, and on cooling to 15°C a solid layer crystallizes out. The two fractions of the winterized oil do not differ radically in most fatty acids, but the polyunsaturated fatty acids are higher in the liquid layer (Ackman, 1980), suggesting that combinations of more than one of 20:5ω3 plus 22:6ω3 or plus 18:4ω3 may be present in a proportion of the molecules. Other molecules may have an excessive proportion of 14:0 plus 16:0, promoting this crystallization (Ackman, 1988). The sharp separation described is hindered if too much 20:1 and 22:1 are present, and the whole mass of the oil may simply solidify. Consideration of these matters and of recent assemblages of fish oil triacylglycerol compositions (Moffat, 1995; McGill and Moffat, 1992), and their stereochemistry (Ando et al., 1992), suggests that there is one basic fatty acid composi-

Stage I

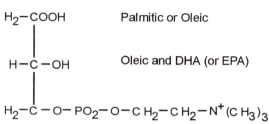

Stage II

FIGURE 11.5. There are literally thousands of combinations of fatty acids possible in triacylglycerols of aquatic organisms. The general biosynthesis through phospholipids leads to a preference for polyunsaturation in the 2-position (stage I), but the final insertion of a fatty acid in stage II can draw on 50 or 60 fatty acids.

tion for cold-water marine fish oils (Ackman et al., 1988). Local or preferred food organisms are probably dominant in modifying these oils.

In body oils from Canadian freshwater fish examined in 1965, the total 22:1 was only 0.3% of fatty acids (Ackman et al., 1967). Although, as already noted, wax esters are not common in freshwater invertebrates, those present must lack the 22:1 alcohol of marine copepods. In other respects, it has been clear since 1965 that freshwater fish "oils" are similar to marine oils in fatty acid composition. Many small lakes and streams in northern latitudes provide insects or insect larvae as food for freshwater fish, presumably presenting a "normal" fatty acid supply (Ackman and Takeuchi, 1986). Insect fatty acids associated with freshwater bodies in Europe (Ghioni et al., 1996; Bell et al., 1994) and in North America (Hanson et al., 1985) show two features differentiating them from fatty acids found in marine fish phospholipids or triacylglycerols. One is the absence of 22:6ω3 and presence of substantial proportions ($\sim$10%) of 20:5ω3, and the other is the occasional presence of 20:4ω6 at levels approaching those of the 20:5ω3. In the analyses of lipids, gammeridae had 40% triacylglycerols and 10% cholesterol, with other lipid classes to bring total neutral lipids to 61%, with polar lipids totaling 36%. An oligochaete analysis showed 55% polar lipid and relatively little triacylglycerol (13%). The fatty acid compositions were rather similar and resembled the insect fatty acid compositions with virtually no 22:6ω3 and 20:5ω3 > 22:6ω3. The chironomidae resembled the gammaridae in lipid classes and fatty acids, except that they had very little 20:4ω6. The 22:6ω3 is also not present in terrestrial arthropods (Uscian and Stanley-Samuelson, 1994). Published literature reports on freshwater fish fatty acid compositions are often either from farmed fish or from those on experimental diets, so publications on natural food chains are often difficult to find.

Tropical marine fish also show the same tendency to contain elevated 20:4ω6, a property discussed by Ackman (1989) for a number of northern Australian fish. This is probably because 20:4ω6 is available from the local photosynthetic organisms of coral reefs. Freshwater fish triacylglycerols do tend to have moderate proportions (~5–10%) of 18:2ω-6 and 18:3ω3 (Table 11.3). These differences in polyunsaturated fatty acids probably account for species-specific different aromas among such fish (Josephson et al., 1984). Lipoxygenases in fish skin or subdermal fats oxidize these fatty acids (Mohri et al., 1992, 1990). The resulting peroxides break down to give mixtures of aldehydes. These are the basis of many distinctive olfactory features of fish when freshly caught. Amounts of lipids or of fatty acids of aquatic invertebrates do not seem to directly contribute to this biochemistry; there is only an indirect influence through the different types of polyunsaturated fatty acids (i.e., ω6 or ω3) contributed to the fish lipids.

In rearing some freshwater fish, notably larval rainbow trout (*Oncorhynchus mykiss*), ω3 fatty acids are deemed "essential." Some confusion exists as to whether the vegetable oil 18:3ω3 by itself is satisfactory. This appears to be settled by a recent feeding study (Wirth et al., 1997). The larvae living off their yolk sacs conserve 22:6ω3, primarily for new phosholipids. When fed 18:3ω3 for four weeks they could utilize these to some extent without elongation, but growth was limited compared to those larvae fed a diet including preformed 22:6ω3. The 22:6ω3 is thus the truly essential fatty acid for larval rainbow trout.

Table 11.4 attempts to show how some fatty acids are phased out of fish lipids during a sequence of feeding steps on intbertebrates in a larval pike food chain. The unusual 18:5ω3 of the invertebrates is seen to pass into the triacylglycerols but not into the phospholipids. Instead, the latter show a drop in saturated fatty acid totals and an increase in the 22:6ω3 characteristic of the fish phospholipids. Fatty acid studies in *Penaeus kerathurus* have confirmed that the conversion of 20:5ω3 to 22:6ω3 follows an elongation to 24:5ω3 and Δ6 desaturation to 24:6ω3, followed by chain shortening to 22:6ω3 (Mourente, 1996). This study confirms that a new metabolic pathway for production of 22:6ω3, replacing the hypothetical Δ4 desaturase (Voss et al., 1992), is valid in a marine crustacean. It depends on the presence of peroxisomes for the chain-shortening step. These bodies have other useful lipid functions in fish (Henderson and Tocher, 1987) but are not well documented in invertebrates. Freshwater *Macrobrachium borellii* is also sensitive to temperature fluctuations affecting mitochondrial oxidation of fatty acids in gill and hepatopancreas (Irazu et al., 1992). Moreover, the 20:4ω6 also increases in the phospholipids. Polyunsaturated fatty acids are thought to be

TABLE 11.4. Sequence of fatty acid profiles in total lipids of feed invertebrates during development of *Esox lucius* larvae in pond culture.[a]

Animal	Cyclopoids	Cladocerans	Simocephalus vetulus	Eurycercus lamellatus	Esox lucius larvae		
Lipid type[b]	TL	TL	TL	TL	TAG	PL	PL
Day	2	2	8	8	8	8	29
Fatty acid							
14:0	5.86	9.57	5.41	3.86	5.80	1.75	tr[c]
16:0	18.90	20.80	23.60	22.81	24.20	27.60	20.80
18:0	8.00	5.53	13.73	17.21	11.16	13.60	8.50
15Br[d]	—[e]	1.93	0.96	1.28	0.40	tr	tr
15:0	1.84	1.82	0.52	1.20	1.22	0.88	—
17:0	1.10	1.69	0.84	0.81	1.35	1.34	—
Σ Saturates	35.70	37.04	46.83	50.07	45.13	46.05	29.30
14:1	tr	0.44	—	tr	—	tr	—
15:1	tr	tr	tr	0.62	0.50	tr	—
16:1	10.23	8.22	7.36	4.66	7.80	5.31	3.10
17:1	0.70	0.60	0.72	tr	tr	0.66	—
18:1	9.72	16.33	12.90	11.50	15.44	13.62	15.67
20:1	tr	tr	0.94	tr	tr	0.47	tr
22:1	tr	1.00	tr	tr	tr	tr	—
Σ Monoenes	20.93	26.59	24.43	18.78	23.74	20.06	18.77
16PUFA[f]	3.84	2.52	2.21	2.67	1.96	0.88	0.47
18:2ω6	4.43	2.86	2.40	2.41	2.70	2.28	3.67
18:3ω6	tr	0.81	1.33	tr	0.44	0.51	0.40

18:3ω3	7.22	7.46	2.81	1.24	1.18	2.41	5.06
18:4ω3	5.26	2.58	2.69	0.50	1.35	0.92	1.07
18:5ω3	1.00	1.00	1.65	1.16	2.75	—	—
20:2ω6	1.01	0.71	1.09	0.82	0.93	0.71	tr
20:4ω6	2.31	1.14	0.43	0.91	1.23	3.98	7.01
20:4ω3	0.60	0.62	0.40	1.14	0.50	0.60	1.20
20:5ω3	2.20	4.59	1.18	3.01	1.83	2.56	6.52
22:2ω6	1.66	4.70	5.67	6.73	2.80	1.01	1.08
22:3ω6	1.40	1.23	1.03	1.71	2.25	0.81	0.50
22:5ω3	2.16	tr	0.50	tr	1.37	2.70	3.40
22:6ω3	4.08	1.02	1.26	0.90	2.80	9.46	17.38
Total ω3	22.52	17.27	10.49	7.95	11.78	19.09	34.56
Total ω6	10.81	11.45	11.95	12.58	10.35	9.71	13.19
Σ PUFA	37.17	31.24	24.65	23.14	24.09	29.68	48.22
Unidentified	5.31	4.01	3.14	6.21	2.67	1.63	0.95

[a]Details between triacylglycerols and phospholipid for one time point are shown for the larvae and the maturation of the phospholipids with time. Data from Desvilettes et al., 1994.

[b]TL, total lipid; PL, phospholipid; TAG, triacylglycerol.

[c]tr, Trace.

[d]Branched-chain fatty acids.

[e]—, Not reported.

[f]Polyunsaturated fatty acids.

critical to some stages of metamorphosis in invertebrates, and the polyunsaturates of these two families (ω6, ω3) may be, to some extent, interchangeable in phospholipids of both freshwater (e.g., *Penaeus monodon;* Merican and Shim, 1996) or marine (e.g., the clam *Chlamys islandica* or the starfish *Ctenodiscus crispatus;* Bell and Sargent, 1985) invertebrates. However, as nonstructural free acids they are important in competing for enzymes to produce either of two families of eicosanoids (Lands, 1986). These biochemicals are powerful agents in all animals. Unfortunately, methods for their analysis are more suitable to larger organisms or cell cultures (Tocher et al., 1996).

Iso- and *anteiso-*fatty acids of Figure 11.3, usually C_{15} and C_{17}, may be of little interest in most situations, although there are exceptions. They are present in all animals, where an amino acid skeleton replaces the first acetate in fatty acid biosynthesis. However, bacteria may use these fatty acids in lipids instead of polyunsaturated fatty acids, and so detritus feeders often show a typically higher proportion of *iso-* and *anteiso-*15:0 and 17:0. Generally, they are ≤0.5% of total fatty acids in fish lipids and may usually be omitted from tables presenting fatty acid compositions, as can the slightly more important straight-chain 15:0 and 17:0 fatty acids. Among the exceptions, 50% of the fatty acids in the lipids of one population of breeding marine amphipods *Pontoporeia femorata* are odd-chain lengths, and these accumulate in the fatty acids of smelt (*Osmerus mordax*), which prey on them (Paradis and Ackman, 1976). Other related species such as *Pontoporeia hoyi* in Lake Michigan (Gauvin et al., 1989; Gardner et al., 1985) or *P. femorata* in the Baltic (Hill et al., 1992) do not have high percentages of odd-chain fatty acids. Cyclopropanoid fatty acids originating from bacteria may be obvious in fish from marshes (Cosper and Ackman, 1983) but are of minor interest except in rare circumstances such as in Lake Baikal in the deep-water organism *Acanthogammarus grewingkii* (Řezanka and Dembitsky, 1994). Such exogenous fatty acids could identify fish or fish fat origins (Ratnayake et al., 1989). The role of bacteria in symbiosis with higher organisms around hydrothermal vents has recently led to several interesting observations on fatty acids in these exotic life forms. The cyclopropane fatty acids, not shown in Figure 11.3 because they are ubiquitous in the bacteria world, were abundant (Fullarton et al., 1995a). The NMID (nonmethylene-interrupted dienoic acids) were plentiful in the unusual tube worms *Ridgeia piscesae,* but the normal *cis-*methylene-interrupted polyunsaturated fatty acids of Figure 11.3 were also well represented (Fullarton et al., 1995b). NMIDs were also found in mussels collected near hydrocarbon seeps (Fang et al., 1993) and are probably more common than suspected in mollusk lipids (Paradis and Ackman, 1977).

A summary of data for the C_{18} and longer chain polyunsaturated fatty acids found in the energy reserves of five commercially important species found in Canada's northern freshwater lakes is provided in Table 11.5. Unpublished data of McLeod and Ackman are supplemented by that for pike from Henderson et al. (1995b). The fish were of commercial size and were processed in Halifax. Usually, two to four fish were headed and the remaining whole bodies were homogenized; subsamples were then extracted in duplicate with $CHCl_3$-MeOH by the

TABLE 11.5. Polyunsaturated fatty acids of triacylglycerol fatty acids of headed whole-body northern Canadian freshwater fish homogenates (duplicate analyses)[a] for comparison and of a European pike liver.[b,c]

Fatty acid	Lake Whitefish *Coregonus clupeaformis*	Tullibee *Coregonus albus*	Burbot *Lota lota*	White Sucker *Catostomus commersoni*	Red Sucker *Catostomus catostomus*	Pike liver *Esox lucius* L.
18:2ω6	3.2	3.4	2.6	4.8	2.2	6.5
20:3ω6	0.2	0.4	0.4	0.3	0.2	0.6
20:4ω6	0.4	1.9	2.0	1.7	0.7	3.0
22:4ω6	0.5	0.4	0.6	0.2	0.1	0.1
22:5ω6	0.8	0.8	1.2	0.3	0.2	0.1
18:3ω3	5.0	4.8	4.1	5.3	3.1	8.6
18:4ω3	2.0	0.7	1.4	1.5	0.8	5.4
20:5ω3	6.2	3.0	5.3	6.1	5.0	8.7
22:5ω3	2.9	2.1	4.2	1.6	1.5	3.7
22:6ω3	6.1	5.0	11.8	3.3	3.1	5.8
Percent lipid	7.44	6.85	2.01	3.58	6.10	—[d]
Wt% TAG	91.64	69.61	40.51	74.85	79.52	—

[a]McLeod and Ackman, unpublished results.
[b]Henderson et al., 1995b.
[c]TAG, triacylglycerols.
[d]—, Not reported.

Bligh and Dyer (1959) procedure. Lipid classes were determined by the Iatroscan TLC-FID techniques. For fatty acid analyses, total lipids were separated by plate TLC; the triacylglycerols were recovered, the fatty acids converted to methyl esters and then analyzed by capillary GLC as described elsewhere (Ackman, 1992; Polvi et al., 1991), except that a DB-23 GLC column was used. All fish were caught in one location (Matheson Island, Lake Winnipeg, Manitoba) in the fall of 1994. The results show strong similarities, presumably related to the sharing of a common food base in a large lake. Exceptions are readily observed, such as the low level of 20:4ω6 in the red sucker. The burbot shows a tendency to elongate both ω3 and ω6 fatty acids to C_{22}, leading to high levels of 22:5ω6 and 22:6ω3. These reproduce the results for tullibee and burbot oils from reduction of commercial fish caught in Lake Winnipeg in the fall of 1965 (Ackman et al., 1967), except that in the 1965 tullibee oil 20:5ω3 was 6.2% and 22:6ω3 was 3.8% of total triacylglycerols. Possibly this could relate to proportions of fatty acids for ovarian maturations as this is a fall spawner. The roe may be rich in phospholipid 22:6ω3. The burbot may spawn in midwinter, but the low total fat of our samples suggests lipid depletion. In 1965 fish, the analysis of oil showed 20:5ω3 as 5.5% and the 22:6ω3 as 7.8%. The European pike liver was included, as this species is also common in northern Canada. The general similarity among the polyunsaturated fatty acids shows that these are ubiquitous in cold freshwater milieus. The differences in fatty acid composition from marine oils are often apparently minor but can have important effects in terms of biochemistry (Innis et al., 1995; Josephson et al., 1984).

11.2.5. Furan and Some Other Unusual Fatty Acids

Omitted from the table and figures are fatty acids containing a central furan ring (Dembitsky et al., 1993; Ishii et al., 1988). Two freshwater phyla and species (respectively, the crayfish *Procambarus clarkii,* the mollusk *Anadonta piscinalis,* and the gastropod *Limnaea fragilis*) are only examples from the recent literature, and furan fatty acids are well known in lipids of both freshwater fish (Sand et al., 1984; Ota and Takagi, 1983) and marine lipids (Ota and Takagi, 1990). They occur in plants as well and are probably generated by lipoxygenases acting on polyunsaturated fatty acids (Guth et al., 1995). Very long-chain polyunsaturated fatty acids such as 24:6ω3, in this case marine in flounder lipids (Ota et al., 1994), are possibly of dietary origin. Such fatty acids are also found for lipids (depot fat) of certain freshwater seals, as well as in a Baltic species (Käkelä et al., 1995). Typical Δ5 marine polyunsaturated fatty acids are reported for sea urchins by Takagi et al. (1986), but the freshwater sponges of Lake Baikal showed various unusual long-chain (C_{28}, C_{29}, and C_{30}) polyunsaturated fatty acids with Δ5 eth-ylenic bonds (Dembitsky et al., 1994). This universal Δ5 desaturation is of interest as the *cis*-methylene interrupted 20:3ω9 has a Δ5 bond and is widely considered a sign of deficiency in polyunsaturated fatty acids in mammals. It can be induced in fish under the same dietary stress (Webster et al., 1994).

11.2.6. Ether Lipids

The "ether lipids" are based on glycerol in which one alcohol oxygen is not esterified to a fatty acid but instead becomes an ether oxygen joining the glycerol with a long chain similar to fatty acids such as hexadecanoic or octadecenoic (Fig. 11.6). They are found in one or more organs of most animals including marine and freshwater life forms. The 1-O-alkyl diacyl glyceryl ethers (DAGE) of Figure 11.6 (II) have been long known because of their frequent occurrence in the liver oils of sharks (Kang et al., 1997; Bordier et al., 1996; Urata and Takaishi, 1996; Bakes and Nichols, 1995). This is not a species characteristic as depot fats of deep-sea squid may also have DAGE (Hayashi and Kawasaki, 1985). There has been much discussion of the reason for DAGE, as they coexist with triacyl-glycerols in marine depot fats, and there is even an entire book on the subject (Mangold and Paltauf, 1983), because the ether lipids are found in mammals as well. Buoyancy control is usually brought forward as the rationale for the coexistence of DAGE and triacylglycerols, but the evidence is not totally convincing that this is the only reason for their existence (Bakes and Nichols, 1995). The bulk of the work has been on marine animal lipids rather than on those of freshwater organisms (Sargent, 1989). The DAGEs (Fig. 11.6, II), if present, are nonpolar and are usually found in depot fats of aquatic organisms. The membrane phospholipids (often loosely called the "polar" lipids) are apt to include to 1-O-alkyl-acyl phospholipids (Fig. 11.6, I), and may well be accompanied by the 1-O-alk-1'-enacyl-alkyl phospholipid (Fig. 11.6, III). The latter structure, often called a plasmalogen, is unstable in mineral acids, cleaving at the ether group to give an aldehyde (Fig. 11.6, V) which in acid preparation of methyl esters becomes a dimethyl acetal. As the dimethyl acetal, this material is sometimes confusing in analyses of methyl esters by GLC, where it may coincide with isoacids (Brosche, 1985). High proportions of the latter with chain lengths corresponding to 16:0, 18:0, or 18:1 are therefore suspect. The result of recent examinations of polar classes of lipids by modern technology is that the proportions among ether phospholipids are often found to be very similar. This is illustrated by the lipids of four different organisms listed in Table 11.6. Many aquatic invertebrates contain the polar phospholipids II, III, and IV of Fig. 11.6 (Sargent, 1989).

The biochemistry of these ether lipids is not linked to any special animal group, although muscle of several sharks has recently been shown to contain all the alk-1'-enyl-acyl and 1-O-alk-1'-enyl-2-acyl-sn-glycero-3-phosphatidyl-ethanolamine and -choline classes, more so in the ethanolamine group of polar lipids than in the choline group (Jeong et al., 1996a). The 1-O-alkyl chains of structure I (Fig. 11.6) were shown by this group to be heavily oriented to 16:0, 16:1, 18:0, and 18:1 structures. The same was true for the ether phospholipids of the ascidian *Halocynthia roretzi* and the sea urchin *Strongylocentrotus intermedius* (Jeong et al., 1996b). The acyl groups in these lipids are interesting in having nearly 50% of 20:5ω3 and much less 22:6ω3. This ratio probably reflects a phytoplankton and macrophyte (epiphyte?) diet. The 2-position acyl chains were highly unsaturated,

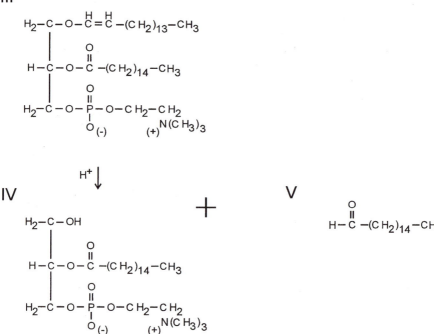
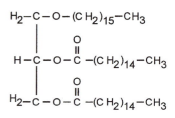

FIGURE 11.6. Ether lipids of aquatic organisms. I and II are stable, but when III is exposed to acids, lysophospholipids can be formed when the 1-O-alkenyl ether group is cleaved. The aldehydes produced (V) may, in turn, be converted to dimethylacetals and appear in gas-liquid chromatography along with methyl esters, often mimicking *iso*-acids.

with up to 50% 22:6ω3. For this reason, the plasmalogens (III) would be expected to be especially important in gills. Comparison among gills of bivalves and fish did not clarify this matter (Nevenzel et al., 1985) but did show that the common farmed channel catfish *Ictalurus punctatus* from fresh water was possibly just as rich in plasmalogens as marine species of fish. Comparisons of the lipids of the

TABLE 11.6. Glycerophospholipid compositions of a freshwater sponge (m/m%), the gills of a crab, a Japanese oyster, and bonito tuna white muscle lipids (w/w%).

Lipid class	Lacustrine sponge[a] Eunapius fragilis	Japanese oyster[b] Crassostria gigas	Pacific crab gill[c] Cancer antennarius	Bonito tuna[d] Euthynnus pelamis
Choline glycerophospholipids	35.5	54.3	54.8	52.2
Alk-1′-enacyl	8.9	3.9	2.1	3.1
Alkylacyl + diacyl	22.8	50.4	52.7	49.1
Ethanolamine glycerophospholipids	38.1	29.9	24.8	24.5
Alk-1′-enacyl	20.6	17.9	13.6	7.1
Alkylacyl + diacyl	15.7	12.0	11.3	17.4
Serine glycerophospholipids	12.4	1.9	2.3	4.5
Inositol glycerophospholipids	3.1	—[e]	1.9	4.4
Lysocholine glycerophospholipids	3.8	3.1	—	8.7
Other	7.1	6.6	5.7	5.7

[a]Early et al., 1996.
[b]Koizumi et al., 1990.
[c]Chapelle et al., 1987.
[d]Ohshima et al., 1989.
[e]—, Not reported.

freshwater crayfish *Procambarus clarkii* and rainbow trout *Oncorhynchus mykiss* showed no great difference, so that salinity was not established as important in respect to the ether lipids. Smolting of salmon is a possible exception to this general view, but the gills adapt in a short time (Takeuchi et al., 1990). In fact, the fish fatty acid pattern for salmon may shift from fresh water to marine on an anticipatory basis (Sheridan, 1994). The reversal of this adaption to allow the salmon to re-enter rivers for spawning suggests that major changes in lipid classes are unlikely. There is little doubt that phospholipids of fish gills are extremely important in salt transport (El Babili et al., 1996; Hansen et al., 1992; Takeuchi et al., 1989), and this applies to crustacea as well (Chapelle, 1986).

A unique report of analysis of these ether lipids is available for the freshwater sponge *Eunapius fragilis* of Lake Michigan; the method applied was ^{31}P nuclear magnetic resonance (Early et al., 1996). Eighteen different phospholipids, including ether lipids, were used to differentiate three habitats on the basis of "indexes" of associated phospholipid types. The differences found in phospholipid profiles (mole %) were relatively minor, but these immobile animals could be differentiated as to origin by powerful statistics. For those more familiar with conventional technology such as HPLC, an extensive thesis has been published in English (Takahashi, 1985) and is applied to polar lipids with comprehensive tabulations of results for western Pacific fish and shellfish.

11.2.7. Prostanoids

The adjustment of phospholipid ratios and fatty acid compositions may be an important response to stress. The model most often tested in fish is thermal adaption (Fodor et al., 1995). In all this elegant work on lipids, the highly sophisticated systems for producing eicosanoids (i.e., prostaglandins) from polyunsaturated fatty acids are often overlooked. Research with freshwater fish indicates that the eicosanoids are produced in all organs and tissues (Henderson and Tocher, 1987). Different tissues of marine fish such as turbot (*Scophthalmus maximus*) are also very active in this respect (Tocher et al., 1996; Henderson et al., 1985). Another example is the effect of dietary fatty acids on eicosanoid production and immune function in Atlantic salmon (Bell et al., 1996, 1993) or rainbow trout (Kiron et al., 1995). For immobile invertebrates such as bivalves (Deridovich and Reunova, 1993), for colonial worms (Toonen and Pawlik, 1996), and even for sea urchins (Kitamura et al., 1993), following metamorphosis, there is evidence that the settling processes are governed by chemical signals based on similar biochemical reactions (Ackman and Kean-Howie, 1995). Prostaglandin activity may also influence production of natural aromas or flavor elements from fatty acids in freshwater or marine fish or mollusks (Lindsay, 1990; Josephson et al., 1984).

The terrestrial world is dominated by ω6 polyunsaturated fatty acids, the aquatic world largely by ω3 polyunsaturated fatty acids. These compete for eicosanoid production (Lands, 1986), but it appears that the ω3-based group may be older and more fundamental, and the ω6 group a terrestrial adaption to capitalize on the more plentiful 18:2ω6 fatty acid of land plants. The colonization of fresh water by diverse organisms of marine origin is only one facet of adaption common even in poikilothermic vertebrates (Sheridan, 1994).

11.3. Conclusions

There is seldom any evidence of one biochemical pathway to exploit lipids for success in the aquatic world. Thus, wax esters may store energy in one species and promote buoyancy in another. Multifunctional roles for lipids seem to be the norm rather than unique to species, and research should focus less on details and more on perspectives.

References

Ackman, R.G. Composition and nutritive value of fish and shellfish lipids. In: Ruiter, A., ed. Fish and Fishery Products. Wallingford, Oxon, U.K.: CAB International; 1995:p. 117–156.

Ackman, R.G. Extraction and analysis of omega-3 fatty acids: procedures and pitfalls. In: Drevon, C.A.; Baksaas, I.; Krokan, H.E., eds. Omega-3 Fatty Acids: Metabolism and Biological Effects. Basel: Birkhäuser Verlag; 1993:p. 11–20.

Ackman, R.G. Application of gas-liquid chromatography to lipid separation and analysis: qualitative and quantitative analysis. In: Chow, C.K., ed. Fatty Acids in Foods and Their Health Implications: New York: Marcel Dekker; 1992:p. 47–63.

Ackman, R.G. Application of thin-layer chromatography to lipid separation: detection methods. In: Perkins, E.G., ed. Analyses of Fats, Oils and Lipoproteins. Champaign, IL: American Oil Chemists' Society; 1991:p. 97–121.

Ackman, R.G. Nutritional composition of fats in seafoods. Prog. Food Nutr. Sci. 13:161–241; 1989.

Ackman, R.G. The year of the fish oils. Chem. Ind. (March 7): 139–145; 1988.

Ackman, R.G. Fish lipids I. In: Connell, J.J., ed. Advances in Fish Science and Technology. Farnham, U.K.: Fishing News Books; 1980:p. 87–103.

Ackman, R.G.; Kean-Howie, J. Fatty acids in aquaculture: are ω-3 fatty acids always important? In: Lim, C.E.; Sessa, D.J., eds. Nutrition and Utilization Technology in Aquaculture. Champaign, IL: American Oil Chemists' Society; 1995:p. 82–104.

Ackman, R.G.; Takeuchi, T. Comparison of fatty acids and lipids of smolting hatchery-fed and wild Atlantic salmon *Salmo salar.* Lipids 21:117–120; 1986.

Ackman, R.G.; Ratnayake, W.M.N.; Olsson, B. The "basic" fatty acid composition of Atlantic fish oils: potential similarities useful for enrichment of polyunsaturated fatty acids by urea complexation. J. Am. Oil Chem. Soc. 65:136–138; 1988.

Ackman, R.G.; Linke, B.A.; Hingley, J. Some details of fatty acids and alcohols in the lipids of North Atlantic copepods. J. Fish. Res. Bd. Can. 31:1812–1818; 1974.

Ackman, R.G.; Eaton, C.A.; Bligh, E.G.; Lantz, A.W. Freshwater fish oils: yields and composition of oils from reduction of sheepshead, tullibee, maria, and alewife. J. Fish. Res. Bd. Can. 24:1219–1227; 1967.

Ahlgren, G.; Blomqvist, P.; Boberg, M; Gustafsson, I-B. Fatty acid content of the dorsal muscle—an indicator of fat quality in freshwater fish. J. Fish Biol. 45:131–157; 1994.

Akimoto, M.; Ishii, T.; Yamagaki, K.; Ohtaguchi, K.; Koide, K.; Yazawa, K. Production of eicosapentaenoic acid by a bacterium isolated from mackerel intestines. J. Am. Oil Chem. Soc. 67:911–915; 1990.

Albers, C.S.; Kattner, G.; Hagen, W. The compositions of wax esters, triacylglycerols and phospholipids in Arctic and Antarctic copepods: evidence of energetic adaptions. Mar. Chem. 55:347–358; 1996.

Ando, Y.; Nishimura, K.; Aoyanagi, N.; Takagi, T. Stereospecific analysis of fish oil triacyl-*sn*-glycerols. J. Am. Oil. Chem. Soc. 69:417–424; 1992.

Arts, M.T.; Robarts, R.D.; Evans, M.S. Energy reserve lipids of zooplanktonic crustaceans from an oligotrophic saline lake in relation to food resources and temperature. Can. J. Fish. Aquat. Sci. 50:2404–2420; 1993.

Bakes, M.J.; Nichols, P.R. Lipid, fatty acid and squalene composition of liver oil from six species of deep-sea sharks collected in southern Australian waters. Comp. Biochem. Physiol. 110B:267–275; 1995.

Bautista, M.N.; del Valle, M.J.; Orejana, F.M. Lipid and fatty acid composition of brackish-water and freshwater-reared milkfish (*Chanos chanos* Froskal). Aquaculture 96:241–248; 1991.

Bell, J.G.; Ashton, I.; Secombes, C.J.; Weitzel, B.R.; Dick, J.R.; Sargent, J.R. Dietary lipid affects phospholipid fatty acid compositions, eicosanoid production and immune function in Atlantic salmon (*Salmo salar*). Prostaglandins Leukotrienes Essential Fatty Acids 54:173–182; 1996.

Bell, J.G.; Ghioni, C.; Sargent, J.R. Fatty acid compositions of 10 freshwater invertebrates which are natural food organisms of Atlantic salmon parr (*Salmo salar*): a comparison with commercial diets. Aquaculture 128:301–313; 1994.

Bell, J.G.; Dick, J.R.; McVicar, A.H.; Sargent, J.R.; Thompson, K.D. Dietary sunflower, linseed and fish oils affect phospholipid fatty acid composition, development of cardiac lesions, phospholipase activity and eicosanoic production in Atlantic salmon (*Salmo salar*). Prostaglandins Leukotrienes Essential Fatty Acids 49:665–673; 1993.

Bell, M.V.; Sargent, J.R. Fatty acid analyses of phosphoglycerides from tissues of the clam *Chlamys islandica* (Muller) and the starfish *Ctenodiscus crispatus* (Retzius) from Balsfjorden, northern Norway. J. Exp. Mar. Biol. Ecol. 87:31–40; 1985.

Bligh, E.G.; Dyer, W.J. A rapid method of total lipid extraction and purification. Can. J. Biochem. Physiol. 37:911–917; 1959.

Bordier, C.G.; Sellier, N.; Foucault, A.P.; Le Goffic, F. Purification and characterization of deep sea shark *Centrophorus squamosus* liver oil 1-*O*-alkylglycerol ether lipids. Lipids 31:521–528; 1996.

Brosche, T. Methyl enol ethers as artifacts in capillary gas chromatographic profiles of aldehyde dimethyl acetals. J. Chromatogr. 345:219–227; 1985.

Bruner, K.A.; Fisher, S.W.; Landrum, P.F. The role of the zebra mussel, *Dreissena polymorpha,* in contaminant cycling; I. The effect of body size and lipid content on the bioconcentration of PCBs and PAHs. J. Great Lakes Res. 20:725–734; 1994.

Castell, J.D.; Mason, E.G.; Covey, J.F. Cholesterol requirements of juvenile American lobster (*Homarus americanus*). J. Fish. Res. Bd. Can. 32:1431–1435; 1975.

Cavaletto, J.F.; Nalepa, T.F.; Dermott, R.; Gardner, W.S.; Quigley, M.A.; Lang, G.A. Seasonal variation of lipid composition, weight, and length in juvenile *Diporeia* spp. (Amphipoda) from Lakes Michigan and Ontario. Can. J. Fish. Aquat. Sci. 53:2044–2051; 1996.

Cavaletto, J.F.; Vanderploeg, H.A.; Gardner, W.S. Wax esters in two species of freshwater zooplankton. Limnol. Oceanogr. 34:785–789; 1989.

Chandumpai, A.; Dall, W.; Smith, D.M. Lipid-class composition of organs and tissues of the tiger prawn *Penaeus esculentus* during the moulting cycle and during starvation. Mar. Biol. 198:235–245; 1991.

Chapelle, S. Aspects of phospholipid metabolism in crustaceans as related to changes in environmental temperatures and salinities. Comp. Biochem. Physiol. 84B:423–439; 1986.

Chapelle, S.; Hakanson, J.L.; Nevenzel, J.C.; Benson, A.A. Ether glycerophospholipids of gills of two Pacific crabs, *Cancer antennarius* and *Portunus xantusi.* Lipids 22:76–79; 1987.

Christie, W.W. Detectors for high-performance liquid chromatography of lipids with special reference to evaporative light-scattering detection. In: Christie, W.W., ed. Advances in Lipid Methodology—One. Ayr, U.K.: The Oily Press; 1992:p. 239–271.

Christie, W.W. High-Performance Liquid Chromatography and Lipids. Oxford: Pergamon Press; 1987.

Cosper, C.I.; Ackman, R.G. Occurrence of *cis*-9,10-methyleneoctadecenoic acids the lipids of immature and mature *Fundulus heteroclitus* (L.) and its roe. Comp. Biochem. Physiol. 75B:649–654; 1983.

D'Abramo, L.R.; Bordner, C.E.; Conklin, D.E.; Baum, N.A. Sterol requirement of juvenile lobsters, *Homarus* sp. Aquaculture 42:13–25; 1984.

de Koning, A.J.; Evans, A.A. Phospholipids of marine origin. The lantern fish (*Lampanyctodes hectoris*). J. Sci. Food Agric. 56:503–510; 1991.

Dembitsky, V.M.; Řezanka, T.; Kashin, A.G. Comparative study of the endemic freshwater fauna of Lake Baikal—VI. Unusual fatty acid and lipid composition of the endemic sponge *Lubomirskia baicalensis* and its amphipod crustacean parasite *Brandtia (Spinacanthus) parasitica.* Comp. Biochem. Physiol. 109B:415–426; 1994.

Dembitsky, V.M.; Řezanka, T.; Kashin, A.G. Fatty acid and phospholipid composition of freshwater molluscs *Anadonta piscinalis* and *Limnaea fragilis* from the river Volga. Comp. Biochem. Physiol. 105B:597–601; 1993.

Deridovich, I.I.; Reunova, O.V. Prostaglandins: reproduction control in bivalve molluscs. Comp. Biochem. Physiol. 104A:23–27; 1993.

Desvilettes, C.; Bourdier, G.; Breton, J-C. Lipid class and fatty acid composition of planktivorous larval pike *Esox lucius* living in a natural pond. Aquat. Living Resour. 7:67–77; 1994.

Early, T.A.; Kundrat, J.T.; Schorp, T.; Glonek T. Lake Michigan sponge phospholipid variations with habitat: a ^{31}P nuclear magnetic resonance study. Comp. Biochem. Physiol. 114B:77–89; 1996.

El Babili, M.; Brichon, G.; Zwingelstein, G. Sphingomyelin metabolism is linked to salt transport in the gills of euryhaline fish. Lipids 31:385–392; 1996.

Exler, J., ed. Composition of Foods: Finfish and Shellfish Products. Human Nutrition Information Service Agriculture Handbook 8–15 (Revised). Washington, DC: U.S. Department of Agriculture; 1987.

Fang, J.; Comet, P.A.; Brooks, J.M.; Wade, T.L. Nonmethylene-interrupted fatty acids of hydrocarbon seep mussels: occurrence and significance. Comp. Biochem. Physiol. 104B:287–291; 1993.

Fodor, E.; Jones, R.H.; Buda, C.; Kitajka, K.; Dey, I.; Farkas, T. Molecular architecture and biophysical properties of phospholipids during thermal adaptation in fish: an experimental and model study. Lipids 30:1119–1126; 1995.

Fraser, A.J. Triacylglycerol content as a condition index for fish, bivalve, and crustacean larvae. Can. J. Fish. Aquat. Sci. 46:1868–1873; 1989.

Fullarton, J.G.; Dando, P.R.; Sargent, J.R.; Southward, A.J.; Southward, E.C. Fatty acids of hydrothermal vent *Ridgeia piscesae* and inshore bivalves containing symbiotic bacteria. J. Mar. Biol. Assn. U.K. 75:455–468; 1995a.

Fullarton, J.G.; Wood, A.P.; Sargent, S.R. Fatty acid composition of lipids from sulphur-oxidizing and methylotrophic bacteria from thyasirid and lucinid bivalves. J. Mar. Biol. Assn. U.K. 75:445–454; 1995b.

Gardner, W.S.; Nalepa, T.F.; Frez, W.A.; Cichocki, E.A.; Landrum, P.F. Seasonal patterns in lipid content of Lake Michigan macroinvertebrates. Can. J. Fish. Aquat. Sci. 42:1827–1832; 1985.

Gauvin, J.M.; Gardner, W.S.; Quigley, M.A. Effects of food removal on nutrient release rates and lipid content of Lake Michigan *Pontoporeia hoyi*. Can. J. Fish. Aquat. Sci. 46:1125–1130; 1989.

Ghioni, C.; Bell, J.G.; Sargent, J.R. Polyunsaturated fatty acids in neutral lipids and phospholipids of some freshwater insects. Comp. Biochem. Physiol. 114B:161–170; 1996.

Giese, A.C. Lipids in the economy of marine invertebrates. Physiol. Rev. 46:244–298; 1966.

Gonzalez-Baro, M.D.R.; Pollero, R.J. Lipid characterization and distribution among tissues of the freshwater crustacean *Macrobrachium borellii* during an annual cycle. Comp. Biochem. Physiol. 91B:711–715; 1988.

Goodnight, S.H. The fish oil puzzle. Sci. Med. 3(5):42–51; 1996.

Goulden, G.E.; Henry, L.L. Lipid energy reserves and their role in Cladocera. In: Meyers, D.G.; Strickler, J.R., eds. Trophic Interactions within Aquatic Ecosystems. American Association for the Advancement of Science Selected Symposium, Washington, D.C., 85, 1981 January 3–8, Toronto, Ontario. 1984:p. 167–185.

Graeve, M.; Kattner, G. Species-specific differences in intact wax esters of *Calanus hyperboreus* and *C. finmarchicus* from Fram Strait—Greenland Sea. Mar. Chem. 39:269–281; 1992.

Grigor, M.R.; Sutherland, W.H.; Phleger, C.F. Wax-ester metabolism in the orange roughy *Hoplostethus atlanticus* (Beryciformes: Trachichthyidae). Mar. Biol. 105:223–227; 1990.

Gurr, M.I.; Harwood, J.L. Lipid Biochemistry—An Introduction, 4th ed. London: Chapman and Hall; 1991.

Guth, H.; Zhang, Y.; Laskawy, G.; Grosch, W. Furanoid fatty acids in oils from soybeans lacking lipoxygenase isoenzymes. J. Am. Oil Chem. Soc. 72:397–398; 1995.

Hansen, H.J.M.; Olsen, A.G.; Rosenkilde, P. Comparative studies on lipid metabolism in salt-transporting organs of the rainbow trout (*Oncorhynchus mykiss* W.). Further evidence of monounsaturated phosphatidylethanolamine as a key substance. Comp. Biochem. Physiol. 103B:81–87; 1992.

Hanson, B.J.; Cummins, K.W.; Cargill, A.S.; Lowry, R.R. Lipid content, fatty acid composition, and the effect of diet on fats of aquatic insects. Comp. Biochem. Physiol. 80B:257–276; 1985.

Hayashi, K.; Kawasaki, K-I. Unusual occurrence of diacylglyceryl ethers in liver lipids from two species of gonatid squids. Bull. Jpn. Soc. Sci. Fish. 51:593–597; 1985.

Hazel, J.R.; Williams, E.E. The role of alterations in membrane lipid composition in enabling physiological adaptation of organisms to their physical environment. Prog. Lipid Res. 29:167–227; 1990.

Henderson, R.J.; Tocher, D.R. The lipid composition and biochemistry of freshwater fish. Prog. Lipid Res. 26:281–347; 1987.

Henderson, R.J.; Millar, R-M.; Sargent, J.R. Effect of growth temperature on the positional distribution of eicosapentaenoic acid and *trans* hexadecenoic acid in the phospholipids of a *Vibrio* species of bacterium. Lipids 30:181–185; 1995a.

Henderson, R.J.; Park, M.T.; Sargent, J.R. The desaturation and elongation of 14C-labelled polyunsaturated fatty acids by pike (*Esox lucius* L.) in vivo. Fish Physiol. Biochem. 14:223–236; 1995b.

Henderson, R.J.; Bell, M.V.; Sargent, J.R. The conversion of polyunsaturated fatty acids to prostaglandins by tissue homogenates of the turbot, *Scophthalmus maximus* (L.). J. Exp. Mar. Biol. Ecol. 85:93–99; 1985.

Herbes, S.E.; Allen, A.P. Lipid quantification of freshwater invertebrates: method modification for microquantitation. Can. J. Fish. Aquat. Sci. 40:1315–1317; 1983.

Hill, C.; Quigley, M.A.; Cavaletto, J.F.; Gordon, W. Seasonal changes in lipid content and composition in the benthic amphipods *Monoporeia affinis* and *Pontoporeia femorata*. Limnol. Oceanogr. 37:1280–1289; 1992.

Innis, S.M.; Rioux, F.M.; Auestad, N.; Ackman, R.G. Marine and freshwater fish oil varying in arachidonic, eicosapentaenoic and docosahexaenoic acids differ in their effects on organ lipids and fatty acids in growing rats. J. Nutr. 125:2286–2293; 1995.

Irazu, C.E.; González-Baró, M.D.R.; Pollero, R.J. Effect of environmental temperature on mitochondrial β-oxidation activity in gills and hepatopancreas of the freshwater shrimp *Macrobrachium borellii*. Comp. Biochem. Physiol. 102B:721–725; 1992.

Ishii, K.; Okajima, H.; Koyamatsu, T.; Okada, Y.; Watanabe, H. The composition of furan fatty acids in the crayfish. Lipids 23:694–700; 1988.

Jeong, B-Y.; Ohshima, T.; Ushio, H.; Koizumi, C. Lipids of cartilaginous fish: composition of ether and ester glycerophospholipids in the muscle of four species of shark. Comp. Biochem. Physiol. 113B:305–312; 1996a.

Jeong, B-Y.; Ohshima, T.; Koizumi, C. Hydrocarbon chain distribution of ether phospholipids of the ascidian *Halocynthia roretzi* and the sea urchin *Strongylocentrotus intermedius*. Lipids 31:9–18; 1996b.

Josephson, D.B.; Lindsay, R.C.; Stuiber, D.A. Variations in the occurrences of enzymatically derived volatile aroma compounds in salt- and freshwater fish. J. Agric. Food Chem. 32:1344–1347; 1984.

Käkelä, R.; Ackman, R.G.; Hyvärinen, H. Very long chain polyunsaturated fatty acids in the blubber of ringed seals (*Phoca hispida* sp.) from Lake Saimaa, Lake Ladoga, the Baltic Sea, and Spitsbergen. Lipids 30:725–731; 1995.

Kang, S-J.; Lall, S.P.; Ackman, R.G. Digestion of the 1-O-alkyl diacylglycerol ethers of Atlantic dogfish liver oils by Atlantic salmon *Salmo salar*. Lipids 32:19–30; 1997.

Kattner, G.; Graeve, M.; Ernst, W. Gas-liquid chromatographic method for the determination of marine wax esters according to the degree of unsaturation. J. Chromatogr. 513:327–332; 1990.

Kean, J.C.; Castell, J.D.; Boghen, A.G.; D'Abramo, L.R.; Conklin, D.E. A re-evaluation of the lecithin and cholesterol requirements of juvenile lobster (*Homarus americanus*) using crab protein-based diets. Aquaculture 47:143–149; 1985.

Keweloh, H.; Heipieper, H.J. *Trans* unsaturated fatty acids in bacteria. Lipids 31:129–137; 1996.

Kiessling, A.; Larsson, L.; Kiessling, K-H.; Lutes, P.B.; Storebakken, T.; Hung, S.S.S. Spawning induces a shift in energy metabolism from glucose to lipid in rainbow trout white muscle. Fish Physiol. Biochem. 14:439–448; 1995.

Kiron, V.; Fukuda, H.; Takeuchi, T.; Watanabe, T. Essential fatty acid nutrition and defence mechanisms in rainbow trout *Oncorhynchus mykiss*. Comp. Biochem. Physiol. 111A:361–367; 1995.

Kitamura, H.; Kitahara, S.; Koh, H.B. The induction of larval settlement and metamorphosis of two sea urchins, *Pseudocentrotus depressus* and *Anthocidaris crassispina,* by free fatty acids extracted from the coralline red alga *Corallina pilulifera.* Mar. Biol. 115:387–392; 1993.

Koechlin, N.; Polonsky, J.; Varenne, J. Accumulation of cholesterol and cholesterol esters in the nephridia of a polychaeta annelida (*Sabella pavonina* Savigny). Comp. Biochem. Physiol. 68A:391–397; 1981.

Koizumi, C.; Jeong, B-Y.; Ohshima, T. Fatty chain composition of ether and ester glycero-phospholipids in the Japanese oyster *Crassostrea gigas* (Thunberg). Lipids 25:363–370; 1990.

Lands, W.E.M. Fish and Human Health. Orlando, FL: Academic Press; 1986.

Lee, R.F.; Patton, J.S. Alcohol and waxes. In: Ackman, R.G., ed. Marine Biogenic Lipids, Fats, and Oils, vol. 1. Boca Raton, FL: CRC Press; 1989:p. 73–102.

Lie, Ø.; Lambertsen, G. Lipid digestion and absorption in cod (*Gadus morhua*), comparing triacylglycerols, wax esters and diacylalkylglycerols. Comp. Biochem. Physiol. 98A:159–163; 1991.

Lindsay, R.C. Fish flavors. Food Rev. Int. 6:431–455; 1990.

Linko, R.R.; Rajasilta, M.; Hiltunen, R. Comparison of lipid and fatty acid composition in vendace (*Coregonus albula* L.) and available plankton feed. Comp. Biochem. Physiol. 103A:205–212; 1992.

Linko, R.R.; Kaitaranta, J.K.; Vuorela, R. Comparison of the fatty acids in Baltic herring and available plankton feed. Comp. Biochem. Physiol. 82B:699–705; 1985.

Litchfield C. Analysis of Triglycerides. New York: Academic Press; 1972.

Mangold, H.K.; Paltauf, F., eds. Ether Lipids: Biochemical and Biomedical Aspects. New York: Academic Press; 1983.

McGill, A.S.; Moffat, C.F. A study of the composition of fish liver and body oil triglycerides. Lipids 27:360–370; 1992.

Merican, Z.O.; Shim, K.F. Qualitative requirements of essential fatty acids for juvenile *Penaeus monodon.* Aquaculture 147:275–291; 1996.

Moffat, C.F. Fish oil triglycerides: a wealth of variation. Lipid Tech. 7:125–129; 1995.

Mohri, S.; Cho, S-Y.; Endo, Y.; Fujimoto, K. Linoleate 13 (S)-lipoxygenase in sardine skin. J. Agric. Food Chem. 40:573–576; 1992.

Mohri, S.; Cho, S-Y.; Endo, Y.; Fujimoto, K. Lipoxygenase activity in sardine skin. Agric. Biol. Chem. 54:1889–1991; 1990.

Mourente, G. In vitro metabolism of ^{14}C-polyunsaturated fatty acids in midgut gland and ovary cells from *Penaeus kerathurus* Forskål at the beginning of sexual maturation. Comp. Biochem. Physiol. 115B:255–266; 1996.

Mourente, G.; Rodríguez, A. Variation in the lipid content of wild-caught females of the marine shrimp *Penaeus kerathurus* during sexual maturation. Mar. Biol. 110:21–28; 1991.

Muje, P.; Ågren, J.J.; Lindqvist, O.V.; Hänninen, O. Fatty acid composition of vendace (*Coregonus albula* L.) muscle and its plankton feed. Comp. Biochem. Physiol. 92B:75–79; 1989.

Nakagawa, H.; Takahara, Y.; Nematipour, G.R. Comparison of lipid properties between wild and cultured ayu. Nippon Suisan Gakkaishi 57:1965–1971; 1991.

Napolitano, G.E.; Ackman, R.A.; Silva-Serra, M.A. Incorporation of dietary sterols by the sea scallop *Placopecten magellanicus* (Gmelin) fed on microalgae. Marine Biology 117:647–654; 1993.

Napolitano, G.E.; Ackman, R.G.; Parrish, C.C. Lipids and lipophilic pollutants in three species of migratory shorebirds and their food supply in Shepody Bay (Bay of Fundy, New Brunswick). Lipids 27:785–790; 1992.

Napolitano, G.E.; Ratnayake, W.M.N.; Ackman, R.G. Fatty acid components of larval *Ostrea edulis* (L.): importance of triacylglycerols as a fatty acid reserve. Comp. Biochem. Physiol. 90B:875–883; 1988.

Navarro, J.C.; Bell, M.V.; Amat, F.; Sargent, J.R. The fatty acid composition of phospholipids from brine shrimp, *Artemia* sp., eyes. Comp. Biochem. Physiol. 103B:89–91; 1992.

Nettleton, J.A.; Allen, W.H.; Klatt, L.V.; Ratnayake, W.M.N.; Ackman, R.G. Nutrients and chemical residues in one- to two-pound Mississippi farm-raised channel catfish (*Ictalurus punctatus*). J. Food Sci. 55:954–958; 1990.

Nevenzel, J.C.; Gibbs, A.; Benson, A.A. Plasmalogens in the gill lipids of aquatic animals. Comp. Biochem. Physiol. 82B:293–297; 1985.

Nichols, D.S.; Hart, P., Nichols, P.D., McMeekin, T.A. Enrichment of the rotifer *Brachionus plicatilis* fed an Antarctic bacterium containing polyunsaturated fatty acids. Aquaculture 147:115–125; 1996.

Nicol, J.A.C.; Arnott, H.J.; Mizuno, G.R.; Ellison, E.C.; Chipault, J.R. Occurrence of glyceryl tridocosahexaenoate in the eye of the sand trout *Cynoscion arenarius.* Lipids 7:171–177; 1972.

Ohshima, T.; Wada, S.; Koizumi, C. 1-*O*-Alk-1′-enyl-2-acyl and 1-*O*-alkyl-2-acyl glycerophospholipids in white muscle of bonito *Euthynnus pelamis* (Linnaeus). Lipids 24:363–370; 1989.

Ohtsuru, M.; Fujii, M.; Ishinaga, M.; Kito, M. Fatty acid composition of fish (fatty acid compositions of fish caught in seas around Yamaguchi Prefecture). J. Agric. Chem. Soc. Jpn. 58:35–42; 1984.

Ota, T.; Takagi, T. Changes in furan fatty acids of testis lipids of chum salmon *Oncorhynchus keta* at spawning season. Nippon Suisan Gakkaishi 56:153–157; 1990.

Ota, T.; Takagi, T. Furan fatty acids in the lipids of kokanee *Oncorhynchus nerka* f. *adonis.* Bull Fac. Fish. Hokkaido Univ. 34(2):88–92; 1983.

Ota, T.; Chihara, Y.; Itabashi, Y.; Takagi, T. Occurrence of all-*cis*-6,9,12,15,18,21-tetracosahexaenoic acid in flatfish lipids. Fish. Sci. 60:171–175; 1994.

Otwell, W.S.; Rickards, W.L. Cultured and wild American eels, *Anguila rostrata:* fat content and fatty acid composition. Aquaculture 26:67–76; 1981.

Padrón, D.; Lindley, V.A.; Pfeiler, E. Changes in lipid composition during metamorphosis of bonefish (*Albula* sp.) leptocephali. Lipids 31:513–519; 1996.

Paradis, M.; Ackman, R.G. Potential for employing the distribution of anomalous non-methylene-interrupted dienoic fatty acids in several marine invertebrates as part of food web studies. Lipids 12:170–176; 1977.

Paradis, M.; Ackman, R.G. Localization of a source of marine odd chain length fatty acids. I. The amphipod *Pontoporeia femorata* (Kroyer). Lipids 11:863–870; 1976.

Parrish, C.C. Separation of aquatic lipid classes by Chromarod thin-layer chromatography with measurement by Iatroscan flame ionization detection. Can. J. Fish. Aquat. Sci. 44:722–731; 1987.

Parrish, C.C.; Yang, Z.; Lau, A.; Thompson, R.J. Lipid composition of *Yoldia hyperborea* (Protobranchia), *Nephthys ciliata* (nephthyidae) and *Artacama proboscidea* (Terebellidae) living at sub-zero temperatures. Comp. Biochem. Physiol. 114B:59–67; 1996a.

Parrish, C.C.; Bodennec, G.; Gentien, P. Determination of glycoglycerolipids by Chromarod thin-layer chromatography with Iatroscan flame ionization detection. J. Chromatogr. A741:91–97; 1996b.

Parrish, C.C.; McKenzie, C.H.; MacDonald, B.A.; Hatfield, E.A. Seasonal studies of seston lipids in relation to microplankton species composition and scallop growth in South Broad Cove, Newfoundland. Mar. Ecol. Prog. Ser. 129:151–164; 1995.

Pellerin-Massicotte, J.; Martineu, P.; Desrosiers, G.; Caron, A.; Scaps, P. Seasonal variability in biochemical composition of the polychaete *Nereis virens* (Sars) in two tidal flats with different geographic orientations. Comp. Biochem. Physiol. 107A:509–516; 1994.

Phleger, C.F.; Grigor, M.R. Role of wax esters in determining buoyancy in *Hoplostethus atlanticus* (Beryciformes: Trachichthyidae). Mar. Biol. 105:229–233; 1990.

Pollero, R.J.; Gros, E.G.; Brenner, R.R. Sterol composition of two freshwater molluscs of genus *Diplodom*. Lipids 18:100–102; 1983.

Polvi, S.M.; Ackman, R.G.; Lall, S.P.; Saunders, R.L. Stability of lipids and omega-3 fatty acids during frozen storage of Atlantic salmon. J. Food Proc. Preserv. 15:167–181; 1991.

Popov, S.; Stoilov, I.; Marekov, N.; Kovachev, G.; Andreev, S.T. Sterols and their biosynthesis in some freshwater bivalves. Lipids 16:663–669; 1981.

Ratnayake, W.N.; Ackman, R.G. Fatty alcohols in capelin, herring and mackerel oils and muscle lipids: 1. Fatty alcohol details linking dietary copepod fat with certain fish depot fats. Lipids 14:795–803; 1979a.

Ratnayake, W.N.; Ackman, R.G. Fatty alcohols in capelin, herring and mackerel oils and muscle lipids. II. A comparison of fatty acids from wax esters with those of triglycerides. Lipids 14:804–810; 1979b.

Ratnayake, W.M.N.; Olsson, B.; Ackman, R.G. Novel branched-chain fatty acids in certain fish oils. Lipids 24:630–637; 1989.

Řezanka, V.M.; Dembitsky, T. Identification of unusual cyclopropane monounsaturated fatty acids from the deep-water lake invertebrate *Acanthogammarus grewingkii*. Comp. Biochem. Physiol. 109B:407–413; 1994.

Roose, P.; Smedes, F. Evaluation of the results of the QUASIMEME lipid intercomparison: the Bligh and Dyer total lipid extraction method. Mar. Pollut. Bull. 32:674–680; 1996.

Saether, O.; Ellingsen, T.E.; Mohr, V. The distribution of lipid in the tissues of Antarctic krill, *Euphausia superba*. Comp. Biochem. Physiol. 81B:609–614; 1985.

Saito, H.; Murata, M. The high content of monoene fatty acids in the lipids of some midwater fishes: family Myctophidae. Lipids 31:757–763; 1996.

Sand, D.M.; Glass, R.L.; Olson, D.L.; Pike, H.M.; Schlenk, H. Metabolism of furan fatty acids in fish. Biochim. Biophys. Acta 793:429–434; 1984.

Sargent, J.R. Ether-linked glycerides in marine animals. In: Ackman, R.G., ed. Marine Biogenic Lipids, Fats, and Oils, vol. 1. Boca Raton, FL: CRC Press; 1989:p. 175–197.

Sargent, J.R.; McIntosh, R.; Bauermeister, A.; Blaxter, J.H.S. Assimilation of the wax esters of marine zooplankton by herring (*Clupea harengus*) and rainbow trout (*Salmo gairdnerii*). Mar. Biol. 51:203–207; 1979.

Shantha, N.C.; Ackman, R.G. Advantages of total lipid hydrogenation prior to lipid class determination on Chromarods-SIII. Lipids 25:570–574; 1990.

Sheridan, M.A. Regulation of lipid metabolism in poikilothermic vertebrates. Comp. Biochem. Physiol. 107B:495–508; 1994.

Sigurgisladóttir, S.; Pálmadóttir, H. Fatty acid composition of thirty-five Icelandic fish species. J. Am. Oil Chem. Soc. 70:1081–1087; 1993.

Smedes, F.; Thomasen, T.K. Evaluation of the Bligh and Dyer lipid determination method. Mar. Pollut. Bull. 32:681–688; 1996.

Steiner-Asiedu, M.; Julshamn, K.; Lie, Ø. Effect of local processing methods (cooking, frying and smoking) on three fish species from Ghana: Part I. Proximate composition, fatty acids, minerals, trace elements and vitamins. Food Chem. 40:309–321; 1991.

Taghon, G.L.; Prahl, F.G.; Sparrow, M.; Fuller, C.M. Lipid class and glycogen content of the lugworm *Abarenicola pacifica* in relation to age, growth rate and reproductive condition. Mar. Biol. 120:287–295; 1994.

Takagi, T.; Kaneniwa, M.; Itabashi, Y.; Ackman, R.G. Fatty acids in echinoidea: unusual *cis*-5-olefinic acids as distinctive lipid components in sea urchins. Lipids 21:558–565; 1986.

Takahashi, K. A novel approach for the identification of lipid molecular species. Mem. Fac. Fish. Hokkaido Univ. 32:245–330; 1985.

Takeuchi, T.; Sampekalo, J.; Nomura, T.; Watanabe, T. Lipid contents and classes in gill of masu salmon *Oncorhynchus masou* during the Parr-Smolt transformation. Nippon Suisan Gakkaishi 56:1527; 1990.

Takeuchi, T.; Kang, S-J.; Watanabe, T. Effects of environmental salinity of lipid classes and fatty acid composition in gills of Atlantic salmon. Nippon Suisan Gakkaishi 55:1395–1405; 1989.

Teshima, S-I.; Kanazawa, A.; Koshio, S.; Horinouchi, K. Lipid metabolism in destalked prawn *Penaeus japonicus:* induced maturation and accumulation of lipids in the ovaries. Nippon Suisan Gakkaishi 54:1115–1122; 1988.

Tocher, D.R.; Sargent, J.R. Studies on triacylglycerol, wax ester and sterol ester hydrolases in intestinal caeca of rainbow trout (*Salmo gairdneri*) fed diets rich in triacylglycerols and wax esters. Comp. Biochem. Physiol. 77B:561–571; 1984.

Tocher, D.R.; Bell, J.G.; Sargent, J.R. Production of eicosanoids derived from 20:4n-6 and 20:5n-3 in primary cultures of turbot (*Scophthalmus maximus*) brain astrocytes in response to platelet activating factor, substance P and interleukin-1β. Comp. Biochem. Physiol. 115B:215–222; 1996.

Tocher, D.R.; Fraser, A.J.; Sargent, J.R.; Gamble, J.C. Lipid class composition during embryonic and early larval development in Atlantic herring (*Clupea harengus,* L.). Lipids 20:84–89; 1985.

Toonen, R.J.; Pawlik, J.R. Settlement of the tube worm *Hydroides dianthus* (Polychaeta: Serpulidae): cues for gregarious settlement. Mar. Biol. 126:725–734; 1996.

Urata, K.; Takaishi, N. Ether lipids based on the glyceryl ether skeleton: present state, future potential. J. Am. Oil Chem. Soc. 73:819–830; 1996.

Uscian, J.M.; Stanley-Samuelson, D.W. Fatty acid compositions of phospholipids and triacylglycerols from selected terrestrial arthropods. Comp. Biochem. Physiol. 107B:371–379; 1994.

Vanderploeg, H.A.; Gardner, W.S.; Parrish, C.C.; Liebig, J.R.; Cavaletto, J.F. Lipids and life-cycle strategy of a hypolimnetic copepod in Lake Michigan. Limnol. Oceanogr. 37:413–424; 1992.

Vaskovsky, V.E. Phospholipids. In: Ackman, R.G., ed. Marine Biogenic Lipids, Fats, and Oils, vol. 1. Boca Raton, FL: CRC Press; 1989:p. 199–242.

Voss, A.; Reinhart, M.; Sprecher, H. Differences in the interconversion between 20- and 22-carbon (n-3) and (n-6) polyunsaturated fatty acids in rat liver. Biochim. Biophys. Acta 1127:33–40; 1992.

Waellert, C.; Babin, P.J. Age-related, sex-related and seasonal changes of plasm lipoproteins in trout. J. Lipid Res. 35:1619–1633; 1994.

Wainman, B.C.; McQueen, D.J.; Lean, D.R.S. Seasonal trends in zooplankton lipid concentration and class in freshwater lakes. J. Plankton Res. 15:1319–1332; 1993.

Webster, C.D.; Lovell, R.T.; Clawson, J.A. Ratio of 20:3(n-9) to 20:5(n-3) in phospholipids as an indicator of dietary essential fatty acid sufficiency in striped bass, *Morone saxatilis,* and Palmetto bass, *M. saxatilis* x *M. chrysops.* J. Appl. Aquacult. 4:75–90; 1994.

Wirth, W.; Steffens, W.; Meinelt, T.; Steinberg, C. Significance of docosahexaenoic acid for rainbow trout (*Oncorhynchus mykiss*) larvae. Fett/Lipid 99:251–253; 1997.

Yang, Z.; Parrish, C.C.; Helleur, R.J. Automated gas chromatographic method for neutral lipid carbon number profiles in marine samples. J. Chromatogr. Sci. 34:556–568; 1996.

Yazawa, K. Production of eicosapentaenoic acid from marine bacteria. Lipids 31:S-297–S-300; 1996.

Young, N.J.; Quinlan, P.T.; Goad, L.J. Cholesteryl esters in the decapod crustacean, *Penaeus monodon.* Comp. Biochem. Physiol. 102B:761–768; 1992.

Zhou, S.; Ackman, R.G. Interference of polar lipids with the alkalimetric determination of free fatty acid in fish lipids. J. Am. Oil Chem. Soc. 72:1019–1023; 1996.

Glossary

acetone-mobile polar lipids (AMPL): operational definition used in thin-layer chromatography for lipids (e.g., pigments, glycolipids, and monoacylglycerols) that are separated from other compounds by acetone as the sole mobile phase.

adiposity: fat (lipid) character of an organism or tissue.

allochthonous: materials formed elsewhere than in situ; or formed outside of the lake basin.

autochthonous: materials formed within the lake basin.

bioaccumulation: accumulation of contaminants by biota from all sources (e.g., water and food).

biochemical indicator: chemical compound characteristic of an organism, which can be used to infer the presence of the organism in the environment and to estimate its biomass.

bioconcentration: accumulation of contaminants by aquatic organisms from water.

biofouling: encrustation by fouling organisms of materials immersed in aqueous systems.

biogenic lipid classes: those lipid classes produced by living matter (e.g., wax esters, steryl esters, triacylglycerols, sterols, glycolipids, and phospholipids).

biomagnification: significant increase in contaminant concentration with each succeeding trophic level above that explained by changes in organism characteristics (e.g., changes in lipid content).

biomarker: see biochemical indicator.

biotransformation: metabolism of contaminants transforming them into different compounds.

Chromarod: quartz rod coated with a thin layer of silica sintered with soft glass.

cyclopropanoid fatty acids: fatty acids that have a ring structure of three carbon atoms, usually near the middle of a fatty acid chain, where the ethylenic bonds are normally found. They are frequently found in bacterial membrane phospholipids, but if found in depot fats of aquatic organisms, they have no known function.

derivatization: preparation of a substance from another substance in such a manner that the general structure of the parent substance is retained.

dietary deficiency: condition in animals resulting from the absence or inadequate amount of a required dietary compound such as essential amino acids, essential fatty acids, vitamins, and several metals.

endogenous lipid: lipid produced by the organism's own metabolism, as opposed to that incorporated from the diet.

energy allocation strategy: biochemical processes by which organisms partition or allocate available (assimilated) energy into the main functions of somatic (protein) growth, basic maintenance, reproductive effort, and energy (lipid) storage. The proportion of energy allocated to each process is dependent on environmental food availability, temporal processes, and physiological condition.

ether lipids: glycerol molecules with a hydrocarbon chain linked by an ether (single oxygen) bond in one outside position. The other two positions may both be occupied by esterified fatty acids (diacylglycerol ethers), or the other outside position may contain a phosphoric acid–amino acid group as in ordinary phospholipids. They may then be called alkylacylphospholipids. (*Note:* the structure of ether lipids is given in Ackman, this volume, Fig. 11.6.)

fasting endurance: ability of an animal to survive and remain active during a short time period when the level of food resources, or its ingestion and assimilation, is insufficient to support the metabolic and growth needs of the animal.

foam: emulsion-like two-phase system in which the dispersal phase is a gas or air.

fugacity: escaping tendency of a contaminant from a compartment.

fugacity capacity: capacity of a compartment to hold a particular contaminant.

gametogenesis: production of gametes (i.e., eggs and sperm).

glacial lake: lake formed by the repeated scouring of the earth's surface during the advancing and receding of glaciers.

gravimetry: determining the amount of a substance from its weight.

growth refuge: condition or process used by some species of fish, particularly in their first year of life, to maximize growth rate or body size to minimize the risk of predation. High growth rates thus provide a refuge from predation.

habitat constriction: ecological term describing a situation in which species are forced, primarily by physicochemical factors, to be constricted or squeezed into a narrow zone of the water column during certain times of the year. For example, low dissolved oxygen in the lower parts of the water column and high temperatures in the upper areas of the water column can cause some fish to be constricted into a narrow band within the middle parts of the water column, where oxygen and temperature conditions are more favorable.

holometabolous: in insects, development with complete metamorphosis (i.e., egg, larva, pupa, adult).

homeostatic capacity: ability of an organism to control or regulate certain critical biological functions or processes within set biochemical and physiological boundaries for that species. If homeostatic boundaries or physiological tolerance limits are exceeded, then the organism becomes increasingly vulnerable to mortality from various sources of environmental stress.

Iatroscan: instrument used for scanning Chromarods through a flame ionization detector.

instar: interval between successive molts of exoskeletons in arthropods. In organisms with an exoskeleton, molting is required for the animal to be able to increase in size.

lamella: thin aqueous film that provides the structure of a foam at its plateau borders.

lipase: enzyme involved in the hydrolysis of lipids.

lipid normalization: concentration in a compartment, organism, or tissue, divided by the fractional weight of lipid of that organism or tissue.

lipophilicity: lipid-like character of a contaminant. It is generally related to the octanol-water partition coefficient of the compound.

lipolytic enzymes: lipases and esterases that hydrolyze ester bonds in lipids.

microlayer: (water microlayer) pertaining to the region of the surface water extending from the air–water interface to a depth of approximately 100 μm.

MUFA: monounsaturated fatty acids; those fatty acids containing a single double bond.

myctophids: sometimes called lanternfish; small (5–20 cm in length) marine fish usually found between 100 m at night and 500 m in daylight hours. Their lipids often include wax esters; this and their small size make them unattractive for food use.

narcosis: nonspecific toxicity of a contaminant that is generally expressed through membrane disruption.

nectobenthic: animals that swim in the water column at deep depths in the hypolimnion where light is limited (e.g., some amphipods found in deep zones of Lake Baikal).

neritic: region of shallow water adjoining the sea coast and extending from the low-tide mark to a depth of about 200 m.

octanol-water partition coefficient: equilibrium distribution constant for partitioning between n-octanol and water. This is generally designated K_{ow} and describes the hydrophobicity or lipophilicity of a contaminant.

osmotic fragility: condition of the cell relative to changes in membrane permeability and electrolyte imbalance that can result in the cell acquiring a hypo- or hyperosmotic state.

petrogenic: relative to rocks and by extension, also used as relative to petroleum.

plasmalogens: almost invariably phospholipids similar to the ether phospholipids except that the ether oxygen is attached directly to an ethylenic bond of the alkyl chain. They are found in most animal membranes and may be called alkenylacylphospholipids.

plateau border: connection of the lamellae within the foam structure, forming an angle of 120°.

profundal: in deep lakes where stratification occurs, it is the region of sediments below the thermocline, where temperature is nearly uniform throughout the year, currents are minimal, and light penetration is much reduced.

PUFA: polyunsaturated fatty acids; those fatty acids containing two or more double bonds.

rod TLC: thin-layer chromatography performed on a Chromarod.

stenothermic: organisms that dwell within a narrow temperature range.

surfactant: any compound that reduces surface tension when dissolved in water or water solutions or that reduces interfacial tension between two liquids or between a liquid and a solid.

track autoradiography: visualization of incorporated radioactivity (e.g., ^{14}C in microscopic algae) through mounting in a photographic emulsion, followed by development and counting of "tracks" of adjacent silver grains, each track denoting emission of a nuclear particle.

trophic marker: chemical compound characteristic of an organism that can be used to trace predator–prey relationships.

Index

Page numbers for entries occurring in figures are followed by an f; those for entries occurring in tables, by a t.